Thermodynamic Dissipation Theory of the Origin and Evolution of Life

Salient characteristics of RNA, DNA and other fundamental molecules suggest an origin of life driven by UV-C light

Karo Michaelian

Alpha Version: 2.0
First printing, Beta Version 1.0, June 5, 2016.
Second printing, Beta Version 2.0, June 20, 2016.
Third printing, Beta Version 3.0, June 30, 2016.
Fourth printing, Beta Version 4.0 July 10, 2016.
Fifth printing, Beta Version 5.0, August 2, 2016.
Sixth printing, Beta Version 6.3, August 16, 2016.
Seventh printing, Beta Version 7.1, September 15, 2016.
Eighth printing, Beta Version 8.0, October 12, 2016.
Ninth printing, Beta Version 9.0, November 22, 2016.
Tenth printing, Beta Version 10.0, December 27, 2016.
Eleventh printing, Alpha Version 1.0, Decembe 28, 2016.
Twelfth printing, Alpha Version 2.0, March 24, 2017.

DOI: 10.13140/RG.2.1.3222.7443
ISBN: 9781541317482 color
ISBN: 9781541366718 black and white
Also available in ebook format.

Cite as: Michaelian, K. (2016) Thermodynamic Dissipation Theory of the Origin and Evolution of Life: Salient characteristics of RNA and DNA and other fundamental molecules suggest an origin of life driven by UV-C light, Self-published. Printed by CreateSpace, Mexico City, ISBN: 9781541317482, DOI: 10.13140/RG.2.1.3222.7443

In memory of Sonia Michaelian (Tataryn)

Contents

List of Figures

List of Tables

About the Author:

Karo Michaelian was born in Wakefield, County Yorkshire, England in 1960, the second of four children, to a Dutch mother and an Armenian father. At the age of 6 the family moved to Canada, settling first near Niagra Falls, Ontario and finally in Edmonton, the capital of the province of Alberta. After a number of years of frog and gopher catching during the summers and cross country skiing during the harsh prairie winters, Karo coursed the program of Honors in Geophysics at the University of Alberta, receiving a gold medal from the *Association of Professional Engineers, Geologists and Geophysicists of Alberta* for graduating in 1981 with the highest marks in the province in his field of study. He followed with a doctorate degree in Experimental Nuclear Physics in 1987, with a thesis on the off-shell effects in the nucleon-nucleon interaction at the meson facility TRIUMF in Vancouver. He then did a 4 year postdoctoral stance at the Paul Scherrer Institute (formally Swiss Institute for Nuclear Research) in Switzerland studying pion absorption in nuclei before obtaining a position as a Research Scientist at the Institute of Physics at the National Autonomous University of Mexico (UNAM) in 1991. He has been in Mexico at the UNAM ever since, working in nuclear physics, scintillation physics, nanoparticle physics, non-equilibrium thermodynamics, ecosystem dynamics and the origin of life. In 2002 Karo was awarded the "Jorge Lomnitz Adler prize" by the *Mexican Academy of Sciences* for his work in complex systems. Karo is a dedicated father of three children. He is a social activist, working to end the alienation of fathers from the lives of their children in western societies, cofounder of the *Missionaries of Science* program for fomenting science education in the schools of Mexico and founder of *One Just World* which aims to bring attention to grave injustices committed against individuals or states, and campaigns to end war mongering practiced by advanced nations against developing nations.

Preface

The *Thermodynamic Dissipation Theory of the Origin and Evolution of Life* celebrates 7 years since it was first published on the Cornell *ArXiv* in 2009 [207]. This theory was the first to recognize the fundamental importance of photon dissipation to the origin and evolution of life and the first to suggest that the fundamental molecules of life were microscopic dissipative structures which "self-organized"[1] under the early Archean (3.9-2.9 Ga) solar UV-C photon flux with the thermodynamic objective of dissipating this flux into heat. Although the conception and incubation of the original idea and its development into a mature theory encountered few major conceptual difficulties, it had a difficult birth, being rejected by many of the leading journals devoted to the origin of life before finally receiving an official adoption two years later in the thermodynamically literate journal *Earth System Dynamics* [210] published by the European Union of Geoscientists.

Despite an enthusiastic response from the reviewers and readers [208], the theory has received little attention from the prebiotic chemistry and origin of life communities. However, no logical or conceptual deficiencies have yet been attached to the theory and new evidences and experiments are accumulating in its favor. The lack of attention appears to have more to do with the unfamiliarity of non-equilibrium thermodynamic concepts among biologist and the origin of life community than with their rejection of the theory based on scientific analysis. An important motivation for writing this book has therefore been to provide a lucid and complete description of the theory and specific concepts of non-equilibrium thermodynamic formalism necessary for a critical understanding of the theory.

The overriding motivation for writing this book, however, was that of providing an accessible introduction to the theory for an educated public which has remained fascinated with the problem of the origin of life ever since Charles

[1] The term "self-organized" appears in quotes, because in reality the material of any dissipative structure or process organizes as a result of an externally imposed generalized chemical potential. It is therefore not strictly correct to include the adverb "self" but here, and in the remainder of this book, we use the phrase "self-organized" as it has traditionally been used in the literature dealing with the thermodynamics of dissipative structures, but always keeping in mind that organization is driven by an imposed *external* potential.

Darwin challenged the embarrassingly naive religious prerogative on this fundamental question. A plausible physics and chemistry based theory for the origin of life is long overdue for this educated public who have remained faithful, for over one and a half centuries, to the belief in an eventual scientific resolution of this mystery. It is this educated public that keeps science funded and therefore to which every individual on the globe owes their very existence. Scientists alone are incapable of assuring the survival of their profession or in bringing science to bear on the progress of society. One of the scientists most important labours is therefore to make his/her work accessible to the public and thereby foment an efficient symbiosis with this scientifically literate public to keep science alive and to keep the human species progressing while becoming acutely aware of the ever changing threats of extinction.

The *Thermodynamic Dissipation Theory of the Origin and Evolution of Life* has at its foundation the irrefutable observation that *all* irreversible processes arise, persist, proliferate, and even evolve (in the most general senses of these words) as a response to a well characterized thermodynamic imperative of generating entropy through dissipating an imposed generalized thermodynamic potential. The phrase "generating entropy through dissipating a generalized thermodynamic potential" means to distribute the conserved thermodynamic quantities; mass-energy, momentum, and angular momentum, charge, etc., over an ever larger number of microscopic degrees of freedom inherent in the quantum particle (or string) nature of the material of the cosmos. Contrary to what is commonly believed, the production of entropy is not incidental, but, in fact, the sole reason for the origin, persistence, proliferation and evolution of any irreversible process. Nature attains this entropy production by constructing processes or structures, known as *dissipative structures*, to more effectively distribute these conserved quantities over ever more microscopic degrees of freedom.

Hurricanes, convection cells, winds and currents, and the water cycle are common dissipative structures, otherwise known as self-organized irreversible processes, which arise and persist to dissipate an inhomogeneous energy distribution manifest in temperature differences. For example, in the case of hurricanes, the temperature difference is that resulting between the hot surface of the sea and the cold upper atmosphere. Lightning strikes and electrical currents in general arise and persist to dissipate the electrical potential due to an inhomogeneous charge distribution established across a conductor, and, a stream of water arises to dissipate a gravitational potential difference between different heights on the surface of Earth due to an inhomogeneous mass distribution[2].

[2] This assertion, that the flow of material due to a gravitational potential, arises from a non-homogenous distribution of material seems, at first sight, to be counter to the other examples given in which the flow of material leads to a more homogeneous distribution of the conserved quantity (for example, diffusion leads to a more homogeneous distribution

The *thermodynamic dissipation theory of the origin and evolution of life* proposes that the dissipative process known as *life* arose, proliferates, and evolves, as a response to the thermodynamic imperative of dissipating the photon potential arising from an energy and volume anisotropy of light, in particular that established between the highly directed and high energy photon beam arriving at the surface of the Earth from the hot surface of the sun, and the homogeneous and isotropic spectrum of low energy photons of the cold cosmic background radiation of outer space. The material of Earth's atmosphere and surface exposed to this light has spontaneously self-organized into dissipative structures, one of which is known as "life", to dissipate this photon potential.

Specifically, the theory presented in this book suggests that life, as we know it, began as a material "self-organizing" process of Nature under the driving force of the prevailing Archean solar photon potential, to produce microscopic dissipative structures known as chromophores (pigments) to absorb and dissipate the solar spectrum in the UV-C region (that photon wavelength region of high enough free energy that covalent bonds between atoms can be broken and re-made directly, without requiring complex biosynthetic pathways, but not of high enough energy to cause ionization of the atoms and thereby destroy the molecular structures themselves).

Chapter 1 provides a general historical sketch of the construction of prior paradigms for understanding the origin of life. In chapters 2 to 4 I describe the intimate relation between dissipation and the organization of material, and how the dissipation of photons in pigments is essential to understanding the origin and evolution of life. In chapter 5 I give the evidence for the ubiquity of organic pigments throughout the cosmos, including in interstellar space, and suggest that there exists a non-equilibrium thermodynamic imperative for proliferating chromophores anywhere in the cosmos where there exists a photon flux of relevant energy and material constituted of the organic elements (H, C, N, O, P, S, etc.).

Chapter 6 describes how non-equilibrium thermodynamic principles can explain why, during the Archean when a solar UV-C flux penetrated to Earth's surface, there existed a natural tendency to produce and proliferate UV-C absorbing chromophores and to form symbiotic dissipative relationships among these, arguing that this follows from the autocatalytic nature of these pigments in dissipating the same prevailing UV-C solar photon potential that produced them. It is a corroborating fact that most of the fundamental molecules of life, those common to all three domains of life, from DNA and RNA to amino acids, phos-

of matter). However, in the case of gravity, this has to do with the fact that, contrary to that arising from the other forces, the density of gravitational energy states increases exponentially with energy (or mass) while for the other forces, the density of energy states increases at most as a power law with energy. The explanation is beyond the scope of this book but a manuscript by the author will be forthcomming.

pholipids, vitamins and other enzymatic cofactors, pigments such as carotenoids and porphyrins, etc., all have chromophores which absorb and dissipate in the UV-C region (Michaelian, 2013; Michaelian and Simeonov, 2015)[214, 217].

Chapters 7 through 14 describe the accumulation of life's salient characteristics through thermodynamic dissipation, while chapters 15 through 17 describe how evolution is driven by dissipation. Chapter 16, in particular, describes how almost all biological evolution can be succinctly described as the evolution of the increase in the photon dissipating capacity of Earth's biosphere through, i) evolving ever more efficient chromophores at capturing and dissipating efficiently the solar photon flux, ii) coupling of these chromophores to other molecules to form complexes of ever greater dissipative efficacy, iii) augmenting the wavelength region of dissipation to cover ever more of the solar spectrum, iv) evolving mechanisms, such as mobile insects and other animals, to spread these chromophores, and the essential nutrients required for their production, over ever more of Earth's surface, v) coupling organic irreversible photon dissipative structures and process to other abiotic dissipative processes such as the water cycle, hurricanes, and atmospheric and oceanic currents.

Chapter 18 and 19 compare our proposed theory with prevailing scenarios for the origin of life and suggests how some traditional and cherished paradigms (including the Darwinian) are in need of reform. This latter assertion will provoke severe criticism of our non-equilibrium thermodynamic theory, but this controversy is vital for fomenting paradigm shifts in favor of advancing our understanding of Nature.

The thermodynamic dissipation theory of the origin of life suggests that every twist and turn, every new direction in evolution taken by life on Earth, or on any planet in any part of the cosmos, be it the formation and proliferation of pigments, the association of different molecular pigments in complexes, the encapsulation of these into a lipid vesicle, the conversion of RNA and DNA into information storing molecules, the symbiosis of different components to form new organisms, the invention of oxygenic photosynthesis, the invention of multicellular plants and animals, the formation of a integrated biosphere coupling biotic and abiotic components, and even the space program of intelligent beings with the ultimate goal of terra forming other planets, all happen for one, and only one, fundamental reason which is thermodynamic in character; to increase the global entropy production of the stellar system under the imposed photon potential of the local star.

The proposition that life and evolution are driven by the dissipation of the solar photon potential is not new. In 1886, only 27 years after the publication of "On the Origin of Species", Boltzmann proposed precisely this idea to describe Darwin's observation of the struggle for existence among organisms. What is surprising, however, is that this remarkable insight remained largely ignored for almost a century until Lars Onsager and Ilya Prigogine began to develop an elegant mathematical formalism to treat non-equilibrium situations, and, until 7 years ago, beginning with the publication of the theory presented here in the Cornell ArXiv in 2009 (Michaelian, 2009)[207], Boltzmann's insight had never been considered as providing a fundamental foundation for the construction of a consistent physical theory for the origin and evolution of life.

Viewing the origin of life from this thermodynamic perspective brings the answers to other important questions into sharper focus. Is there a purpose to life and a direction to evolution, or is life just an accident of circumstance, wandering aimlessly through the epochs, as modern biology contends? Was the origin of life on Earth a unique event and are we therefore alone in the universe? The theory presented here emphatically argues the contrary; that the organization of material into dissipative chromophore structures, such as DNA and RNA and other fundamental molecules of life, and their later symbiotic associations and evolution into ever greater dissipative systems, what we consider today as life, under the thermodynamic imperative of dissipating the solar photon flux, should occur within material surrounding almost every second (or younger) generation star emitting in the UV-C wavelength region. These later generation solar systems are made up of the remnants of previous stars which ended their lives as red giants or supernovas and thus have a sufficient proportion of heavy elements, particularly carbon, necessary for building complex pigments and sustaining life as the dissipative process that we know it. Since the great majority of stars in the cosmos are of second (or younger) generation and since most emit at least somewhat in the UV-C wavelength region, the universe should be teaming with life similar to ours. Perhaps, also with life very different from ours, but still recognizable, since it is known that photon dissipation occurs equally well when the same fundamental pigments are in solvents for organics other than water and thus this kind of organic based dissipation is viable under very different physical conditions such as temperature and pressure. This speculation will be developed in some detail in chapter 20.

The thermodynamic perspective presented here implores us to consider as more general forms of living ecosystems all irreversible processes dissipating in whatever photon wavelength region, or, even more generally, dissipating whatever generalized chemical potential, from the photon absorbing chromophores immersed in the liquid methane and ethane lakes and puddles on the surface of Titan fomenting the methane rain cycle, to the great southern and northern vortices on Venus fed by the heat generated by photon dissipation in chromophores floating in the Venusian atmosphere (see figure 0.1), to the heavily pigmented and water transpiring great Redwood forests on Earth.

The directional nature of covalent bonding in carbon based organic molecules, and therefore the practically unlimited number of distinct variations on these, suggests that the favored photon wavelength region for starting life similar to our own would be that of the UV-C and UV-B regions inclusive, where there is enough energy to make and break covalent bonds, but not enough energy to ionize and therefore destroy the molecule.

Chapter 20 considers the evolutionary stages of planetary dissipation that incipient life similar to ours would necessarily go through, from the autocatalytic photochemical formation of pigments during stage 1, to the speculative, but thermodynamically consistent, suggestion of intelligent construction of cosmic worm holes of stage 16. From the consideration of this list of dissipative stages in the evolutionary history of life, it must be emphatically claimed that *we have already discovered extraterrestrial life* at stages one (that of the autocatalytic photochemical formation of pigments) and two (that of the irreversible coupling of dissipation in pigments to abiotic irreversible processes), and probably stage three (that of a dissipative symbiosis of distinct molecular complexes), on a number of planets and moons throughout our solar system.

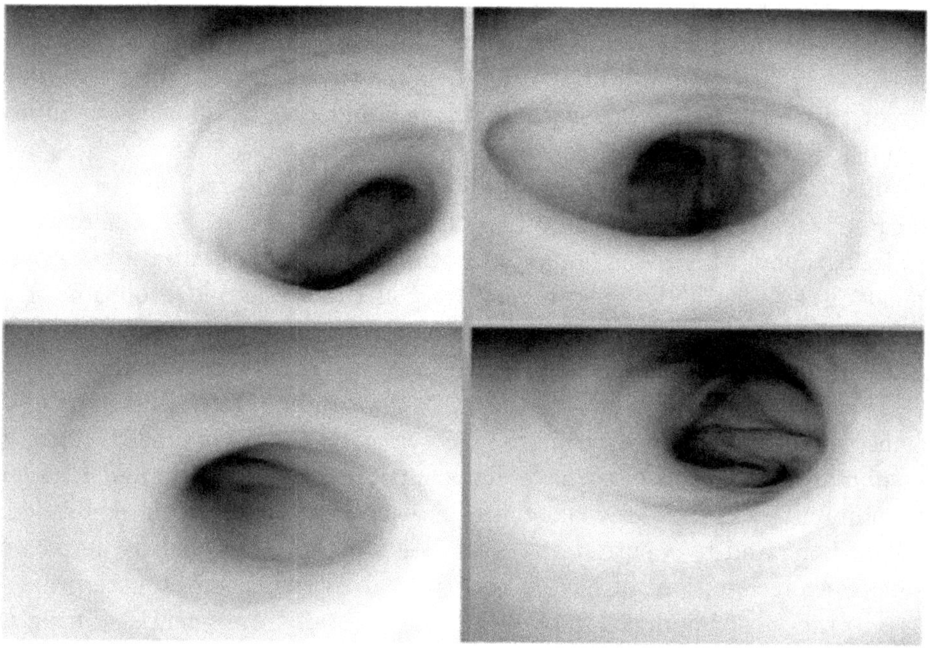

Fig. 0.1. Four views of the great southern vortex on Venus. The vortex consists of a great convection cell fed by the heat generated by photon dissipation in pigments floating in the clouds of the planet. This image was taken in the infrared. The darker regions correspond to the hotter regions in the cloud cover at lower altitude. In the most general conception of life, that of a dissipative system employing molecular pigments at its foundations, the southern vortex must be considered as a manifestation of a living Venusian biosphere. There is also a northern vortex of lesser strength. Similar vorticies also exist at the poles of Saturn. Image credit: ESA, Venus Express.

I believe, in fact, that we are now at the threshold of being able to discover the existence of a great ubiquity of extraterrestrial life at each of the 16 stages listed in chapter 20 having finally understood how material organizes into dissipative structures or processes under a generalized chemical potential and having access to detection technologies which are constantly becoming more sensitive. Chapter 20 of this book also provides thermodynamically grounded speculation on what incredible varieties of dissipative life may await discovery and discusses how we should most effectively design a cosmic search for these.

This book is being published in a rather unconventional form for a number of reasons. The author has been made painfully aware of inherent prejudices in the origin of life and prebiotic chemistry communities, in particular, the resistance of peers to the consideration for publication ideas foreign to their particular traditional and established lines of research[3]. The author is also adamant that the information presented in this book remains accessible to anyone interested in the topic of the origin of life. For these two reasons, this book is being published on a popular web server and can be downloaded in .pdf or EPub format for a minimal cost, or can be ordered in standard paper book print format (see page after title page for the ISBN of different formats). Although this book has not gone through the usual editorial peer review, the published articles on which most of this book is based have, and that should give it sufficient credibility within the scientific establishment.

The author would like to express his admiration and gratitude to all those dedicated scientists that are cited herein, or cited in the articles referenced in the bibliography, whose long hours of careful labour were essential to the development of this thermodynamic perspective on the origin and evolution of life presented here which I first called the "Thermodynamic Theory for the Origin of Life" published in the Cornell ArXiv (Michaelian, 2009)[207], and then as the "Thermodynamic Dissipation Theory of the Origin of Life" published under the auspices of the European Union of Geoscientists (Michaelian, 2011)[210]. Of course, many others not cited have also contributed enormously and I implore the community to forgive my economizing, oversight, or just plain ignorance of these authors, and to ensure that these scientists are given credit where credit is due. Peer recognition, of having contributed to the scientific enterprise, is, after all, the most important motivation that sustains the scientist during the long hours of self-imposed exile from the rest of humanity while attending to the arduous task of unravelling Nature's well kept secrets while at the same time foregoing her delights that abound all around them.

I would like to acknowledge the work of my collaborators Norberto Santillán Padilla and Aleksandar Simeonov who's contributions have, respectively, been

[3] Most of the published papers on which this book is based suffered an exaggerated editorial process of on average two years, most of that time sitting in limbo on the editors desk.

instrumental in the experiments concerning UV-C induced DNA denaturing and in the demonstration that the fundamental molecules of life are pigments in the UV-C and abundant throughout the cosmos. I also thank these two collaborators for their critical review of, and comments and suggestions on, the manuscript. I am indebted as well to my students Julian González, Vasthi Alonso, Noemi Candía, Patricia Jácome, Jessica Gatica, Norberto Santillán, Jorge Arroyo, Zulema Armas, Julian Mejía, Adriana Reyna, Oscar Rodríguez, and Eduardo Cano, whom have taken it upon themselves to understand and consider as plausible the basic premise of this theory and to have taken the dangling threads and helped weave them into a smart fabric. I have no doubt that many of them will have more to say about many aspects of this theory, and about the thermodynamics of life and evolution in general, in the future.

Much of the work presented in this book has been made possible thanks to generous grants from the Direción General de Asuntos del Personal Academico (DGAPA), grants IN118206, IN112809, and IN102316, of the Universidad Nacional Autonoma de México (UNAM). I am also indebted to Jose Leonel Torres of the Universidad Michoacana de San Nicolas de Hidalgo (UMSNH) and to Luis Manuel Gaggero of the Universidad Autonoma del Estado de Morelos (UAEM) for providing a congenial environment for pursuing these themes while on sabbatical leave.

The theory presented in this book and the information presented in support of it, although carefully filtered and consistent with thermodynamic law and physical principles, has yet to be thoroughly verified experimentally and accepted by the scientific community at large. It may be, therefore, that details of the hypothesis may eventually need revision and the author will take it upon himself to make timely the necessary revisions and experimental updates to future editions of this book.

Basic knowledge of thermodynamics and biochemistry would be helpful to the reader, but is not essential. I have endeavored to leave most of the detailed analysis and mathematical formalisms to either boxes which can be skipped without great loss of continuity or understanding, or to the references cited in the bibliography, and instead have relied heavily on figures, graphs, and some analogies to clarify concepts. A glossary defining the technical terms used throughout the book, located at the end of the book just before the bibliography, can also be consulted to help clarify concepts. My hope is that the reader will not struggle with understanding the theory, but instead will enjoy a rapid reading of the book and close it with, not only new insight into one of the most challenging problems of contemporary science, that of explaining the origin and evolution of life in physical-chemical terms, but with a resolve to take an active part in this fascinating new line of non-equilibrium thermodynamic research which, if headed in the right direction, seems to be teaching us once again humility, since we are left with no other choice but to recognize our tiny thermodynamic niche within the incomprehensibly vast arena of a dissipative universe.

Mexico City, April, 2016 Karo Michaelian

The following abstract can be skipped by those without a thermodynamic background. Its contents will become clear after studying the book.

ABSTRACT

Any theory addressing the origin of life must take into full account the fact that life is an irreversible thermodynamic process and, like all irreversible processes, its origin, persistence, and evolution as a "self-organized" system is due to the dissipation of an imposed external generalized chemical potential, i.e., to the production of entropy. Entropy production is not incidental to the process of life, but rather the fundamental reason for its existence. Present day life augments the entropy production of Earth in its solar environment by dissipating ultraviolet and visible photons into heat through organic pigments in water. This heat provides a secondary potential which drives a host of secondary dissipative processes such as the water cycle, ocean and wind currents, hurricanes, the carbon cycle, the sulfur cycle, etc.

If the thermodynamic function of life today is to produce entropy through photon dissipation, then this probably was its function at its very beginnings. It turns out that both RNA and DNA when in a water solvent are very strong absorbers and extremely rapid dissipaters of UV light within a part of the Sun's spectrum that penetrated the prebiotic atmosphere and had enough energy to make and break covalent bonds, the 230–290 nm (UV-C) wavelength region. The amount of ultraviolet light reaching parts of the Earth's surface within this spectral region during the whole of the early Archean (3.9-2.9 Ga) could have been as high as 5 W/m^2, or some 30 orders of magnitude greater than it is today at 260 nm (where RNA and DNA absorb most strongly) due to the lack of oxygen and ozone in the Archean atmosphere.

In fact, not only RNA and DNA, but many fundamental molecules of life (those common to all three domains of life; archea, bacteria, and eukaryote) are also pigments that absorb in the UV-C, and many of these also have chemical affinity to RNA and DNA. Nucleic acids may thus have acted as acceptor quencher molecules to these UV-C-excited antenna pigment donor molecules by providing an ultrafast channel for dissipation through their conical intersections.

It will be shown in this book that there exists a non-linear, non-equilibrium thermodynamic imperative to the abiogenic UV-C photochemical synthesis and proliferation of these pigments over the entire Earth surface due to their autocatalytic activity in augmenting the solar photon dissipation rate. The implication is that the fundamental molecules of life and their complexes were thus microscopic self-organized dissipative structures.

A simple mechanism to explain enzymeless replication of RNA and DNA at the beginnings of life can be given within the same dissipative thermodynamic framework by assuming that life arose when the temperature of the primitive seas had cooled to somewhat below their denaturing temperature. The ratio of $^{18}O/^{16}O$ found in cherts of the Barberton greenstone belt of South Africa indicates that the Earth's surface temperature was around 80 °C at 3.8 Ga, falling to 70 ±15 °C about 3.5 to 3.2 Ga, suggestively close to RNA or DNA denaturing (uncoiling and separation) temperatures. During the night, the surface water temperature would drop below the denaturing temperature and single strand RNA or DNA could act as an extension template for the formation of a complimentary strand, leading to double strand RNA or DNA. During the daylight hours, RNA and DNA would absorb UV-C light and convert this energy directly into heat, thereby raising the local ocean surface temperature enough to allow for denaturing of RNA or DNA. Direct experimental evidence for UV-C light induced denaturing of DNA has now been obtained and will be given in chapter 12.2 of this book.

The reproduction process would have been repeated with each diurnal cycle. This Ultraviolet and Temperature Assisted Replication (UVTAR) mechanism bears similarity to polymerase chain reaction (PCR), a routine laboratory procedure employed to multiply DNA segments, but where the heating and cooling cycle is replaced by a UV-C light cycle. Since denaturation would be most probable in the late afternoon when the Archean sea surface temperature would be highest, and since late afternoon submarine sunlight is somewhat circularly polarized, the homochirality of the organic molecules of life can also be explained within the proposed dissipative thermodynamic framework.

The fact that the aromatic amino acids also absorb strongly in the UV-C region and have been shown to have chemical affinity to their codons, or anti-codons, suggests that they might have originally acted as antenna pigments to increase dissipation and thereby provide more local heat for the UVTAR mechanism operating on RNA and DNA as the sea surface temperature cooled further. Many fundamental molecules of life also have affinity to the grooves in RNA and DNA, making these molecules more hydrophobic and thus propense to stick to the water surface, thereby assuring maximal UV-C exposure. The accumulation of information in RNA or DNA, e.g., coding for the aromatic amino acids or for molecules affecting the hydrophobicity, can thus be related to reproductive success derived from dissipative efficacy under this UVTAR mechanism. This ultraviolet and temperature assisted replication mechanism thereby associates replication and selection with dissipation, and thus evolution with thermodynamics.

Evolution is driven by *thermodynamic selection* of greater global entropy production based on this dissipation-replication relation. The irreversible process of life is coupled with other abiotic dissipative processes and thermodynamic selections acts on the hierarchy of these processes to increase the entropy production of Earth in its solar environment. The Darwinian paradigm of natural selection operating at the individual level is not an accurate representation of thermodynamic selection and its embracement as a theory of evolution has led to irresolvable paradoxes.

It is suggested that the traditional origin of life research, which expects to describe the emergence of life as an fortuitous autocatalytic event which somehow evolved to fall under the dominion of the Darwinian paradigm, without overwhelming reference to the dissipation of an external potential, is erroneous. Traditional evolutionary theory based on an implicit metaphysical "will to survive" of the individual and the tautology of "survival of the survivors" has little explicative value. This becomes particularly apparent when it is realized that no new insights on the origin of life come from applying the Darwinian paradigm to life's incipient molecular stages. The imposed environmental potentials, particularly the solar photon potential, and the entropy production resulting from their dissipation, are crucial to understanding the emergence, proliferation, and evolution of life.

Part I

Introduction

1. The Origin of Life

1.1 Ideas from Antiquity

Nammu, goddess of the primeval sea, the ultimate origin of all things, gave birth to An, god of the sky, and Ki, goddess of Earth. Out of the union of An and Ki all the gods were born. Humans and the animals were created by the gods to serve them and give them sustenance. So began life, according to the Sumerians living in Mesopotamia around 4000 B.C. (Willis, 1993)[408].

For the ancient Greeks of 1000 B.C., the gods of sea, sky and earth retained their prominence, portrayed by the figures of Poseidon, Zeus, and Gaia respectively. Under the Greeks, the gods began to proliferate in number and to take on responsibility for endowing human traits and follies. Athene was the goddess of wisdom, Aphrodite the goddess of love, and Ares the god of war.

Within this uncompromising world of mischievous gods protagonizing lavish myths, the beginnings of scientific enquirery and objective reasoning were brewing. The great library at Alexandria included a work by Aristotel in which he gave what Darwin would, two millennia later, consider as the first written account of natural selection;

> "Therefore, all things together (that is all parts of one whole) happen like as if they were made for the sake of something, these were preserved, having been appropriately constituted by an internal spontaneity; and whatsoever things were not thus constituted, perished, and still perish."(Aristotle, 350 B.C.)[6]

The Romans adopted the Greek gods under new names, with Neptune, Jupiter, and Terra representing the sea, sky and earth respectively. Under the Romans, the gods began to be represented in human form while at the same time were liberated from many of their dishonorable human exploits and moral failings. Religion was becoming more sophisticated but the incipient scientific revolution started by the Greeks was not to last. The ancient library at Alexandria was deliberately burnt to the ground in 391 A.D. and one of its most famous curators and scholars, Hypatia, was brutally murdered in 415 A.D. by an angry mob in a frenzied defense of the mono-theist Christian archbishop Cyril in his

Fig. 1.1. Aristotle (384 – 322 BC) provided the first written account of what Darwin would 2000 years later consider as the first description of natural selection. Image credit: Roman copy after a Greek bronze original by Lysippos from 330 BC. Public Domain.

public dispute over political power with pagan Orestes, governor of Alexandria, who frequently called on Hypatia for her council.

Monotheism, although dating perhaps from the 14th century B.C. in the embodiment of the Sun god Aten of the Egyptians under Pharaoh Akhenaten and later, between the 8th to 6th century B.C., in the embodiment, for example, of figures such as Ahura Mazda of the Indo-Iranian religions, by the fourth century A.D. had spread rapidly throughout the Middle East and Europe, due in large part to its sophistication and efficiency of idea, that of an all powerful deity responsible for, and in control of, everything. A tight hierarchal structure adopted by practitioners of the monotheist religion, with a select few human intermediaries between the *All Mighty* and the people, served the protagonists of the emerging brutal feudal societies in attaining ever greater accumulation of political power and wealth (Engels, 1884)[94]. In this new pragmatic and decisively patriarchal religion, life began when man was created by God, in his image, out of clay for his amusement and vanity, while woman and the animals were created for the utility, amusement and vanity of man[1].

[1] "For a man indeed ought not to cover his head, forasmuch as he is the image and glory of God: but the woman is the glory of the man. For the man is not of the woman; but the woman of the man. Neither was the man created for the woman; but the woman for the man." Bible, 1 Cor 11:3-10.

"Wives, submit yourselves unto your own husbands, as unto the Lord." Bible Eph 5:18, 22-24.

Fig. 1.2. Hypatia (c. AD 370 - 415), philosopher and astronomer who taught at the Platonist school at Alexandria. Daughter of the mathematician Theon Alexandricus (c. AD 335 – c. 405). She was stripped of her clothes and dragged through the streets until death by an angry Christian mob in revenge over her defense of pagan Orestes, governer of Alexandria, in his public dispute for power with the Christian archbishop Cyril. Image credit: Raphael (1483-1520). Public Domain.

After many centuries of intellectual dark ages in Europe and most of the rest of the world, during which continual wars were waged over who's gods were the relevant ones, and within tribes witches and other religious dissidents were regularly burned at the stake (see figure 1.3) or brutally tortured until death by some other innovative mechanism, a scientific renascence emerged beginning in the late 15th century Europe. The renascence threatened the prevailing monotheistic religions by questioning their ideological jurisdiction over the historical truth and over many of Nature's mysteries. Courageous individuals appeared every now and again to question the veracity of the natural philosophy espoused in the "holly" books, usually with dire consequences for the skeptical individual.

During the dark ages, with no light of science to guide the individual, there was extreme social pressure to conform to the prevailing religion and to uphold an institutionalized set of beliefs, no matter how incredible or naive these seemed, since these offered the only hope for salvation from disease and other poorly understood dangers. How comforting it must have felt in these times, for example, to have someone to pray for your safe journey home, or to be convinced that your child's death was not in vain.

The pressure to conform was severe. A rebellious but courageous individual Domenico Scandella (1532–1599), also known as Menocchio, immortalized in the book "The Cheese and the Worms: The Cosmos of a Sixteenth-Century Miller" by Carlo Ginzburg (1980)[106], was a literate miller with the cognitive ability to draw far reaching logical conclusions from the small amount of literature that he had access to in his home town of Montereale, Italy. He very publicly insisted that the soul expired on the death of an individual, that Jesus had a human father and that Mary was definitely not a virgin, that the Pope had no special power given to him from God but merely exemplified the qualities of a good man, and that Christ had not died to "redeem humanity". To those of his village he preached that the only sin was to harm one's neighbor and that to blaspheme caused no harm to anyone but the blasphemer (Ginzburg, 1980)[106]. Eventually, however, Menocchio was betrayed by the local priest. Forsaking the advice of his children and friends, during his inevitable inquisition he criticized the opulent and abusive nature of the Catholic church and its highest practitioners in times of severe generalized poverty. After years of imprisonment and forced testimony during which Menocchio captivated his catholic inquirers, he was "burned at the stake" for his beliefs in 1599.

Fig. 1.3. Until the end of the European inquisitions around the middle of the 18th century, religious dissidents who questioned the veracity of the "holly" books were brutally tortured until death. A common practise was "burning at the stake", considered as a kind of batism by fire. This painting is by Grigoriy Myasoyedov (1835-1911) of the burning at the stake of the rebelious Orthodox protopope Avvakum of Russia. Image credit: Public Domain.

Galileo, another rebellious and courageous individual who insisted on presenting the evidence for the Copernican view of a Sun centered solar system, in obvious contradiction to the holly scriptures which put the Earth decisively at the center of the universe, was tried by the Roman inquisition in 1633 and found guilty but met with a less harsh punishment of public ridicule and house arrest. Throughout these dark ages, the origin of life, particularly human life, remained the strict ideological property of the church, which saw it firmly as the result of God's all mighty power, providence and kindness.

It was not until the publication of "On the Origin of Species" by Charles Darwin in 1859 (Fig. 1.4) that the prevailing dogmatic ignorance, jealously guarded by the western church for over one and a half millennia, finally surrendered some ground to scientific enquirery practised by a small but well educated segment of western society who began to consider the possibility of a material origin of life based on new physical and chemical discoveries. It was also around this time that it became understood that the Earth was much older than the official ecclesiastical institutionalized age of 6000 years decreed by the catholic church based on a "scholarly reading" of the Bible by archbishop James Ussher of Ireland in the seventeenth century. The gradual acceptance of the geological evidence in favor of a much older Earth, along with the evidence accumulated and published by Darwin for the slow but incessant change of species, made evolutionary theory even more attractive to the enlightened European mind which was by now also beginning to reap the benefits of scientific enquirery into almost every sphere of life.

Even Darwin, however, could not free himself from the overbearing censorship of conscience inspired by the church. In "On the Origin of Species" Darwin appears to have been obliged to write, "The Creator originally breathed life into a few forms or into one. Then evolution took over.". However, in 1871, twelve years after publishing these words in his best selling book, in a private letter to his friend Joseph Dalton Hooker, Darwin admitted that life may have had a chemical origin. He wrote that life may have started in a "warm little pond, with all sorts of ammonia and phosphoric salts, lights, heat, electricity, etc. present, so that a protein compound was chemically formed ready to undergo still more complex changes" (Pereto, 2009) [276].

Although Darwin is now generally recognized as the first important investigator and popularizer of the idea of species evolution, notions of the mutability of species through some form of natural selection actually have a history predating Darwin. In the Historical Sketch of the sixth edition of his book *On the Origin of Species*, in a clear effort to quell recriminations by others that they had preceded him in discovering the theory of evolution, Darwin mentions the previous descriptions of species evolution, including Aristotel's given above, and an account derived from his grandfather's, the physician Erasmus Darwin, reflections on the subject. Lamarck's (1744-1829) theory of evolution through the "use and disuse"[2] is also acknowledged and critically reviewed. Recognizing not only these specific predecessors, Darwin goes on to admit that the essence of his

[2] Lamarck, a profesor of botany at the Muséum d'Histoire Naturelle published an influential work *Philosophis Zoologique* (1809) in which he expounded a theory asserting that physical characteristics of animals evolved according to their use or disuse. For example, a giraffe's

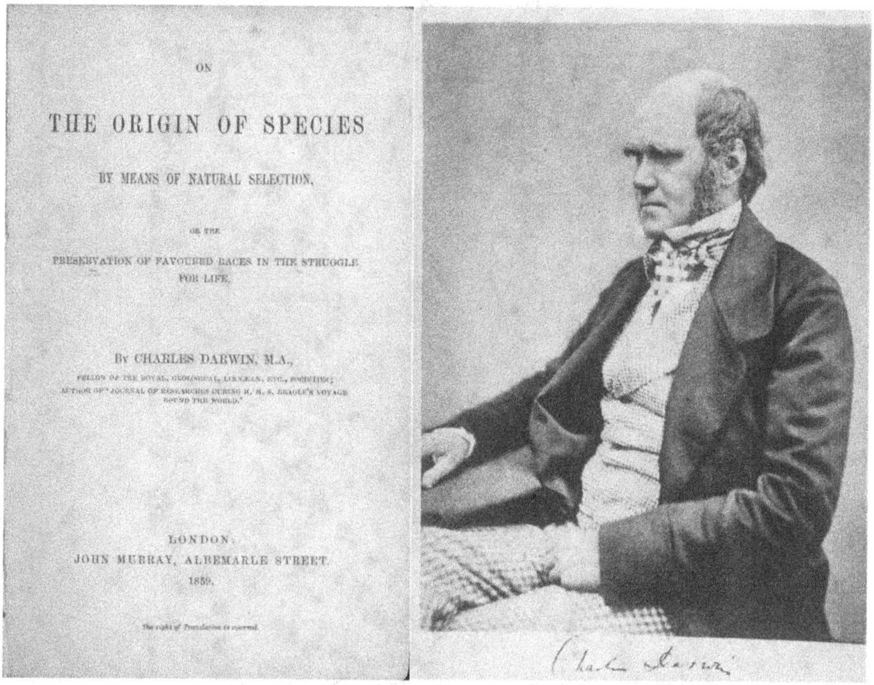

Fig. 1.4. Charles Darwin in a photo taken around 1854 (five years before the publication of On the Origin of Species) and the front page of his book which finally ended the jeausoly guarded perrogative of the church on the origin of life and paved the way from a naive dogmatic view to a scientific and materialistic perspective on life and evolution. Image credit: Public Domain.

theory was already published by Dr. W.C. Wells in 1813, 46 years before the publication of his book, and then again by Patrick Matthew in 1831, as well as in numerous other, less specific, descriptions of evolution[3].

The idea of the mutability of species and evolution was "in the air" so to speak, it only needed a methodical mind to collect and tie together all the loose ends that floated about. Darwin's main contribution was in weaving these facts into a consistent framework, including an explicit description of a mechanism for natural selection, and in the collection of a large amount of empirical data which provided undeniable evidence for evolution. Darwin, in contradistinction to Alfred Wallace (who unwittingly forced Darwin into rapidly publicizing their

neck gradually became elongated as it stretched for grazing on ever higher branches. These aquired traits were assumed to be inherited to future generations.

3 Darwin's recognition of prior theories of evolution through natural selection, left until the 6th edition of his book, appears to have been a direct response to an angry letter written to him by the English mathematician Baden Powell who claimed plagerism of his own writings on the subject of evolution.

Fig. 1.5. Portrait of Erasmus Darwin, grandfather of Charles Darwin, by Joseph Wright of Derby (1792). A natural philosopher, physician, slave-trade abolitionist, inventor and poet. Erasmus's writings on species evolution had an important influence on Charles. Image credit: Public Domain.

similar theories together[4]) believed that humans were no exception and had also evolved through natural selection along with all other species.

The church was vehemently opposed to Darwin's theory at the time of its publication, even in its benign version meant for public consumption in which an omnipotent creator was still required and credited for the initiation of life. As soon as Darwin's book appeared in print, the inevitable attacks began. In one of the most notable public debates arranged to "expose Darwin's folly", Samuel Wilberforce, Archbishop of Cantebury, turned to Thomas Henry Huxley, Dar-

[4] Alfred Russell Wallace (1823-1913), working as a naturalist in Indonesia, sent Darwin a letter asking him for his help in publishing his theory of evolution through natural selection which was identicle in form, but lacking in substance and empirical evidence, to that which Darwin had been working on for 15 years prior but had still not published. After consultation with his academic friends Sir Charles Lyell and Sir Joseph Hooker, Darwin was able to convince Wallace that both works should be presented together at the Linnaean Society Meeting in London on July 1, 1858.

Fig. 1.6. Samuel Wilberforce, Archbishop of Cantebury, on the cover of Vanity Fair February 1869. The Archbishop arranged a debate with Darwin to "expose Darwin's folly". Darwin never attended the debate for failing nerves, but Darwin's colleague and trusted friend Tomas Henry Huxley, later to become known as "Darwin's bulldog", volunteered instead. It was widely reported that after Huxleys reply to Wilberforce, indicating that he would perfer the lineage of the monkey to that of a man of magnificant traits but who would use those traits to ridicule a scientific debate, "at that point in the debate a number of fair ladies in the audience lost their composure and fainted". Image credit: Carlos Pellegrini (1839-1889), Public Domain.

win's appointed representative in the debate[5] and his most faithful supporter, later to be both adversely and affectionately referred to as "Darwin's bulldog", and asked him if he considered himself as a descendant of the monkey from his grandfathers or his grandmothers lineage[6]. Huxley, after an equally eloquent but rigorously scientific rebuttal contested "If I were able to choose as ancestors between a defenseless monkey and a man magnificently blessed by nature, and of great influence, but who used those traits to ridicule a scientific discussion

[5] Wilberforce's well known outspoken and over burdening character provoked anxiety in the timid Darwin and the thought of a confrontation led Darwin to ask his closest friend and scientific ally Huxley to substitute for him in the debate. Darwin did not attend.

[6] Grandparents were most respected in Huxley's days, maximizing the impact of the insult.

and discredit those who sought the truth, I wouldn't hesitate in deciding for the lineage of the monkey."

Fig. 1.7. Thomas Henry Huxley (1825-1895) was both adversely and affectionately referred to as "Darwin's Bulldog". This picture was taken in 1874. Image credit: Unknown photographer, Public Domain.

One hundred and fifty years later, the Catholic church has come to accept evolution through natural selection and now permits its followers to believe in what they may. The church still insists, however, that the human "soul" can not be derived from the material but rather was placed into each individual by the *All Mighty*. In 1996, a scientifically enlightened, but still dogmatic, Karol Józef Wojtyła (Pope John Paul II) wrote;

> "In his encyclical Humani generis (1950), my predecessor Pius XII has already affirmed that there is no conflict between evolution and the doctrine of the faith regarding man and his vocation, provided that we do not lose sight of certain fixed points.... Theories of evolution which, because of the philosophies which inspire them, regard the spirit either as emerging from the forces of living matter, or as a simple epiphenomenon of that matter, are incompatible with the truth about man."

The leadership of the catholic church today has intellectually evolved to the point of Darwin's public posture of 1859, with the acceptance of evolution but still attributes the essence of life to "the Creator". Denying a role for "the Creator" in the origin of life would certainly call into question the very need for belief in an "All Mighty" and would undermine the theoretical foundations of the church. Although, at least for the present time, the unresolved complex scientific questions such as the human "soul", human "free will", and the "origin of life" leave spaces for a "god of the gaps"[7], in western society the teachings of the church on natural philosophy have become almost completely superseded by the enlightenment born of scientific enquirery. However, although many contemporary societies have been liberated from the prerogative of the gods, a difficult battle lies ahead for many others that are still struggling with the recognition of the scientific method as the only viable route to understanding and progress.

Religion is in decline in both eastern and western societies, but science has not, and cannot, make religion obsolete since the majority of humans are still, and will probably always remain stubbornly scientifically illiterate. Humans have a spiritual need to feel connected to the universe by aspiring to understand the fundamental reason for their existence and that of the universe. Religion has simple answers to complex questions to satisfy the intellectually lazy, but beyond that, the institution of religion has the power to unite and to provide a psychological safe haven for humans during difficult times. The gods have become socially ingrained within the collective human psyche and from time to time, when under undue stress (for example, that provoked by the devastating and relentless neo-imperialist wars inflicted by western powers on the developing world, particularly on the Muslim countries of the Middle East), humans return to their deities for consolation, strength, and unity and will take measures to obliterate their own science and history, and even the history of their gods, if these are seen as challenges to their unity under the difficult circumstances.

Over the long run, however, scientific deduction and free enquirery seem destined to displace religious myth and blind faith. Progress can be measured today in scientifically oriented western societies which have evolved from religious feudalism to scientific capitalism or scientific socialism. Today in England, over 50% of male adults declare themselves to be atheist[8]. However, even in these scientifically advanced societies, religion has been conserved as a practical means of coercion and control over the aspirations of a large and often uneducated

[7] The phrase "god of the gaps" can be traced to Henry Drummond, a 19th-century evangelist lecturer, from his Lowell Lectures on The Ascent of Man. He criticized Christians who emphasize aspects of nature that science has yet to explain as evidence for the existence of God, "gaps which they will fill up with God".

[8] Among females in England, only 37% declare themselves as athiests. This gender difference probably has to do with socialized lower levels of interest in science among females and with a socialized mind set of vulnerability and dependence on the male figure.

labouring class while facilitating the endowment of special power and privilege to a certain ambitious few.

This trend towards the consciousness of a knowable, material, and objective reality is not necessarily irreversible as the examples of ancient Greece and contemporary Middle East demonstrate. In the opinion of the present author, the route to a stable and pleasant society in which humans may live in relative peace and liberty, a society worthy of the human intellect, is to ensure that this trend towards objective thinking and the use of the scientific method continues and becomes ever more popular. A plausible, and generally accepted, physical-chemical description of the origin of life will help in this regard, and, to this end, the remainder of this book is dedicated.

1.2 Modern Ideas

The example of the theory of evolution through natural selection mentioned above shows how advances in science, although often being inspired by particular novel insights, do not occur randomly in a vacuum, but instead occur when the scientific environment is ripe for their development. One of the most challenging questions still to be answered by science, and now ripe for the picking, is; exactly how did life arise on Earth?[9] Darwin, avoided publicly stating his beliefs on the origin of life but suggested privately that "life may have had a chemical origin".

However, no detailed theory for the origin of life was ever formulated by Darwin, and essentially no new insights can be obtained from applying his theory of evolution through natural selection to the incipient stages of life. In fact, it will be argued in Chapter 17, that Darwin's framework in general should not be considered as being complete, even concerning the later stages of biological evolution of complex organisms, since it is not based on physical-chemical law, and this fact alone leads to a certain amount of unavoidable tautology and ambiguity which brings to the forefront inherent but completely unfounded assumptions. An example of the tautology is the suggestion of the "survival of the fittest", and "the fittest", because of the complexity and interconnectedness of biotic and abiotic processes, must be clothed in terms of "those that survive" so we are left with the circular argument "survival of the survivors" (Popper, 1978)[287]. An example of a related unfounded assumption in the Darwinian paradigm is the *inherent drive (or will) to survive* of the individual. Without an

[9] There are those that argue that it is more probable that life arose elsewhere in the Universe and was transported to Earth via comets or meteroites. This theory, known as Panspermia, has been promoted extensively by the English astrophysicist Fred Hoyle and the Indian astrobiologist Chandra Wickramasinghe. However, the problem of the origin of life is not solved by this scenario, it is only transported to another location and another, albeit earlier, time (some time as large as the age of the Universe, perhaps 15 Ga). More about this theory will be presented in chapter 19.16.

inherent organismal drive to survive there can be no Darwinian evolution and no continuity of life. Survival of the fittest and the drive to survive must first be founded on physical-chemical law if Darwinian theory is to make sense and we will consider how to do this from within a non-equilibrium thermodynamic framework in chapter 6.

Even though there appears to be no route to addressing the origin of life directly from within the theory of evolution through natural selection itself, Darwin's theory is, however, of fundamental importance since a new theory purporting to be more encompassing would certainly have to be at least consistent with all that already established as being correct.

1.2.1 Oparin

Discounting the metaphysical explanations (such as, for example, spontaneous creation[10]), the first materialistic and scientifically sophisticated theory of the origin of life was proposed in 1924 by the Russian biochemist Aleksandr Ivanovich Oparin (figure 1.8) in his book "The Origin of Life" [258] first published in Russian in 1924 with an English translation in 1936 (Oparin, 1936)[258].

Benefiting from the scientific-materialist school of Frederich Engels (1820-1895) prevailing during his adolescence in Russia, Oparin considered what was suspected in his day concerning the composition of the early Earth's atmosphere (water vapor H_2O, carbon dioxide CO_2, methane CH_4, and ammonia NH_3) and suggested that, under the action of electric discharges from storm clouds and ultraviolet light from the sun, these primordial and common compounds would break up and re-form into the basic compounds of life such as the amino acids. According to Oparin, these compounds would collect and concentrate in shallow pools at the shores of the oceans. Oparin believed that on increasing their concentration, they would somehow become subject to the evolutive paradigm proposed by Darwin in 1859. Making an analogy to the Darwinian mechanism, Oparin speculated that competition for basic organic molecules would lead prebiology to diversify and increases in complexity. However, the most difficult question, that of how the inanimate mixture of chemicals could somehow resolve

[10] An entertaining example of spontaneous creation was that espoused by the renowned French medical practitioner Ambroise Paré (1510 -1590), famous for his numerous battle field implimentations, such as ties to stop hemoraging. Paré asserted that he had dug up a hollow rock in his wineyard which contained a large frog that "could only have been generated spontaneously out of the wreched humidity of the rock" since it was apparently closed on all sides.

It wasn't until 1859 (the same year that "On the Origin of Species" was published) that Louis Pasteur put an end to the spontaneous creation theories by demonstrating, through careful experiments with sterilization, that all life was generated out of some seeds of life (germs) that even floated in the air. His work provided the basis of modern sterilization techniques.

Fig. 1.8. Aleksandr Ivanovich Oparin (1894-1980) suggested in 1924 that the the fundamental molecules of life could have been produced by the action of lightning, ultraviolet light, and temperature on a reducing atmosphere. His predictions were verified by experiments carried out by Stanley Miller and Harold Urey in 1953. Image credit: Unknown, Fair Use.

itself into organisms with a will to survive and that began to compete for resources and thereby become subjected to Darwin's paradigm, or, in other words, how chemical selectivity and "competition" for primordial molecules could be linked with natural selection, was not considered by Oparin.

1.2.2 Schrödinger

Erwin Schrödinger, one of the founders of quantum theory, wrote an interesting book first published in 1944 entitled "What is Life?"[325] based on a series of lectures given under the auspices of the Dublin Institute for Advanced Studies at Trinity College in February, 1943. The purpose of the lectures were, in Schrödinger's own words, "an attempt to address the question, in a way understandable to both the biologist and the physicist, of; How can the events in space and time which take place within the spatial boundary of a living organism be accounted for by physics and chemistry?". Schrödinger's book analyzed much of the experimental work by Max Delbruk and Hermann Joseph Muller concerning the mutation of fruit flies using high temperatures and ionizing radiations. Schrödinger directed attention to Boltzmann's deep insight into how life was "living off entropy production", suggesting that energy flow through organic material could lead to material structuring since the energy flow also included an inherent flow of low entropy. In his book, written ten years before the discovery of the structure of DNA and RNA by Watson and Crick, Schrödinger discusses how genetic material giving rise to inheritance must be of a quasiperiodic nature in order to provide a reservoir for information.

Other than presenting some novel perspectives and reviving Boltzmann's deep insight, Schrödinger's book did not really offer anything new, but it was however very important in stimulating the interest of many physicists as it led them to consider the origin of life as a viable topic for study under physical and chemical law. One of those physicists who openly acknowledged having been inspired by Schrödinger's book and who would go on to discover the structure of DNA was Francis Crick. Before Schrödinger's book, life was considered as a subject too complex to be treatable at the level of fundamental law. Eminent scientists, such as Eugene Wigner, even suggested that the known laws of physics and chemistry were insufficient for describing life and that new laws would eventually have to be found.

Fig. 1.9. Erwin Rudolf Josef Alexander Schrödinger (1887-1961). One of the founders of quantum mechanics and writer of an engaging book entitled "What is Life" which inspired many physicists to take up the study of life from a chemical-physical basis. Image credit: Nobel foundation, Public Domain.

1.2.3 Urey and Miller

In the early 1950's, the American biochemist Harold Urey and his doctoral student Stanley Miller decided to put to experimental test the theory of Oparin.

With an arrangement consisting of a spherical glass container filled with a mix of what was considered during Millers day as the best hypothesis for the early atmosphere, not differing much from the assumption of Oparin, but now containing some hydrogen sulfide H_2S gas which would have been available from frequent volcanic eruptions on early Earth, Miller subjected the mixture to continuous electric discharges of 60,000 V. In less than a week, Miller was identifying within his glass sphere amino acids, formyl, urea, acetic acid, hydrocyanic acid, and even more complex organic molecules such as sugars, alcohols and lipids (Miller, 1955)[223]. Somewhat later, Miller showed that the mechanism for amino acid formation in his experiments was the so called Strecker condensation in which HCN and certain aldehydes formed by the electrical discharges on his mixture of primitive gases could condense out together with some ammonia to form a large set of amino acids (some of which were from the 22 used by life, but others were only found in carbonaceous chondrite meteorites). Miller's results provided strong evidence in support of Oparin's proposition, thereby providing the foundations for an unprecedented number of experimental and theoretical investigations into the origin of life, a line of research that has since come to be known as *prebiotic chemistry*.

Fig. 1.10. Stanley Miller (1930-2007), co-originator, along with his doctoral thesis advisor Harold Urey, of prebiotic chemistry, next to the apparatus in which fundamental molecules of life were obtained from electrical discharges (which provided free energy in both the UV light given off and in the pressure fronts produced) on primordial gases. Image credit: NASA, Public Domain.

The Miller experiments have been criticized on the grounds that Earth's atmosphere during the early Archean was probably not as reducing (containing as much hydrogen) as simulated by Miller. In similar experiments with a more neutral atmosphere, it was shown that the production of amino acids was much less efficient. However, more recent experiments performed by Cleaves et al. (2008)[50] in neutral atmospheres containing oxidation inhibitors, such as ferrous iron, the yield of amino acids greatly increases.

Investigations into prebiotic chemistry, however, until relatively recently, were restricted to near equilibrium conditions (except for some use of periodic electrical discharges or UV light to provide sufficient energy for breaking and making covalent bonds or for favorably biasing important reaction pathways (Powner et al., 2009)[288]), and even though the fundamental molecules of life have appeared from time to time with small concentrations in these experiments, it is still not at all clear how a dynamical processes such as life could spontaneously emerge out of a free energy minimizing isolated system of chemical reactions. These systems always move towards the thermodynamic equilibrium state in which state the large Gibb's free energy required for formation of the complex molecules of life give them very low probabilities of existence[11].

Life, as we now know it, however, is a dynamic, non-equilibrium process which uses DNA and RNA to code for and produce proteins which are the basic units that determine life's physical and chemical characteristics. Particular proteins called "enzymes" are needed to denature the two strands of the DNA double helix, to hold the strands apart long enough for the process of extension, to synthesize new DNA nucleobases, and to locate and fix these bases on the separated single strands. Still other "proof reading" enzymes correct most errors along the new complimentary DNA strand. Without these proteins, DNA cannot be faithfully synthesized or copied, but without DNA, these proteins cannot be manufactured because the information for their complex structure must come from DNA. This enduring paradox of "What came first, the protein or the DNA?" or "the chicken or the egg" most succinctly summarizes the impasse that scientific enquirery had arrived at in trying to understand the origin of life in the decade after the discovery of the structure of RNA and DNA.

[11] The Gibb's free energy, defined by $G = E + PV - TS$, (where E is the internal energy, P the pressure, V the volume, T the temperature and S the entropy) is the quantity that Nature minimizes for an isolated system under conditions of constant temperature and pressure (normal laboratory conditions). In equilibrium thermodynamics, the probability of finding a given molecular structure depends exponentially on its Gibb's free energy. The lower the Gibb's free energy, the more probable the molecule. Non-equilibrium thermodynamics in the non-linear regime offers a solution to the problem of predicted low probabilities for complex molecules. If the molecule acts as a catalyst for its own production, or for the dissipation of a coupled thermodynamic potential, then its concentration may grow to values many orders of magnitude greater than those predicted by equilibrium thermodynamics. This will be described in detail in chapter 6.

1.2.4 RNA World

An avenue which offered a potential resolution to this dilemma was found in the early 1960's when Carl Woese, Francis Crick (codiscoverer of the structure of DNA), and Leslie Orgel showed that, although less stable and easily degraded at high temperatures, RNA was easier to synthesize than DNA, and that RNA could act as an enzyme to catalyze its own reproduction. So began a fruitful line of research which was to become known as the "RNA World". Although many interesting discoveries were made that will undoubtedly help in the final resolution of the problem of the origin of life, in 1997, after almost 40 years of experimenting within the RNA World, Leslie Orgel admitted, "After years of trying, however, we have been unable to achieve the second step of replication, copying of a complementary strand to yield a duplicate of the first."

Work has continued however within the RNA World and although it is no longer considered the panacea it once was by prebiotic chemists (Bernhardt, 2012)[19], Orgel (2004) [260] is now cautiously optimistic with regard to the abiogenic synthesis of RNA, suggesting that other, undiscovered, routes to these molecules may eventually be found. The most difficult current problems, according to Orgel, with abiogenesis are; (1) the production and stability of ribose competing with other more easily synthesized and more stable sugars, (2) the difficulty of the polymerization of nucleotides to form polynucleotides, (3) the problem of the racemic mixture of chiral nucleotides frustrating the template-directed copying of polynucleotides and, perhaps the most difficult, (4) the replication of RNA without the assistance of enzymes.

Some of the problems related with the RNA World proposal, and why the non-equilibrium thermodynamic perspective presented here suggests instead an RNA + DNA + Fundamental Molecules World are discussed in chapters 9 and 19.4.

1.2.5 Panspermia

One of the most important questions, the answer to which would help significantly in determining how life arose, is that of where, or more particularly, in what environment, the fundamental molecules of life were produced. Oparin, Miller and others suggested that photochemical reactions in Earth's atmosphere were the primary source of the organic material which contributed to the primordial soup. However, posterior indications from surviving sediments of the era suggest that the atmosphere was less reducing than simulated in the Miller experiments. A reducing atmosphere is basically one in which hydrogen is present in quantities and can donate its electron to a shared covalent bond. In less reducing atmospheres, it is much more difficult, but not impossible, to make the fundamental molecules.

Hoyle and Wickramesinghe and other Panspermia proponents, and many astrobiologists not connected with Panspermia, have suggested that the original organic material was produced in space and delivered to Earth via comet and asteroid collisions (see chapter 19.16 for more details). Their argument is based on the fact that fly-by missions to comets have demonstrated large amounts of organics on their surface and analysis of meteorites also indicates that a large amount of organic material must be available in space. It is also a fact that early Earth was impacted by comets and meteorites at a much higher rate than that of today until at least the end of the period known as the "late lunar bombardment era[12]" (~ 3.9 Ga).

Hoyle and Wickramasinghe have further proposed the possibility that bacteria-like organisms could actually be alive and be growing on comets, and that a meteoritic impact could free rock containing bacteria from one planet and transport them to another. Although it is most probable that some of the original organic material was delivered to Earth via comets or asteroids, and that somewhere in the universe life has been transported from one planet to another, our proposed autocatalytic production mechanism for the fundamental molecules of life (pigments in the UV-C) based on dissipation, which I present in chapter 6, would have been a mechanism sufficient to produce large quantities of such molecules in the atmosphere or on the ocean surface of the Archean Earth, without the need to resort to extraterrestrial seeding.

1.2.6 Deep Sea Hydrothermal Vent, or an Ocean surface Origin?

More recently, it has been suggested that submarine hydrothermal vents would have been ideal nurseries for incipient life since in this case early life would have been protected by water from the supposedly dangerous ultraviolet light arriving at the early Earth's surface. It is argued that incipient life could have used the reducing conditions and the free energy available in temperature and chemical gradients at these vents to produce the necessary fundamental molecules. However, there has been little experimental confirmation of routes to any of the fundamental molecules of life through this form of high pressure and high temperature chemistry. This proposal is, however, consistent with a generally accepted thermophilic, or hyperthermophilic, origin of life (see chapter 19.11 and figure 19.3) as determined from a comparative analysis of the genomes from the three domains of life; the archea, bacteria, and eukaryote (Schwartz and Lineweaver, 2004)[329].

Stanley Miller and Jeffrey Bada, however, have performed experiments on the stability of the fundamental molecules of life at high temperatures and

[12] The "late lunar bombardment era" refers to the observation that there was a last important period of cratering of the moon (and thus undoubtedly also of Earth) corresponding to intense asteroid or coment bombardment, which occured at approximately 3.9 Ga.

this has led them to conclude that hydrothermal vents are places of molecular destruction rather than molecular production, since all sea water is recycled through hydrothermal vents at temperatures of over 350 °C every 10 million years (Miller and Bada, 1988)[225]. Miller further suggests that this high temperature recycling would also argue against the idea of gradual accumulation of organic material through comet or asteroid delivery.

From the thermodynamic perspective of life as a dissipative process, it also does not make much sense to attach the origin of life to such an insignificant free energy source as hydrothermal vents compared to the free energy which would have been available in sunlight at the Earth's surface. In this sense, probability would favor an atmospheric or surface abiotic synthesis of the primordial molecules involved in the origin of life. Today, even with very few high energy photons available to make and break covalent bonds, the ocean surface is known to be an important biomolecular factory. The most probable location for the origin of the fundamental molecules of life, and of life itself, would thus appear to be the ocean surface.

There are other reasons besides ample free energy available for dissipation that favor an ocean surface origin. Molecules adhering to the air-water interface would have been spared from the cycling through submarine hydrothermal vents and the destruction that Miller referred to. The ocean surface is subjected to large diurnal (day/night) variations of its physical characteristics which could have helped foment the incipient molecular reproduction processes utilizing UV-C light (see chapter 12). The ocean surface is also a region of high concentration of metal ions and other catalyst cofactors. Furthermore, at the surface there is an important component of circularly polarized submarine light which may have been responsible for engineering the homochirality into life (see chapter 14). More evidence for the ocean surface being the cradle of life will be given in chapter 10.

During the past 60 years since the first experiments of Miller, much understanding of the abiotic synthesis of the fundamental molecules has been obtained, such that there is now a consensus that all of the fundamental molecules of life could have been produced abiotically on Earth, or in the local space environment of Earth and delivered to Earth via comets or asteroids. However, little progress has been made on understanding how the "vitality" of life arose, that is, the kinetic or dynamical aspects of life such as reproduction, proliferation and evolution. The main obstacles to advancement, in the estimation of the present author, has been the scant appreciation of the non-equilibrium nature of life, the fact that life is inherently dependent on dissipating an external potential, and also the failure to recognize the incompleteness of the Darwinian paradigm. The deficiencies of this paradigm become particularly evident when one attempts to apply it at the molecular level. Unless the implicit drive (will) to survive of living organisms can be understood in physical-chemical terms, and understood also at the microscopic level of the fundamental molecules of life, there can be no honest scientific licence for accepting as complete, or even correct, the Darwinian paradigm. It is the intention of this book to show that life can be described as *microscopic* dissipative structuring leading to the fundamental molecules (or pigments) and their support structures, such as the cell and all other complexifications up to the biosphere, and that this perspective leads to a much more profound understanding of life and evolution based on physical and chemical principles.

1.2.7 Gaia Theory

In support of our non-equilibrium thermodynamic perspective, an important advance in understanding the origin of life comes from the realization of just how strongly living systems interact with, and alter, their environment. Pioneering work in this regard was carried out by James Lovelock through validation studies of his theory of Gaia (named after the Greek goddess of the Earth). The revelation of strong and complex interactions of living organisms with their physical environment led Lovelock to the suggest that the entire Earth was one great living organism, capable of auto-regulation of its physical attributes such as; the global temperature, the amount of salinity in the oceans and the oxygen content of the atmosphere, as well as many other attributes, all in the supposed interest of increasing the suitability of the environment for life.

During the early years of Gaia, the early 1970's, the theory was unfairly discredited by neo-Darwinists who insisted on defending the deficient Darwinian paradigm and who maintained a very narrow view of the nature of life (see, for example, Richard Dawkins criticism of Lovelock's ideas in chapter 17.4 and (Dawkins, 1989)[72] as well as Lovelock's response in *Gaia: Medicine for an ailing planet* [192]). Although it is now generally accepted that Lovelock's whole Earth organism Gaia may be more metaphoric than was originally conceived, there is no question that the strong interaction between the biotic and the abiotic, as evidenced in the data collected for establishing the theory of Gaia, is an accurate description of Nature.

Gaia has awakened life scientists and system scientists to the extraordinary influence and interdependence of life on the physical environment of Earth. Once ridiculed, and his articles shunned from mainstream journals, Lovelock's talks at invited conferences are now preceded, as well as postceded, by standing ovations.

In our interpretation, the one great "living" organism of Gaia can be identified with the great non-equilibrium dissipative structure, involving the coupling of both biotic and abiotic dissipative structures and processes, known as the *biosphere*. The biosphere is "living" off solar photon dissipation, but it is the photon dissipation which takes precedence in Nature, not the particular organisms nor the particular nature of the biosphere.

On Earth, however, thermodynamic evolution (see following chapter) has fashioned the biosphere into a very efficient and stable dissipater with apparently few distinct competitive alternatives given Earth's initial conditions. For many thousands of millions of years, Nature has maintained the biosphere as the relevant dissipating system on Earth, even against strong external perturbation (such as a 30% increase in solar luminosity since the beginnings of life – see chapter 4.1). Many of the mechanisms through which Nature has maintained and enhanced the stability of the biosphere were discovered by Lovelock and collaborators during the elaboration of his theory of Gaia, and these mechanisms will be shown in chapter 17.4 to have led to evolutionary increases in global dissipation. In this respect, the work of Lovelock and collaborators should be considered as pivotal in laying the empirical foundations for a chemical-physical description of the origin and evolution of life.

Fig. 1.11. James Lovelock next to a statue of Gaia, the Greek godess of Earth. "Life does more than adapt to Earth. It changes Earth to its own purposes." Image credit: Bruno Comby, CC 1.0.

The theory of Gaia, of the strong interaction between biotic and abiotic systems or processes, together with the thermodynamic dissipation theory of life presented here, suggests that a *thermodynamic selection* based on entropy production through photon dissipation (to be detailed in Part II, but specifically in chapter 17.3) permeates all levels of the biosphere in a set of nested hierarchies from the fundamental molecules all the way up to the highest level, the biosphere itself. In chapter 17 we take a look at the global repercussions of the dissipative theory for living systems at the higher hierarchal levels of the ecosystem and the biosphere, suggesting a thermodynamic dissipative explanation for the stability of the biosphere and propose a new indicator of ecosystem health for assessing endangered ecosystems based on photon dissipation efficacy (see chapter 17.5). Furthermore, viewing these higher levels of life, together with life at the molecular level, from this thermodynamic perspective as dissipative systems, allows a clear identification of the deficiency in the Darwinian paradigm and suggests just how this deficiency should be remedied in order to provide a more complete physical-chemical theory of life and evolution.

Chapter 17.4 considers how the evidence collected in establishing the theory of Gaia supports the basic premise of this book; that the origin of life and its subsequent evolution is, in essence, the microscopic dissipative structuring and proliferation of ever more efficient and widely spread organic pigments which carry out the non-equilibrium thermodynamic imperative of dissipating the externally impressed solar photon potential.

1.2.8 Autocatalytic Chemial Reactions

Autocatalytic chemical reactions are known to play fundamental roles in the chemistry of life (Nicolis and Prigogine, 1977)[248]. Such reactions have also been suggested to have been necessary at the origin of life because autocatalytic reactions provide exponential amplification even at the molecular level and such amplification is one of the characteristics of life, for example as depicted in the logistic equation describing the Malthusian population catastrophe[13] which greatly influenced Darwin.

Orgel (1995)[259] has reviewed some of the experimental autocatalytic chemical reactions pertaining to RNA and DNA single strands acting as templates for the construction of a similar strand, a route to replication first realized by Watson and Crick shortly after their discovery of the DNA structure. Perhaps the best known autocatalytic chemical reaction involving nucleic acids is the polymerase chain reaction (PCR – see figure 1.12) discovered by Mullis (1990)[239] and now a routine and indispensable laboratory procedure.

[13] Thomas Robert Malthus (1766-1834) wrote an influential paper entitled "An Essay on the Principle of Population" in which he suggested that the world's population, under ideal conditions, would increase without bounds and would thus neccesarily be brought under control by wars, famines and epidemics. Darwin appealed to the Malthusian hypothesis of population increase without bounds and the ensuing catastrophe in order to explain the struggle for existence and natural selection. However, there is nothing in the Darwinian paradigm that explains the proliferation of a population under ideal conditions from a fundamental physical-chemical basis. This is one of the important deficiencies of the Darwinian paradigm.

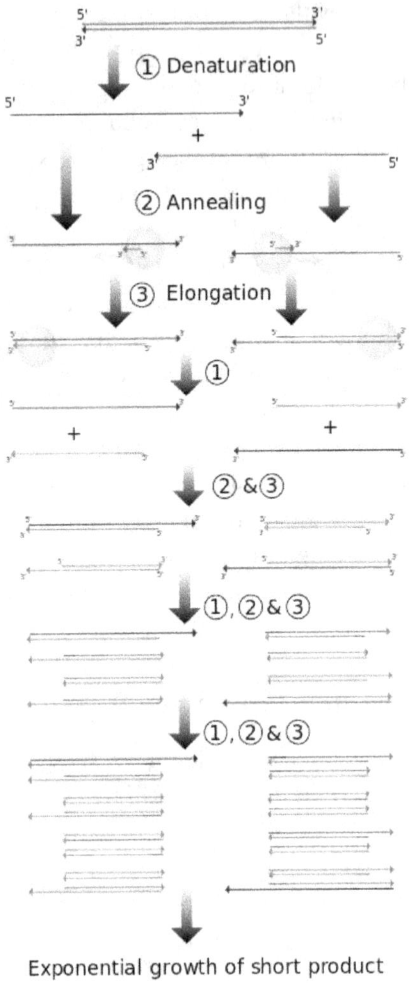

Fig. 1.12. Polymerase chain reaction (PCR). A laboratory process discovered by Mullis (1990)[239] to amplify exponentially an original fragment of DNA or RNA using activated nucleotides, primers, the enzyme polymerase, and temperature cycling. The fragment to be amplified is delineated by oligos known as primers which give polymerase (spheres in diagram) a starting point for extension. Amplification is exponential in the desired fragment, but only arithmetic in the original DNA. This is an example of an autocatalytic template directed process which is common in life today and most probably was also comon at the dawn of life. Image credit: Madprime, Public Domain.

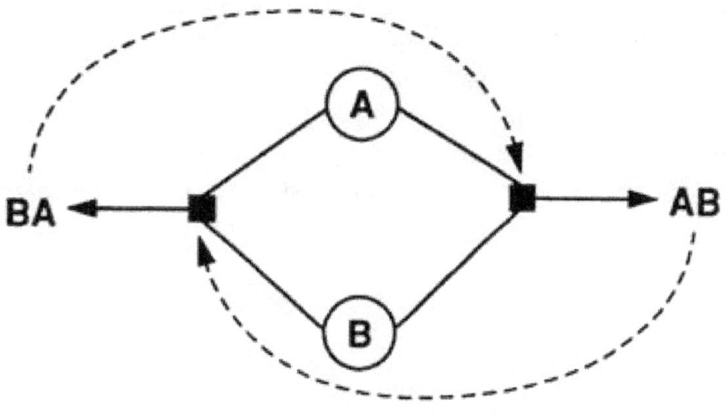

Fig. 1.13. Kauffman suggested that the origin of life 3.85 billion years ago could have arisen as a set of autocatalytic chemical reactions in which the product of one reaction could have acted as the catalyst for another, and so on in a closed cirlce. However, such chemical perpetual motion theories, without explicit consideration of the dissipation of an external generalized chemical potential, are not viable and are now considered as defunct theories.

The biochemist Stuart Kauffman developed autocatalytic theory for the origin of life further, suggesting that a set of imbedded autocatalytic chemical reactions involving enzymes (biological catalysts) could be what bootstrapped life to vitality from inorganic substances (Kauffman, 1986; 1993)[152, 153]. Wächtershäuser (1988)[388] and later Morowitz et al. (2000)[235] proposed an autocatalytic reductive citric acid cycle as a possible mechanism for producing molecules such as sugars, lipids and amino acids on the early Earth. However, the suggestion of a primordial autocatalytic chemical reaction has so far not enjoyed empirical confirmation because, in the opinion of the present author, the relevant dissipation of an imposed potential has not hitherto been correctly identified nor emphasized as the driving force behind the kinetics. Any autocatalytic scheme, without explicit consideration of the dissipation of an imposed generalized chemical potential, an external driving force, would be the chemical equivalent of a perpetual motion machine and therefore non-viable (see figure 1.13).

In chapter 6 we take up the search once again for an autocatalytic reaction scheme, but this time for an autocatalytic *photochemical* reaction which feeds off the dissipation of the solar photon potential. Photochemistry has a number of advantages over normal thermal chemistry, in particular, a much greater amount of energy can be delivered by the photon and channeled into a very free energy intensive reaction such as the breaking and reformation of covalent bonds. Photochemistry also provides many more distinct reaction pathways than thermally induced reactions, and this will be discussed in chapter 6. Finally, the photon potential is, and always has been, the most intense external potential available on Earth and it makes sense that if the thermodynamic function of life today is to dissipate photons, then life would have very early, if not since its beginnings, latched on to this most important potential. In chapter 6, I give one such very specific example of an autocatalytic photochemical reaction leading to the exponential enzymeless replication of specific oligonucleotides.

Part II

Thermodynamic Foundations of the Origin of Life

2. Dissipation; A New Paradigm for the Origin and Evolution of Life

Non-equilibrium thermodynamic formalism makes it clear that all irreversible processes arise and persist to produce entropy, to dissipate a generalized thermodynamic potential. Boltzmann realized this only 27 years after the publication of Darwin's theory of evolution through natural selection, but Boltzmann's insight has never been well understood and therefore unappreciated by those in the biological, evolutionary, and prebiotic chemistry sciences. Without the consideration of such an impressed thermodynamic potential, the system may "organize", for example through the minimization of some system inherent thermodynamic potential such as the Gibb's free energy, as in the formation of crystals, but there can be no *dynamic* stationary states such as those which characterize the vitality of life. The system will eventually reach thermodynamic equilibrium in which all internal flows vanish and the system comes to a dynamic "death" at maximum entropy, known as the equilibrium state.

Non-equilibrium thermodynamics, however, provides the appropriate answer; given an imposed external potential, the system may self-organize into what Ilya Prigogine (1967)[289] termed "dissipative structures" with an inherent internal dynamics that evolves towards increases in the rate of dissipation of this externally impressed potential. The external potential is defined by the environment and it simply becomes meaningless to study the dynamic process, which we call "life", separated from its environment. This will be discussed in detail in chapter 4 on the organization of material through dissipation.

Contemporary conceptions of the origin of life see it as a rather extraordinary and unique event in which a fortuitous distribution of concentrated reactants somehow led to spontaneous autocatalytic replication which endowed, rather mystically, the original complex of these molecules with a kind of inherent "drive (will) to survive and propagate" (for example, Richard Dawkins' not so metaphorically conceived "selfish gene" - Dawkins (1976)[71]). However, the incredulity of such an expectation becomes apparent when it is realized that this sort of spontaneous dynamic organization is never observed isolated in Nature. All dynamic processes associated with the structuring of material that we are familiar with (called irreversible processes in thermodynamic language because they occur in a preferred direction in time) from water currents and winds, to hurricanes, convection cells, and the water cycle, are always associated with the dissipation of an externally imposed generalized potential (e.g. a temperature gradient, a concentration gradient, a chemical potential, an electrical potential, or a gravitational potential). For example, a tropical storm does not spontaneously arise and evolve into a hurricane under its own "inherent will to originate and survive". A storm originates and develops into a hurricane under the thermodynamic imperative of dissipating the external temperature gradient imposed over the system by the warm ocean surface and the cold upper atmosphere (see Fig. 2.1). The atmospheric gases of which the storm is composed are driven to dynamic structuring by this external potential of the temperature gradient. The material of the storm organizes dynamically and apparently "spontaneously" in this manner because doing so the material of the storm system is more efficient at dissipating the external temperature gradient. Macroscopic and dynamic structuring of material, leading to a breaking of symmetry of the system in both space and time, occurs in response to macroscopic dissipation of, in the case of the hurricane, an energy gradient, but more generally, a gradient of any conserved quantity (mass-energy, momentum, angular momentum, charge, etc.[1]).

Another well known example of macroscopic structuring of material in response to the dissipation of a generalized chemical potential is that discovered by Borris Belousov in the early 1950's, and later analyzed in more detail by Anatol Zhabotinsky. Known as Belousov-Zhabotinsky reactions, they consist of a system in which two irreversible processes take place; chemical reactions and diffusion of the compounds. These systems arrange themselves (self-organize) in space and time in order to dissipate more efficiently the two involved generalized chemical potentials driving the two irreversible processes; the imposed chemical affinities of the reactions (obtained from the chemical potentials of the

[1] The "etc." here refers to other conserved variables due to other, lesser-known, symmetries in Nature. One example is guage invariance that leads to conservation of particle number. Another symmetry of Nature leads to the Noether charge which arises from a diffeomorphism invariance. For example, it is now understood that a black-hole has an entropy equal to one quarter the surface area of its event horizon. If it has an entropy, then it must also have a temperature and this lead Stephen Hawking to propose that black-holes emitt particles with a black-body spectrum. These ideas are, in fact, much more general than the results obtained from Einstein's field equations for gravity. The thermodynamic parameters derrived in more general theories can be shown to depend only on the metric and not on the field equations which led to the metric. One must, however, generalize the concept of entropy in these models (which may no longer be one quarter of the area of the event horizon). This could be done by using the first law of black hole dynamics, $TdS = dM$. Since the temperature is known, this equation can be integrated to determine S. Padmanabhan (2010)[266] explains how this was done by Wald (1993)[391] and it turns out that the entropy can be related to a conserved charge called Noether charge which arises from the diffeomorphism invariance of the theory (Padmanabhan, 2010). Dissipation of the Noether charge over ever more microscopic degrees of freedom is also a source of entropy production and similarly leads to dissipative structuring of material.

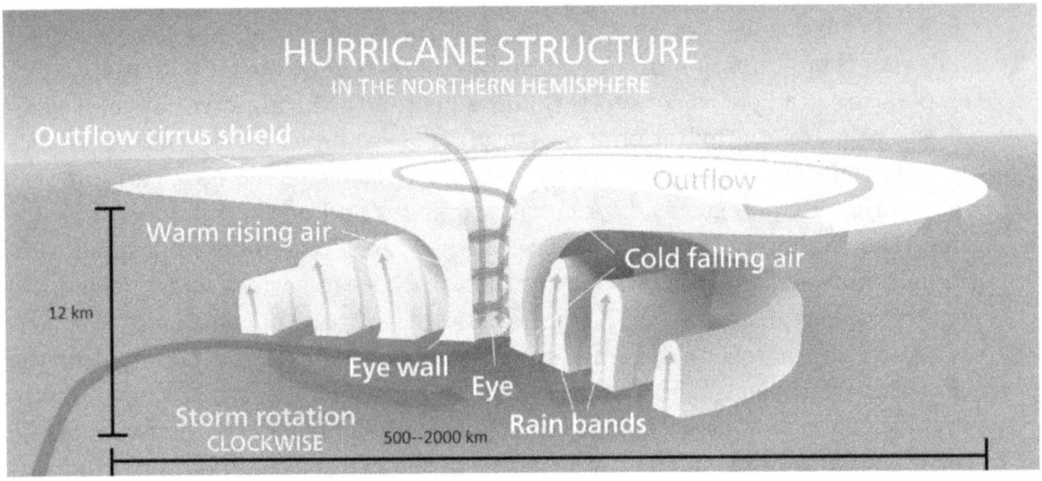

Fig. 2.1. A hurricane originates, persists, and evolves as a non-equilibrium thermodynamic response to dissipating the imposed external temperature gradient over the atmospheric gasses. This dissipation is, in essence, the self-organization of material in such a manner so as to equilibriate the kinetic energy distributions of the molecules in the warm ocean surface with those in the cold upper atmosphere. Image credit: Kelvinsong, Public Domain.

reactants and products) and the concentration gradients of these. The dissipative structures which arise in these systems are particular oscillating patterns of concentration gradients of the reactants and products which not only oscillate in space, but also in time, making the system appear to be alive (Fig. 2.2).

In distinction to *macroscopic* structuring, *microscopic* structuring of material can also occur in response to the dissipation of a generalized chemical potential, but the relevant mechanisms appear to be different from those for macroscopic structuring. Examples of microscopic structuring through dissipation, such as molecules produced through photochemical reactions, will be given in section 3.4. Microscopic dissipative structuring is particularly relevant for life. In particular, it will be suggested that the organic pigments (including the fundamental molecules of life) self-organize and proliferate under an impressed photon potential. In this view, the origin of life was driven by microscopic dissipative structuring of material under the prevailing Archean solar photon potential.

In the case of life, the apparent spontaneity of its origin and its continued persistence, and even its evolution, are overwhelmingly in response to the thermodynamic imperative of dissipating the external *photon potential*; that potential arising from the directed short wavelength solar photon spectrum incident on Earth (similar to a black-body spectrum at a temperature of the surface of the sun, 5800 K) and that of the isotropic, long wavelength, cosmic microwave background spectrum of outer space at a black body temperature of 2.7 K. It is the dissipation of the solar photons which drove the microstructuring of

Fig. 2.2. Turing patterns of concentration gradients of reactants and products formed in a Beluosov-Zhabotinsky system of chemial reactions with diffusion. The patterns oscillate in space and time giving the appearance of a living system. This breaking of space and time symmetry is a non-equilibrium themodynamic result and is driven by the thermodynamic principle of optimizing the rate of entropy production (in this case dissipating a chemical potential and a gradient of concentration). Image credit: Public Domain.

the pigment molecules, directly through UV-C light in the Archean (see figure 2.3), and still does today, although indirectly through visible light. These microscopic dissipative processes are, in turn, at the foundations of the global dissipative process known as the biosphere. More details about the autocatalytic photochemical reaction networks involved in the production of the fundamental molecules of life and other pigments, and the non-equilibrium thermodynamics of this microscopic dissipative structuring process will be given in chapters 6 and 12.

The heat provided by the dissipation of photons in pigments provides a secondary generalized chemical potential for driving other biosphere processes, from the denaturing double helix strands of DNA during the Archean thereby allowing replication (see chapter 12), to the water cycle, hurricanes, and atmospheric and oceanic currents of today. These processes in turn provide a generalized chemical potential for still further processes, such as human planting of crops, etc. A hierarchal nested set of dissipative processes therefore exists, which, as a whole, is known as the "biosphere" (see figure 2.4). There are many positive feedbacks among the different irreversible processes, and among the different hierarchal loops, which increase the overall rate of global entropy production,

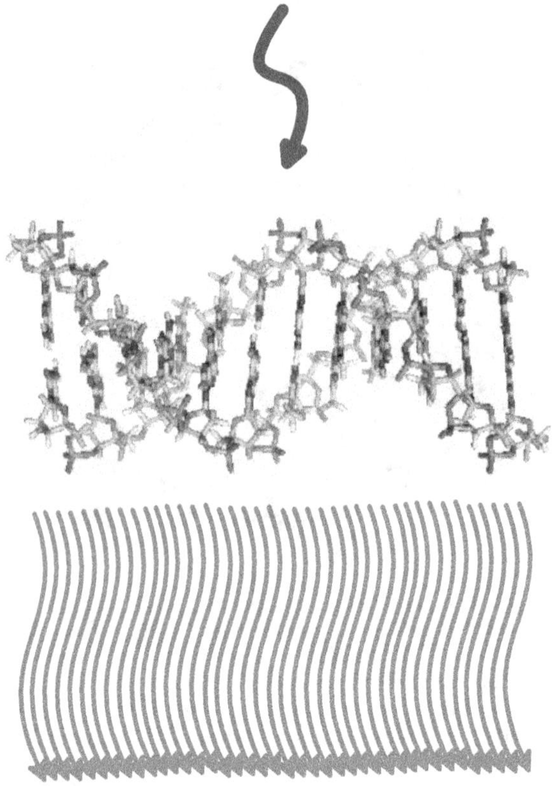

Fig. 2.3. A single UV-C photon in the Archean could be converted into between 30 to 40 infrared photons by the absorption and dissipation on pigments (the fundamental molecules of life – those common to all three domains of life) in a water solvent environment. The figure shows the particular example of dissipation of a single UV-C photon on a nucleobase of DNA. This process produces entropy and, in fact, the dissipation of photons is the principal thermodynamic work performed by life still today. However, absorption and dissipation of photons today on pigments is confined to the visible and near UV regions of the solar spectrum. The infrared photons (heat) produced provide the thermodynamic force (temperature gradient) for a large number of other dissipative processes in the biosphere, such as the water cycle and wind and ocean currents (see figure 2.4).

but at the same time makes the isolation of any one particular component (e.g. the Darwinian organism) for independent analysis a futile exercise.

The non-equilibrium thermodynamic imperative of optimizing dissipation at the global level takes effect at all inferior nested levels, including at the organismal level, and this level has become to be known as the level of *Darwinian selection.* However, the Darwinian perspective of selection acting only at the individual level is short sighted and does not capture the essence of the thermodynamically driven global evolutionary process. For example, the processing of wood products by humans from the Amazon forests provides them an income and therefore greater individual Darwinian "fitness", however, from the global thermodynamic perspective, destroying the Amazon forests has the net effect of reducing the global photon dissipation rates, meaning that at the global level it will be thermodynamically selected against. But, how does this occur? This occurs as an interference of the harvesters tree felling actions with other thermodynamic potentials necessary for their existence. The Amazon harvesters and their vigilant kin in the rest of the world begin to be subjected to negative forces affecting their populations, for example in the affect of their actions on the local water cycle necessary for their crops on which their livelihood also depends.

Apparent fitness at the individual level, on which strict Darwinian theory is based, is really a complex, many level, quantity involving both biotic and abiotic components that by no means can be simply sorted out or even easily imagined on this multidimensional level (hence the tautology in the Darwinian paradigm). It is not individual survivability or reproductive success that is selected, but rather global thermodynamic dissipation efficacy. Rather than individuals or species going extinct, a particular local generalized chemical potential spawning the dissipative organism would simply "dry up" leaving the organism or species, etc., without a necessary external driving potential and the system will eventually disintegrate, as in the case of a hurricane passing over land, or in the example given above of the Amazon harvesters loosing their crops to draught. That thermodynamic selection is indeed operating on a global biosphere level which reaches down to the individual level, can be seen in the increasingly many human programs to reduce global deforestation in all countries. The evolution of the biosphere as a response to optimizing photon dissipation will be discussed in more detail in chapter 17.

Looking at the global dissipation process from space, the material in Earth's atmosphere and on Earth's surface organize to dissipate the high energy incoming photons into many more low energy photons which are emitted isotropically into space (see figure 2.5). The amount of dissipation attributable to life and the other coupled abiotic processes occurring within the biosphere can be determined by comparing the incoming solar radiation spectrum to the red-shifted outgoing emitted spectrum of Earth (see figure 2.6). Contemporary life plays an important role in this global dissipation process and there is little doubt that this was the role played by life ever since its origin in the Archean.

In chapter 7 it will be shown that salient properties of RNA and DNA and of the other fundamental molecules of life (those common to all three domains; archea, bacteria, and eukaryote), all support the conjecture that they were originally microscopic dissipative structures operating in the UV-C and that life arose as an irreversible thermodynamic process to dissipate this part of the solar spectrum prevailing at the surface of Earth during the Archean. These same properties also suggest that RNA and DNA eventually formed a symbiosis, to-

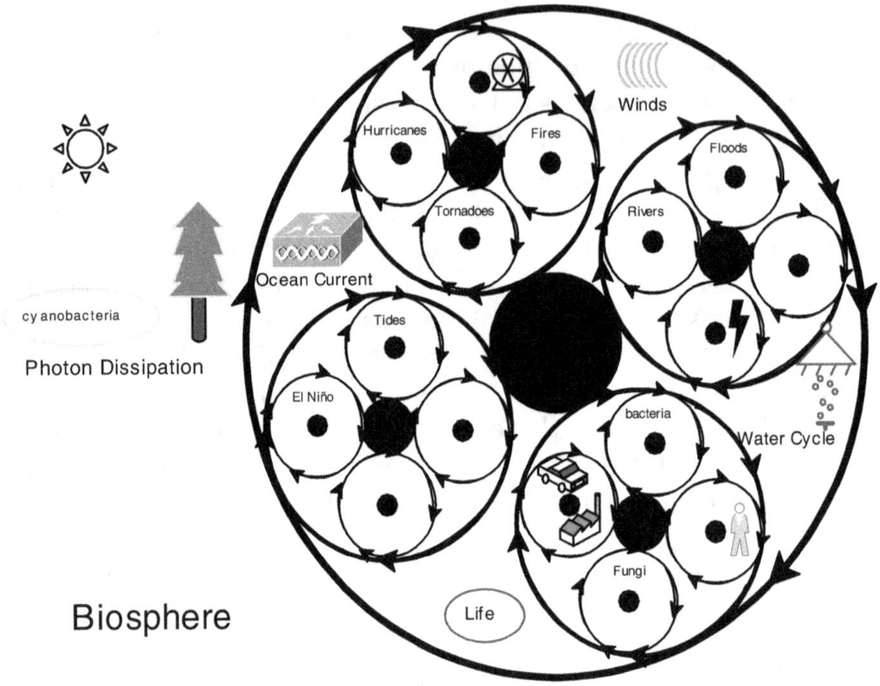

Fig. 2.4. The global dissipative process known as the biosphere consists of many coupled irreverssible processes with positive feedback. The greatest entropy producing process operating in the biosphere is the absorption and dissipation of sunlight in pigments in plants and cyanobacteria. This process gives rise to secondary generalized chemical potentials which spawn other irreversible processes such as the water cycle, winds and ocean currents, and more complex life. Each dissipative inner loop is feeding off the generalized chemical potential produced by an outer loop in the heirarchy. The size of the loops and arrows indicate the relative importance of the dissipative process. Thermodynamic selection towards greater entropy production is operating at each level but is ultimately decided at the level of the biosphere, i.e. the greatest loop, which is concerned with photon dissipation, primarily in organic pigments.

gether and in complexes with other fundamental molecules of life, and later with enzymes and other proteins, to increase photon dissipation.

Much later (perhaps a billion years later), when UV-C light no longer penetrated to Earth's surface because of the emission into the atmosphere of the UV pigment oxygen by living organisms as a result of the invention of oxygenic photosynthesis (oxygen and ozone strongly absorb and dissipate UV light), RNA and DNA made the final transformation from UV-C photon dissipaters into information storing and manipulating molecules for use in single cells or multi-cellular plants and animals. RNA and DNA still retained their thermodynamic function of fomenting the solar photon dissipation, albeit now indirectly, by storing and supplying information for building pigments and their support structures which began to dissipate at the more intense visible wavelengths.

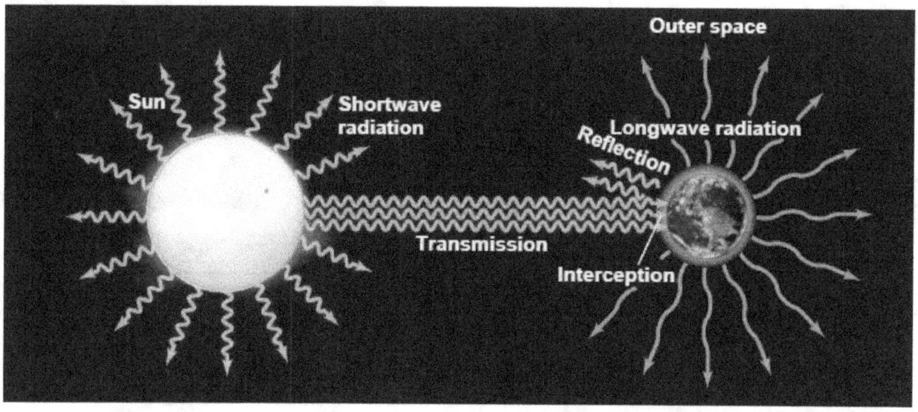

Fig. 2.5. A highly directed and high energy photon beam from the Sun is incident on Earth. Irreversible processes occuring on Earth, life being one of these, dissipate this incoming beam into a flow of photons of much lower energy emitted isotropically into space. Material on Earth organizes into pigments and supporting structures in order to facilitate this dissipation. Both the conversion of the high energy incident beam into a lower energy emitted beam, and its dispersion into a 4π solid angle produce entropy (Michaelian, 2012b)[213]. Image credit: Adapted from Zajeta Slika.

Non-linear systems driven by an imposed external potential, and under the variation of a particular constraint, can pass through an instability point at a bifurcation leading to new locally stable states. It is most probable that the new stable state taken by Nature is one of greater entropy production since fluctuations in the direction of, or in favor of, states with greater dissipation will be reinforced by the external potential (see chapter 3.3). This kind of *thermodynamic selection* acting on external perturbation has evolved a complex biosphere of ever greater global entropy production, principally through photon dissipation by evolving, i) pigments which dissipate over an ever greater region of the solar spectrum available at Earth's surface, ii) greater efficiency of pigments at absorption and dissipation of photons, iii) quenching of pigment radiative decay channels such as fluorescence and phosphorescence, iv) increases in the size of the photon absorption cross section while decreasing the physical size of the pigment molecule (by, for example, increasing the conjugation of molecules[2], or by increasing the strength of molecular dipole moments), v) mobile mechanisms, such as animals, for spreading nutrients and seeds and pollination, thereby allowing pigments to spread over an ever greater surface area of Earth wherever sunlight and water is available, and vi) evolving a coupling of biotic photon dissipation to other abiotic dissipative processes such as the water cycle.

[2] "Conjugation" refers to the formation of consecutive double and single covalent bonds, a situation which increases the non-localllity of the electrons in the molecule, thereby leading to collective electronic excited states with energies in the UV and visible (see chapter 4.3).

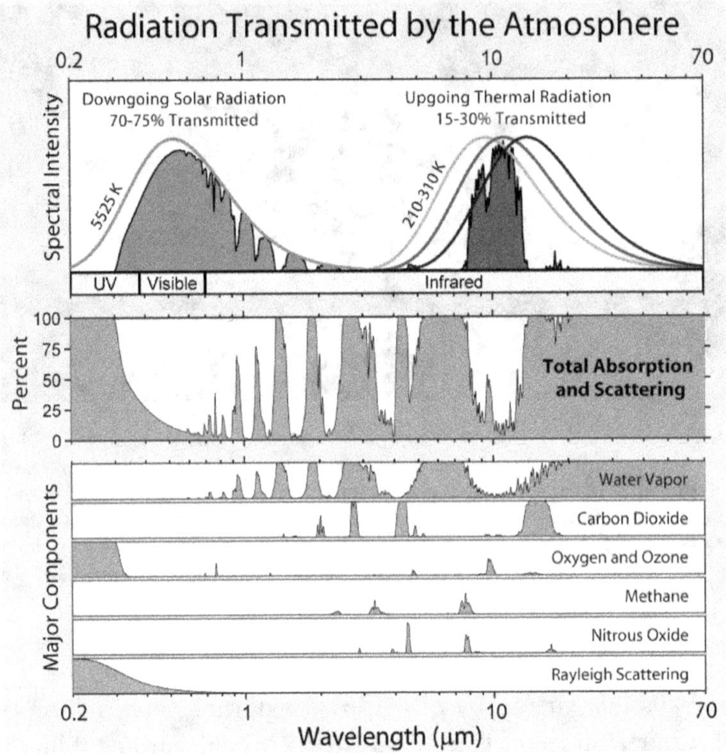

Fig. 2.6. The incoming solar spectrum (in red) and the outgoing Earth emitted spectrum (in blue) after photon dissipation by life and other abiotic irreversible processes occuring in the biosphere. Approximate black-body spectra at the relevant temperatures are overdrawn. The shift in wavelength of the incoming energy spectrum to the outgoing spectrum is a measure of the photon dissipation or entropy production due to material "spontaneously" organizing into dissipative structures (pigments, complexes, life, the water cycle, winds, currents, etc.) in the atmosphere and on the surface of Earth. Total absorption and scattering in the atmosphere of the different wavelengths is given as well as the decomposition into the contribution due to absorption and scattering attributed to each of the different atmospheric gases. Image credit: Robert A. Rohde, Creative Commons.

3. Material Organization through Dissipation

3.1 Non-equilibrium Thermodynamics and Life

In 1886 Boltzmann (Fig. 3.1) realized that the material of life was organizing to dissipate the solar photon potential and that this could be somehow behind the perceived organismal struggle for survival that Darwin wrote about. Boltzmann wrote "The general struggle for existence of animate beings is not a struggle for raw materials – nor for energy which exists in plenty in any body in the form of heat, but a struggle for [negative] entropy, which becomes available through the transition of energy from the hot Sun to the cold earth." (Boltzmann, 1886)[22]. It is curious just how this 130 year old insight into material structuring through dissipation has been re-discovered numerous times in the intervening years, each time being hailed in the popular scientific literature as providing new insight into life and evolution, but how there has been a conspicuous complacency towards taking Boltzmann's insight to its final logical consequences regarding the origin of life.

The formal derivation based on thermodynamic foundations of this principle of material organization under a generalized chemical potential was obtained by Ilya Prigogine (Fig. 3.2) in the 1960's and was lucidly presented in his book entitled "An Introduction to the Thermodynamics of Irreversible Processes" (Prigogine, 1967)[289]. Prigogine, more than anyone else before him, took Boltzmann's ideas further down the road to understanding the origin of life by delineating the conditions under which material will spontaneously organize under an imposed external generalized chemical potential and identifying the possibility that dissipation (entropy production) could increase through this organization. For this work on what he called *dissipative structures*, Prigogine was awarded the Nobel prize in Chemistry in 1977.

Prigogine extended equilibrium thermodynamic formalism to cover non-equilibrium conditions (including an imposed external potential over the system) by assuming local equilibrium in small but still macroscopic regions of the system, which, as a whole, was out of equilibrium. This required the inclusion of a space and time dependence of the usual thermodynamic variables, assuming

Fig. 3.1. Ludwig Eduard Boltzmann (1844 –1906), around 1886, when he first realized that living systems were dissipating the impressed solar photon potential and he attributed this as being behind the struggle for existence. Although Boltzmann eventually became famous for deriving a microscopic statistical mechanics basis of thermodynamics, his ideas were originally rejected by his peers since they required a revolutionary atomistic view of material. This rejection aggravated a condition of severe depressions that he suffered from. While on vacacion, with his wife and daughter swimming in the bay of Duino on the Adriatic sea, Boltzmann, undoubtedly the most influential scientist of all time, hung himself in the bathroom of his hotel room. He was 62. Image credit: Unknown, Public Domain.

always that local equilibrium could be established so that the Gibb's equation[1] remains valid locally, and invoking the fundamental conservation laws (known as the continuity equations) which do the accounting for the production and flows, into or out of the system, of matter, energy, entropy, momentum, etc. (see Fig.

[1] The Gibb's equation is perhaps the most fundamental equation in equilibrium thermodynamics. It is derived from the fact that the entropy of a system is a function of a number of independent variables, in particular $S(E, V, n_\gamma)$ where E is the internal energy of the system, V is its volume, and n_γ are the mole numbers of substance γ. It is then possible to write that $dS = \frac{\partial S}{\partial E} dE + \frac{\partial S}{\partial V} dV + \sum_\gamma \frac{\partial S}{\partial n_\gamma} dn_\gamma$. The partial derrivatives of the entropy have well defined thermodynamic names and so we can write the previous equation as,

$$dS = \frac{1}{T} dE + \frac{P}{T} dV - \sum_\gamma \frac{\mu_\gamma}{T} dn_\gamma$$

which is known as the Gibb's equation. Here, T is the temperature, P the pressure, and μ_γ the chemical potential of substance γ.

Fig. 3.2. Ilya Prigogine (1917-2003) extended equilibrium thermodynamic formalism to non-equilibrium situations. He showed that material will spontaneously organize to dissipate an external potential and coined the term "dissipative structures" to describe this structuring of material in space and time. This structuring occurs spontaneously as a response to a thermodynamic imperative of producing entropy through the dissipation of the imposed external potential. Image credit: Unknown, Public Domain.

3.3). Prigogine showed that if the imposed external potential over the system was constant (fixed external constraints), then the system will eventually arrive at a *stationary state* in which the position dependent thermodynamic variables for the system become constants in time, or, in the case of non-linear systems, may fluctuate about constant average values.

For a linear system, the stationary state was shown to be unique and stable. However, if the system was non-linear then multiple stationary states were available and the states were only locally stable or not stable at all (an instability point). In this latter case, a microscopic fluctuation could be sufficient to lead the system to another state corresponding to a different macroscopic regime.

The following box goes into more detail about stationary states. It can be skipped by those with an adverseness to mathematics without much loss of continuity of argument, but I recommended it to the courageous since dominating the mathematical language of non-equilibrium thermodynamics will lead to a more complete understanding of the thermodynamic dissipation theory of the origin and evolution of life.

For a time relaxed system under constant external constraints, the entropy, S, of a system, due to an external flow of entropy into or out of the system, d_eS/dt, and an internal production of entropy, d_iS/dt, due to irreversible processes occurring within the system itself, becomes constant in time, i.e.

$$\frac{dS}{dt} = \frac{d_eS}{dt} + \frac{d_iS}{dt} = 0.$$

The entropy production d_iS/dt has been shown by Prigogine to be a linear sum of generalized flows J multiplying generalized forces X over all n irreversible processes occurring within the system,

$$\frac{d_iS}{dt} = \sum_{k=1}^{n} J_k X_k. \tag{3.1}$$

Prigogine demonstrated that if there exists a non-linear relationship between the rate (flow J) of the dynamic process and the gradient of the generalized chemical potential (force X) causing the flow, e.g. $J_k = F(X_1, X_2, ...X_n)$, where F is some non-linear function of the forces X_k, then multiple solutions to the stationary state $dS/dt = 0$ exist and these solutions could demonstrate spontaneous symmetry breaking in both space and time (another way of saying dynamic organization of material). Furthermore, different solutions of the equations for the steady state may have different rates of internal entropy production, d_iS/dt.

Prigogine showed that amplification of microscopic fluctuation at an instability would often lead to macroscopically complex organization of material with long range correlation which may increase the dissipation of the imposed external potential. The entropy production of the system (directly related to the rate of dissipation of the external potential) would generally, but not necessarily, increase through this "spontaneous" organization of the material in the system.

Although the existence of bifurcations in non-linear systems (see section 3.3) implies that there cannot exist a potential which could be optimized in order to determine the direction of evolution in time of the system (as there exists in equilibrium thermodynamics), the analysis showed that the variation of the entropy production with respect to the free forces (those non-fixed forces which organize the system internally) had a negative definite sign which could be used as a *general evolutionary criterion* to determine the local direction of evolution of the system in certain regions of its phase space.

Starting from the entropy production as a linear sum of forces times flows of the irreversible processes occurring within a system (Prigogine, 1967)[289],

$$\mathcal{P} = \frac{d_i S}{dt} = \sum_k J_k X_k \geq 0. \tag{3.2}$$

we can write,

$$d\mathcal{P} = d_X \mathcal{P} + d_J \mathcal{P}$$

and the therefore,

$$d_X \mathcal{P} = d\mathcal{P} - d_J \mathcal{P} \leq 0 \tag{3.3}$$

where the final inequality is due to the general evolutionary criterion derived by Prigogine. More on this criterion, and its application to the autocatalytic proliferation of organic molecules which dissipate the solar photon potential, will be given in chapter 6.3.1.

In accordance with the second law of thermodynamics, the overall entropy of the system plus environment had to increase, although that of the system itself could decrease, while the entropy production rate $\mathcal{P}$ generally increases. A summary of the important concepts of Classical Irreversible Thermodynamics is given in figure 3.3.

A well known and instructive example of the spontaneous generation of order (symmetry breaking in space) under an externally imposed generalized chemical potential, in order to dissipate more effectively that potential, is that of the Bénard cell (see figure 3.4).

Prigogine's non-equilibrium thermodynamic formalism has been applied to the population dynamics of ecosystems (Michaelian, 2005)[206] as an alternative to the merely descriptive Lotka-Volterra-type equations commonly used to model ecosystems. This allows extrapolations to regions where no data exist and to predictions concerning the direction of evolution of ecosystems (assuming the evolution towards ever greater global entropy production – see chapter 3.3) which cannot be made within the Lotka-Volterra framework. The optimization of entropy production, although it cannot be derived from within the frame work established by Prigogine, its demonstration from first principles still under much debate, can be intuitively understood through heuristic arguments concerning the relative size of fluctuations at the bifurcation instability for the different phase space paths allowed for the system (see section 3.3). It has also been shown to be a principle validated empirically in numerous physical processes, and these processes will be discussed in section 3.2.

Recently, Phil Attard (2008; 2009)[9, 10] has questioned the use of the ordinary (nominal, or first) entropy variable in optimization principles for predicting system evolution in Nature, arguing that the ordinary entropy variable, our S

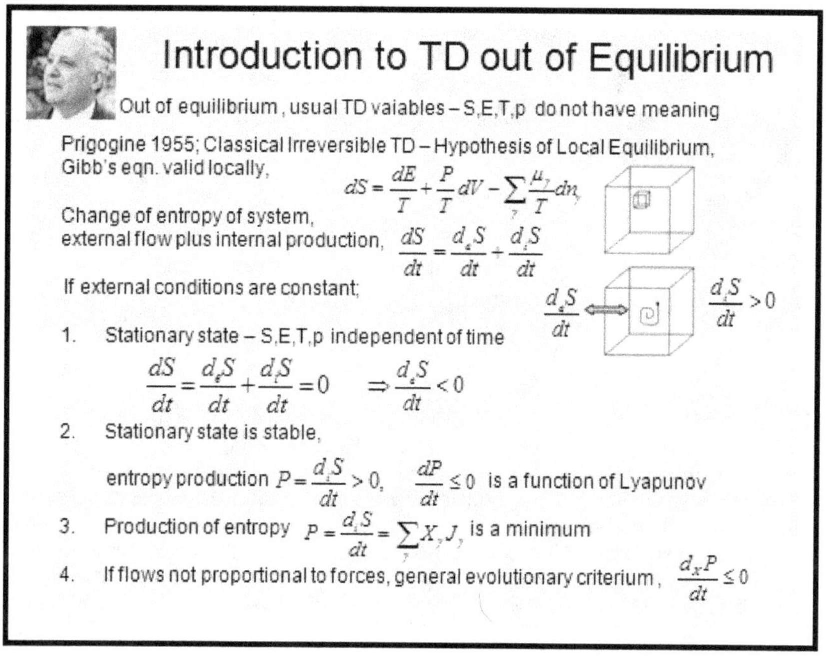

Fig. 3.3. Summary of the concepts of Classical Irreversible Thermodynamics (CIT) as initiated by Lars Onsager and developed by Ilya Prigogine. For systems out of equilibrium, the normal thermodynamic variables S, E, T, p, etc. do not have meaning, however, local equilibrium can be assumed so that the Gibb's equation remains valid and all the normal thermodynamic variables retain their usual meaning but now locally (they become functions of both space and time). With linear relations between the flows J and forces X, there is only one stationary state and it corresponds to a minimum of entropy production with respect to variation of the free forces (or flows). For the non-linear situation, many stationary states become available, each with a different entropy production and stability property. In the general non-linear case, the entropy production is not a minimum, but a "general evolutionary criterium" applies, i.e. the variation of the entropy production with respect to variation of the free forces is negative semi-definite. Given internal or external fluctuations, the general evolutionary trend is towards stable stationary states of greater entropy production. An explanation for this will be given in section 3.3.

above, is explicitly time-independent and therefore could not be used to describe evolution. Attard sees the necessity of ascribing the existence of a new variable, the *second entropy*, and a new *non-equilibrium* thermodynamic law, which together lead to an entropy production optimality principle to explain the evolution of increases in entropy production observed in Nature. The following section giving Attard's arguments in mathematical form can be skipped by those with an adverseness to mathematics.

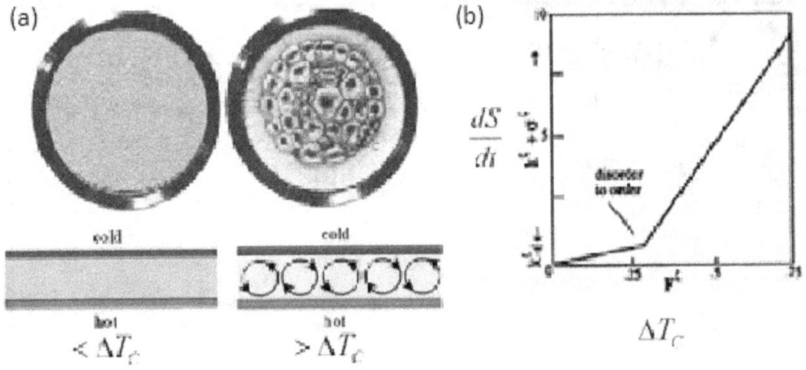

Fig. 3.4. An example of dissipative structuring; the Bénard cell. The onset of structuring of material in space and time occurs as a result of an external potential imposed over the system and a non-linear relation between flows and forces. The left part of the figure (a) describes the linear situation for the relation between the force (temperature gradient) and the flow (heat flow) when the temperature difference over the system is below some critical value ΔT_c and the system remains homogeneous (no spatial symmetry breaking) and there exists only conduction of heat. The right part of (a) describes the non-linear situation when the temperature difference between the two plates is greater than some critical value ΔT_c for which convection cells spontaneously arise (Bénard cells) and spatial symmetry breaking occurs. The direction of liquid circulation in the convection cell is determined by the initial microscopic fluctuation which grew into the cell. Two solutions are allowed at the bifurcation point, one with clockwise rotation and the other anti-clockwise. Which one is chosen is determined by a particular microscopic fluctuation. In (b) the internal entropy production for the system is plotted as a function of the temperature difference. There is an increase in slope, meaning a greter rate of increase in entropy production, at the critical value of the temperature gradient at the onset of convection, describing a disorder to order transition.

3.1.1 Second Entropy

The traditional entropy, or the first entropy $S^{(1)}$ is formally defined as,

$$S^{(1)}(x|E) = k_B \ln \int_E d\Gamma \delta(x - \widehat{x}(\Gamma)),$$

where x represents the values of the macroscopic variables which define the macrostate and $\widehat{x}$ represents the values of the macroscopic variables evaluated at the particular point Γ in the phase space of the system, all of which are equally weighted, k_B is Boltzmann's constant, and the integral is restricted to all regions of phase space on the surface of constant energy E.

This first entropy $S^{(1)}$ is related to the number of microstates compatible with a given macrostate x, and the second law of thermodynamics indicates only a time direction for evolution, that of increasing $S^{(1)}$. Therefore, the definition of the first entropy and the second law taken together say nothing about an optimization of a quantitative rate of entropy production.

Attard maintains that since the ordinary entropy variable and the second law of thermodynamics are time independent formulations, any derivation of an optimality principle in time based on these variables would be ambiguous. In order to include time explicitly, Attard defines a new entropy which he calls the "second entropy" $S^{(2)}$ which he also calls a *transition entropy* since, rather than measuring the number of microstates consistent with a macrostate, it measures the number of molecular configurations (microstates or points in phase space) compatible with the transition between macrostates, $x \longrightarrow x'$, occurring in a specified time τ. Formally it is written as (Attard, 2008)[9],

$$S^{(2)}(x', x|\tau, E) = k_B \ln \int_E d\Gamma \delta(x - \widehat{x}(\Gamma))\delta(x' - \widehat{x}(\Gamma(\tau|\Gamma, 0))),$$

where the second factor $\delta(x' - \widehat{x}(\Gamma(\tau|\Gamma, 0)))$ assures that the integration is only over those phase space points Γ that lead to the transition $x \longrightarrow x'$ in time τ. This new entropy corresponds to the rate of transition form one macrostate to another $x \longrightarrow x'$, and is thus a measure of the number of paths in phase space available for the transition, the greater the second entropy, the greater the number of paths available, and the greater the transition rate.

A second law of *non-equilibrium thermodynamics* is then proposed by Attard as (Attard, 2008;2009)[9, 10],

> *The second entropy increases during spontaneous changes in the dynamic structure of the total system.*

Dynamic structure is defined as a macroscopic flux or rate; a transition between macrostates in a specified time. Attard indicates that "the new law is deliberately analogous to the familiar second law of equilibrium thermodynamics. Instead of a constrained macrostate, however, one has a constrained flux, and instead of the equilibrium state in which the macrostate no longer changes and entropy is a maximum, one has the steady state in which the fluxes cease to change and the second entropy is a maximum".

The second entropy, also called the transition entropy, is the number of molecular configurations associated with a transition between macrostates in a specified time. Attard argues that second entropy is optimized in nature, leading to dynamically organized material which lead to greater transition rates between two macroscopic states of different entropy.

Attard's second entropy and second law of non-equilibrium thermodynamics concepts have been used to derive several non-equilibrium statistical mechanical results (Attard, 2008;2009)[9, 10], giving his formulation some scientific credibility. Attard's criticism regarding the ambiguity of traditional non-equilibrium thermodynamic optimality principles based on a time independent entropy function appear to be valid. However, the correctness and utility of his particular framework of second entropy still has to be thoroughly accessed by the community at large. Attard's and similar theories based on maximizing the Shannon information entropy $S_I = -\Sigma p_\Gamma \ln p_\Gamma$ to determine the unique probability distribution p_Γ of phase space paths Γ for transitions between states, such as Dewar's (2003)[77] utilization of Jaynes' theory, are still being debated and it may be some years before a generally valid and accepted optimality principle for non-equilibrium thermodynamics emerges. Although the existence of a general optimality principle for non-equilibrium systems is highly debated in the contemporary literature, it seems obvious that an optimality principle must indeed exist and should be frameable within some theory of non-equilibrium thermodynamics given the regularities observed in evolution such as increases over time of global dissipation rates and increases in complexity of self-organized material under an imposed external potential.

Attard's optimization principle for second entropy, his second law of non-equilibrium thermodynamics, if correct, would provide a framework for the empirically observed dissipative structuring of conical intersections in organic pigments (which allow rapid conversion of excited molecular electronic energy into heat) since the number of phase space paths through a conical intersection for transitions between the first electronic excited state of pigment molecules and their ground state is large for such molecules in a thermal bath. DNA and RNA and other fundamental molecules of life contain conical intersections. More on conical intersections will be presented in chapter 7.

3.2 Optimallity Principle in Nonequilibrium Thermodynamics

Another way of describing this material organization principle, without resorting to the thermodynamic jargon which may be confusing to those without a background in physics, is the following: Within the constraints dictated by the physical nature of the environment (including the universal conservation laws, the particular values of the physical constants, and the nature of the microscopic degrees of freedom available in the material for carrying the conserved quantities), Nature will "manufacture" dynamical structures (or more correctly said "processes") which aid in distributing the conserved quantities (energy, momentum, angular momentum, charge, etc.) over ever more microscopic degrees of freedom available provided by the material in the environment. A spontaneous self-organization of material to enforce a sort of assertive socialism, if you will, leading to the smallest possible GINI index[2] with respect to the distribution of Nature's most coveted quantities (her conserved quantities) over all her degrees of freedom. In the case of hurricanes and life, these internal degrees of freedom over which energy can be distributed correspond to the translational, vibrational and rotational modes of the environmental atmospheric or liquid water molecules, and finally, to a still greater distribution of the conserved energy over the photons of an infrared spectrum which is emitted by Earth into space.

Although the system itself, under the application of an external potential, may reduce its entropy through organization by dissipative structuring (the conserved quantities become distributed over less microscopic degrees of freedom within the system itself), the entropy of the greater environment increases through the systems interaction with it; dissipation of the external potential imposed over the system by the environment. Taking the system plus environment as a single isolated system, this structuring can be seen to occur in strict observance of the second law of thermodynamics, which states that the conserved quantities of an isolated system must either become more dispersed over the available microscopic degrees of freedom, or remain at maximal dispersal (in equilibrium). The second law of thermodynamics, expressed with the concept of entropy representing the measure of the dispersion of the conserved quantities, must be obeyed for the isolated system as a whole, but not for any part of the system taken as a separate entity and analyzed as if artificially separated from the environment with which it is interacting.

It should be emphasized that entropy is produced not only by distributing the conserved quantity of energy over the many available microscopic degrees of freedom in the environment, but entropy is also produced by distributing over more microscopic degrees of freedom any conserved quantity, for example, energy, angular momentum, linear momentum, charge, etc. These variables which obey conservation laws are precisely those that are derived from the fundamental symmetries in Nature - this fact is known as Noether's theorem after its discoverer Emmy Noether (see figure 3.5).

There is much empirical evidence which argues for an optimality principle in nature suggesting that non-equilibrium systems evolve towards increasing global entropy production (such as Attard's second law of non-equilibrium ther-

[2] The GINI index is a measure from economic theory of the spread of wealth in an economy. The smaller the GINI index, the more uniform the spread of wealth. The spread of wealth is intimately related with general living conditions. Nordic countries have the smallest GINI indices while Latin American and African countries have the largest.

Fig. 3.5. Emmy Noether (1882 – 1935), mathemetician who realized that conserved quantities in Nature were associated with specific symmetries. For example, conservation of energy can be derived from the symmetry of the physical laws of Nature with respect to time translation, conservation of momentum derives from the symmetry of the laws of Nature with respect to space translation, and the conservation of angular momentum can be derived from the symmetry of the laws with respect to rotation. The second law of thermodynamics describes the increase in the distribution of these conserved quantities over ever more microscopic degrees of freedom with time, but does not describe exactly how this increase occurs. Attard's second law of non-equilibrium thermodynamics suggests that this increase occurs in such a manner that the second entropy increases during any spontaneous change of the system (see section 3.1.1). Image credit: Unknown photographer, Public Domain.

modynamics, see section 3.1.1). The following is not meant to be an exhaustive list, but instead attests to the variety of some of the more common phenomena which have been analyzed and found to be consistent with such an optimality principle;

1. Systems of chemical networks tend to follow reaction routes which maximize the entropy production given constant external flows of the reactants and products (Garcia et al., 1996)[99].
2. There is evidence that not only the metabolic rates of organisms have increased over evolutionary time scales, but that the metabolic rate per unit biomass has also increased (Zotin, 1984)[431].
3. Global circulation patterns can be accurately described with simple models of heat transfer by optimizing the entropy production (Paltridge, 1979)[265].

4. Ecosystems pass through a set of stages, known as succession, of ever greater entropy production through optimizing photon dissipation rates (Schneider and Kay, 1994b)[324].
5. Evolution leads to greater entropy production in bacterial metabolic pathways (Unrean and Srienc, 2011)[382].
6. Evolution has led to ever more complex and dissipative biotic, and coupled biotic-abiotic, processes (Zotin, 1984; Michaelian, 2012a)[431, 212].
7. Some systems such as plastic polymers left to age under visible or ultraviolet light tend to go from highly transparent or reflective to highly absorptive and dissipative (Michaelian, 2014)[215].
8. Liquid diffusion problems, corresponding to flows in viscous media, can be resolved by maximizing the dissipation (Onsager, 1945)[257].
9. The experimental chemical reaction rates for a complex set of reactions transforming phosphate P and adenosine diphosphate ADP into adenosine triphosphate ATP were found to be determined well by maximizing the entropy production (Dewar et al., 2006)[78].

Although there is still no consensus with regard to the establishment of a valid theoretical demonstration of an optimality principle in non-equilibrium thermodynamics implying the evolution of increases in entropy production over time (besides the confirmed non-variational general evolutionary criterion of Prigogine, see Eqs. (3.3) and (6.6)) the heuristic explanation given in the following section, attributed to the present author, appears to be plausible. In any case, even if proven incorrect, the list of empirical evidences given above strongly suggests that an optimality principle will one day eventually be confirmed.

3.3 Explanation for the Evolution of Increases in Entropy Production

The following mathematical description of a non-linear system and how it may evolve under non-equilibrium conditions to ever greater entropy production is not put into a box since it is fundamental to the understanding of why Nature evolves towards states of every greater photon dissipation. The derivations are not complicated and it is strongly recommended that the reader make an effort to understand these arguments as they form the physical-chemical basis of what will be termed *thermodynamic selection* in our non-equilibrium thermodynamic paradigm which will replace *natural selection* in the deficient Darwinian paradigm.

The following mathematical example is from Kondepudi and Prigogine (1998)[168] but we develop it further to give an explanation of the optimality principle of probabilistically determined increases in entropy production over time. Consider the stationary (time independent) solutions of the following non-linear partial differential equation describing the dynamics of some variable α,

$$\frac{\partial \alpha}{\partial t} = -\alpha^3 + \lambda\alpha = 0, \tag{3.4}$$

where λ is some parameter (for example, it may be the value of some external potential or constraint over the system). If we are only interested in real solutions (i.e. not complex) this equation has a number of different possible solutions depending on the value of λ. For $\lambda \leq 0$, the only real solution is $\alpha = 0$, but for $\lambda > 0$ there are two solutions, $\alpha = \pm\sqrt{\lambda}$. These solutions as a function of the parameter λ are graphed in figure 3.6,

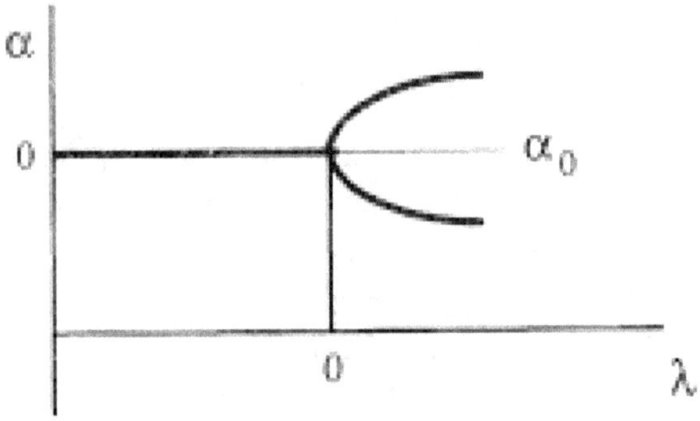

Fig. 3.6. Bifurcation of a stationary state solution of a non-linear partial differential equation describing the dynamics of some variable α as a function of parameter λ.Taken from Kondepudi and Prigogine (1998)[168].

This shows that at a value of the parameter $\lambda = 0$ there is a bifurcation leading to the two new solutions $\alpha = \pm\sqrt{\lambda}$. It is possible to check the stability of the solutions by determining the time variation of a small perturbation δ added to the solution. It is easy to show (see Kondepudi and Prigogine 1998)[168] that the perturbation δ decreases exponentially in time so the solution $\alpha = 0$ is stable until the point $\lambda = 0$ at which it becomes unstable. For $\lambda > 0$ the two solutions $\alpha = \pm\sqrt{\lambda}$ are stable. Any microscopic fluctuation at $\lambda = 0$ will therefore cause the system to take the route of one of these two solutions. Which of the two solutions the system takes in this mathematical example will be arbitrary because each has the same probability. However, in the real world, fluctuations in the direction of the route of greatest dissipation of the external potential will be reinforced and therefore somewhat larger (for example, the thermal fluctuations of a charge in the presence of an external electrical potential will be reinforced in the direction of the pole of opposite sign, i.e. in the direction of current flow, which is the direction of greater dissipation of the external electrical potential). It will then be more probable, but not certain, that the system takes the route of greatest dissipation of the external potential. This is, in fact, what is observed in the real world and some examples of this were given in section 3.2.

The general tendency of the evolution of a driven non-linear system, given the fact that the stationary states are only locally stable, and given the existence of external perturbation which can change the values of some constraint λ, and given the existence of both internal and externally induced fluctuations, is towards stationary state solutions of ever greater entropy production. The non-linear dynamic equations describing the time evolution of the system are accidented with bifurcations and the particular route taken by the system through these bifurcations is determined by the amplification of a particular microscopic fluctuation (see figure 3.4). Fluctuations in the direction of the solution of greater dissipation of the external potential are reinforced and this gives rise to a probabilistic evolution to generally greater entropy production. It is this route through bifurcations, generally leading to greater entropy production, which endows a "history" to the irreversible process or system (see figure 3.7).

This amplification of fluctuation is a particular characteristic of non-equilibrium thermodynamic systems that does not occur in systems close to equilibrium where there is no external potential driving the system. Near equilibrium the response to fluctuation is described by the Le Chatalier-Braun principle which describes how internal forces are generated such that fluctuations are damped and the system always returns to the unique equilibrium state.

3.4 Microscopic Dissipative Structuring

As we have shown, macroscopic spontaneous organization of material in space and time occurs under an externally imposed generalized chemical potential in order to increase the dissipation of the imposed potential or, in other words, to spread the conserved quantities of Nature over ever more microscopic degrees of freedom, which, in other terms, is called the *production of entropy*. For example, in the case of the dissipation of an energy gradient over a material system, through which there will exist a continual flow of energy, internal fluctuations can then lead the system into a completely new state of order through self-amplification of a perturbation, similar to what happens to the concentration of a product catalyst in an autocatalytic chemical reaction. Such a self-amplification or autocatalytic activity requires a non-linear relation between the forces and flows (see box in chapter 3.1) and this non-linearity also allows the system to have numerous solutions (stable states).

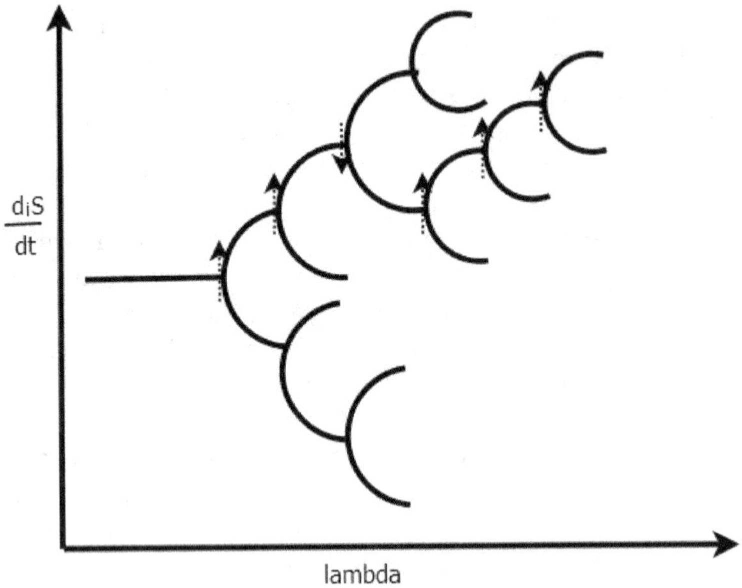

Fig. 3.7. Bifurcation diagram showing how, for non-linear non-equilibrium systems, two or more solutions may suddenly appear for the same value of a particular parameter λ of the system (which may be, for example, the strength of the potential over the system). Which branch solution the system takes depends on microscopic fluctuations (externally or internally induced) and their "reinforcement" by the applied external potential (usually stronger in the direction of greater dissipation d_iS/dt) and this leads to a unique "history" for the system, generally, but not necessarily, through states of ever greater entropy production d_iS/dt. The dotted arrows represent the direction of the fluctuation which prevailed for a particular evolutionary history. This probabilistic selection through the different stable state solutions towards ever greater entropy production is what I have termed *thermodynamic selection*.

An interesting example of this is the formation of Bénard cells (see figure 3.4) in a liquid system held under a temperature gradient, maintained by a lower plate held at high temperature and an upper plate held at low temperature, and subjected to a gravitational field. A particular microscopic fluctuation which produces a small movement of a particular heated volume element of the fluid towards the upper plate will be reinforced because of its lower density with respect to the colder water above it. If the fluctuation is large enough, then this force for the upward movement of the volume element of the fluid will gain over the viscous frictional forces acting against the movement and over the conduction which tends to stabilize the system, and the volume element will continue its upward trajectory pulling in warmer liquid behind it. In this manner, a Bénard cell establishes itself. This cell, who's dimensions depend on the physical characteristics of the material and confining volume defined by the imposed potential, will now provide a strong local perturbation, stimulating neighboring volumes of liquid to also form cells, and very quickly the whole systems fills with Bénard cells. The initial cell, or, in fact, a tiny portion thereof, thus acted as a catalyst for the second cell, and so on in an exponential, explosive manner, all started by a single random fluctuation of a microscopic volume element.

Is there a finite size to these Bénard structures? In a Bénard system, there are two irreversible processes occurring, the first is heat conduction and the second is convection. Heat conduction is the simple diffusion of vibrational energy of the liquid molecules from the lower plate to the upper plate. Covenction is the dynamical flow of macroscopic amounts of liquid due to density gradients resulting from the temperature gradient. Convection spontaneously arises when the physical conditions allow it, when the upward buoyant force due to the lower density of a heated liquid volume element becomes greater than the viscous forces of friction that oppose its movement. It is also necessary that heat flow due to convection becomes predominant with respect to that of conduction which tends to dampen the collective dynamical movement. This, of course, requires a sufficient temperature gradient to be maintained over the system, and the conduction of heat to be sufficiently slow from the hot plate to the cold so that convection outperforms conduction in dissipation. As the distance between the two plates is decreased, heat diffusion (conduction) becomes ever more "instantaneous" and even though the temperature gradient becomes greater, Bénard convection cells eventually cannot compete at heat dissipation and do not interfere with the more efficient dissipative process of conduction. At these small dimensions, only conduction remains. This happens at a minimum cell size of convection, for a water solvent, corresponding to approximately tens of μm (1μm $=10^{-6}$ m).

As another example of the limit to the size of a dissipative structure, consider a macroscopic reaction-diffusion system, such as that giving rise to the Belousov-Zhabotinsky chemical oscillator (see figure 2.2). In such a macroscopic system, diffusion times are large compared with typical chemical reaction times and so the reaction times can be considered to be instantaneous. Alan Turing predicted that in a system of reacting and diffusing chemicals, homogeneity could be broken and macroscopic patterns of products and reactants would spontaneously emerge if there existed non-linearity in the system, for example, an autocatalytic agent (an activator) who's effectivity was limited by an inhibitor. In a reaction-diffusion system, the characteristic size of the macroscopic pattern is, by necessity, greater than that of the average diffusion length (the mean distance a molecule travels between reaction events). For biological systems in a water solvent environment, the minimum size for dissipative structuring of this reaction-diffusion type is also of the order of microns (10^{-6} m) (Hess and Mikhailov, 1994)[131].

Therefore, a dissipative conduction-convection Bénard, or a reaction-diffusion Belousov-Zhabotinsky process could not give rise to structuring at the sub-micron scale. However, dissipative structuring at the sub-cellular level, and even at the nanometer scale, is well known as can be witnessed by the many processes occurring inside a living cell, a clear example being molecular motors or translation in the ribosomes (see figure 3.8). There must, therefore, be other dissipative mechanisms besides convection-conduction and reaction-diffusion operating at the nanoscale. At the nanoscale dimension, particularly for soft (biological) material there is cohesion between atoms and this allows for molecular structural transitions (structural phase transitions, for example between different geometric isomer states, between the ground and excited states, between molecular complexes or excimer states, or phase separation of different components). Such structural transitions to greater molecular order can be initiated by energy flow through the material just as occurs in the case of the Bénard cell, since low entropy can be carried along with the energy flow. Any ordering, however, will occur only if dissipation occurs and ordering will generally only persist if dissipation persists.

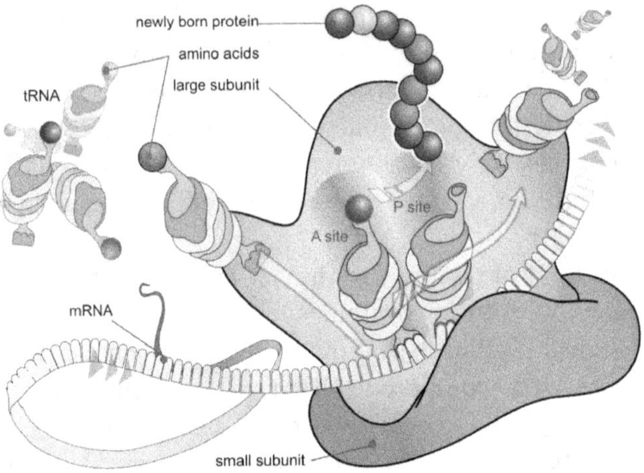

Fig. 3.8. Diagram of the cell organelle ribosome where genetic information is converted into proteins. This process occurs at the submicron scale, implying the existence of *microscopic* structuring through dissipation. Molecular motors are another example of micrscopic dissipative structuring. Image credit: LadyOfHats, Public Domain.

Chemical or photochemical reactions and diffusion, together with structural phase transitions involving internal degrees of freedom of the molecule, can lead to what is known as *microscopic dissipative structuring*. This new field of research, although formally recognized and introduced since the 1960's (Prigogine, 1967)[289] (see below), is only recently receiving a lot of attention for technological reasons since dissipation induced nanoscale structuring can persist even after removal of the external generalized chemical potential, due to atomic and molecular mobility issues, and can thus be used in a large number of novel practical applications where controlled nanostructuring is required. Hess and Mikhailov have studied particular mechanisms for soft matter nanostructuring based on equilibrium phase separation of two molecular species at a surface and a photon induced non-equilibrium de-absorption process (Hess and Mikhailov, 1994) [131].

Photon induced processes are often associated with microscopic dissipative structuring since photons can deliver an intense amount of energy of low entropy locally on a very small amount of material. This leads to very large generalized forces at microscopic dimensions which can lead to microscopic self-organization as a response to dissipate the imposed photon potential. A typical example of this is a lasing system. In a system operating in the lasing mode, photons of a particular chamber resonant wavelength stimulate excited atoms in the material to emit photons of the same wavelength in a coherent manner, forming a self-organized system of atoms emitting coherently from a particular excited state along with a beam of coherent mono-frequency photons. Even though the organized state of the coherent output photon beam is of low entropy (as compared to a black-body spectrum containing the same amount of energy) entropy production is gained by vacating the lower lying vibrational states superimposed on the electronic excited state of the material, allowing the pumping field to be dissipated more effectively than would happen without stimulated emission. A laser heats up under operation and it is this heat emitted to the environment that is a measure of the inherent entropy production of the system (see figure 3.9).

Examples of microscopic self-organized dissipative structures, which are of much relevance to the origin of life, are the fundamental molecules of life and their associations in complexes, in particular the nucleobases and their complexes as single and double RNA and DNA strands. These molecules, which absorb and dissipate UV-C light (see chapter 4) can be formed from simpler precursor molecules by the very same UV-C photons that they eventually dissipate with such efficacy. Although the particular photochemical reaction pathways to the multitude of pigments still need to be delineated, their existence in particular cases has been established and in general is evidenced, for example, by the large quantity of organic pigment molecules found throughout the cosmos where UV light is available (see chapter 5). Today, a similar situation exists on Earth with pigments dissipating in the visible which are organized through more complex biosynthetic pathways but remain, nonetheless, microscopic dissipative structures dissipating the visible light that produced them.

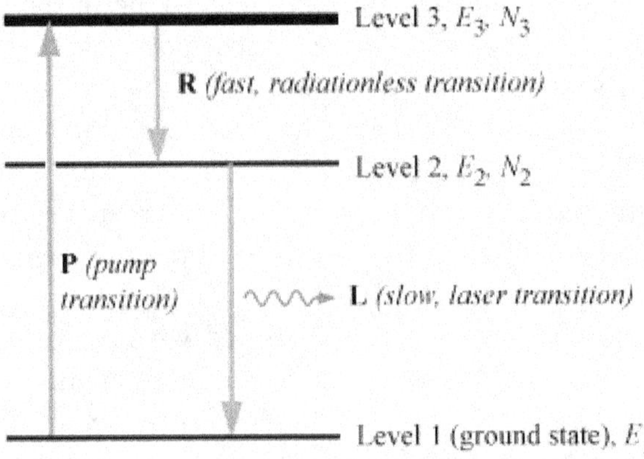

Fig. 3.9. A laser is an example of microscopic dissipative structuring. The system organizes internally under an external pumping field to dissipate this field more effectively. The electric pumping field excites the material which then decays to the lowest energy excited state E_2 through internal conversion (dissipation into heat). Photons from the decay of this lowest energy excited state then stimulate other electrons in the same lowest energy excited state to decay and emit a photon of the the same wavelength. This de-populates the lowest energy excited state E_2, allowing the higher energy excited states E_3 to decay to the lowest energy exicted state E_2 through internal conversion. This non-radiative decay of excited state electronic energy into heat which is expulsed to the environment is the entropy production of the system. A laser is an example of microscopic dissipative structuring involving the degrees of freedom related to the distribution of energy over the internal electronic states of the material. Image credit: Bob Mellish, Creative Commons.

Microscopic self-organization through dissipation is what distinguishes life from other macroscopic dissipative structures that we usually consider as being abiotic in nature, such as hurricanes or convection cells. Such microscopic structuring permits the important concept of organismal memory of the external generalized chemical potential available in the organisms environment due to atomic mobility issues, that is, nanoscale structuring can persist even after removal of the external generalized chemical potential. This persisting structure can therefore become useful for resuming dissipation if the potential reappears in the organisms environment, or may be used as a base for a still more complex dissipative organization.

As an example, the stereo chemical coding in RNA and DNA for another antenna UV-C pigment, the amino acid tryptophan, happened because tryptophan could utilize (in a non-equilibrium thermodynamic sense) the conical intersection of RNA or DNA to rapidly dissipate its electronic excitation energy after the absorption of a UV-C photon. The RNA or DNA codon specifying for tryptophan was therefore part of a microscopic dissipative structure created during the Archean that has remained (and is even utilized) to this day, even though the external generalized chemical potential that created it (the UV-C photon potential) no longer exists in the environment. RNA and DNA thus eventually became information storing molecules, and that information can inevitably be associated with the dissipation of an externally imposed generalized chemical potential at some point in the organisms history. More on this subject of information encoding in relation to dissipation will be given in chapter 13.

The small size of the self-organized pigment is related to the short wavelength of the photons that they dissipate. At these wavelengths there is sufficient free energy in a single photon to break and re-make covalent bonds between atoms. Macroscopic abiotic dissipative structures such as hurricanes generally dissipate long wavelength light in the infrared (temperature gradients). Examples of microscopic dissipative structuring relevant to the origin of life, in particular to the production and proliferation of the nucleotides and their polymerization into strands of DNA or RNA will be given in chapters 6.1 and 6.2.

Such microscopic self-organization through dissipation and the persistence of nanoscale structuring would not be unique to Earth but should be occurring wherever there existed UV-C or UV-B light and the organic elements; on planets, protoplanetary systems, on asteroids, comets, and even in the galactic gas and dust clouds. In fact, evidence for the cosmic ubiquity of organic pigments will be given in chapter 5. On Earth, the increase in photon dissipation which came about through global proliferation of the pigments depended on the further complexification of material into mobile biotic organisms for transporting the pigments and the nutrients required for their proliferation to regions where sunlight exists but where nutrient content is low. This is the thermodynamic relevance of animals and more will be said about this in chapter 16.6. The alternative is also possible, that of diverting sunlight into regions of high organic nutrient content, and this may be a useful definition of an "intelligent" animal. Speculation on this kind of thermodynamic utility of intelligence attributed to humans for fomenting dissipation will be given in chapter 20.

The following box presents the extension to the formalism of Classical Irreversible Thermodynamics necessary for including internal degrees of freedom (inter-atomic or molecular) in the analysis, in order to explain microscopic dissipative structuring. This extension is due to Prigogine (1967)[289]. This box can be skipped by those with an adverseness to mathematical analysis without loosing much continuity of argument.

Irreversible processes can also arise to distribute the conserved quantities of Nature, such as energy, momentum, angular momentum, charge, etc., over internal degrees of freedom of the atom or molecule, and this leads to microscopic dissipative structuring. These internal degrees of freedom may be, for example; excitations, orientation of spin, orientation of electric dipole moment, molecular deformation, isomerization, etc.

The starting point is the Gibb's formula (see footnote on p. 40);

$$\frac{dS}{dt} = \frac{1}{T}\frac{dE}{dt} + \frac{p}{T}\frac{dV}{dt} - \frac{1}{T}\int_{\gamma} \mu(\gamma)\frac{\partial n(\gamma)}{\partial t} d\gamma \tag{3.5}$$

where $n(\gamma)$ is the density of molecules in state γ so $n(\gamma)d\gamma$ is number of molecules for which the internal parameter lies between γ and $\gamma + d\gamma$.

Consider a continuity equation for $\partial n(\gamma)/\partial t$. First assume that the change of γ is discrete, i.e. γ is changed by transformations from or into neighboring states $\gamma - 1$ or $\gamma + 1$. Then

$$\frac{dn_\gamma}{dt} + (v_\gamma - v_{\gamma-1}) = 0 \tag{3.6}$$

where v_γ is rate $\gamma \to (\gamma + 1)$ and $v_{\gamma-1}$ is rate of $(\gamma - 1) \to \gamma$.
If γ is a continuous parameter, then

$$\frac{\partial n(\gamma)}{\partial t} + \frac{\partial v(\gamma)}{\partial \gamma} = 0 \tag{3.7}$$

is a continuity equation in the "internal coordinate space" γ. $v(\gamma)$ is the reaction rate (flow) determining how the molecules change along coordinate γ.
In vector notation

$$\frac{\partial n(\gamma)}{\partial t} = -div\ \boldsymbol{v}(\gamma). \tag{3.8}$$

By partial integration, Eqn. (3.5) can be transformed into

$$\frac{dS}{dt} = \frac{1}{T}\frac{dE}{dt} + \frac{p}{T}\frac{dV}{dt} - \frac{1}{T}\int_{\gamma} \frac{\partial \mu(\gamma)}{\partial \gamma} v(\gamma) d\gamma \tag{3.9}$$

so the internal production of entropy is

$$\frac{d_i S}{dt} = -\frac{1}{T}\int_{\gamma} \frac{\partial \mu(\gamma)}{\partial \gamma} v(\gamma) d\gamma > 0. \tag{3.10}$$

Prigogine then postulated a further refinement to the second law of thermodynamics. In each part of the internal coordinate space, the irreversible processes proceed in a direction such that a positive entropy production results. This implies that

$$\sigma^* = -\frac{1}{T}\frac{\partial \mu(\gamma)}{\partial \gamma} v(\gamma) > 0 \tag{3.11}$$

σ^* is the entropy production per unit volume of the internal configuration space. σ^* has the usual form of a product of an affinity (or force) $-\frac{1}{T}(\partial\mu(\gamma)/\partial\gamma)$ and a rate $v(\gamma)$ of irreversible process.

If there is a potential energy which varies with γ (e.g. if γ is the angle θ of a dipole with respect to an external electric field e, then

$$E_{pot} = -me\cos(\theta) \tag{3.12}$$

where m is the dipole moment, there then appears a corresponding "force" $-\partial E/\partial\gamma$ in the entropy production,

$$\sigma^* = -\frac{1}{T}\left(\frac{\partial \mu(\gamma)}{\partial \gamma} + \frac{\partial E_{pot}}{\partial \gamma}\right) v(\gamma) > 0. \tag{3.13}$$

Such a thermodynamic analysis has advantages over a kinetic analysis, for example, by assuming that there exists a linear relation between the rate and the affinity, minimizing the entropy production leads to a formulation of Debye's theory of the orientation of dipoles in an alternating electrical field in the stationary state.

3.5 Why Irreversibility?

As we have seen in earlier sections of this chapter, material organizes into macroscopic or microscopic dissipative structures or processes under the imposition of an external generalized chemical potential. This is an apparent imperative of Nature, the dissipation of her conserved quantities over ever more microscopic degrees of freedom, the second law of thermodynamics when referring to the isolated global system, i.e. system plus its environment, but, a relevant question is; Why would Nature do this? It is this tendency of Nature which gives rise to irreversibility and to the material structuring seen in dissipative systems, one of which we call "life". This question has been contemplated ever since Boltzmann developed a statistical mechanical interpretation of thermodynamics. So far, however, no consensus on its resolution has been reached.

This question is, of course, intimately related with the question of an optimality principle in non-equilibrium thermodynamics. In fact, it appears to be just a restatement of the question regarding the reason for the existence of an optimality principle and it can be answered on similar grounds; that of probabilistic directions taken at bifurcations in driven non-linear systems (see section 3.3). However, since there is still no consensus, it is considered a still open problem. Some believe that a more complete understanding may require a more profound understanding of the nature of space and time, perhaps even requiring the acknowledgement of the 10 dimensional world invoked in string theory. Here, instead of going to such complex paradigms, we first present the traditional views on the nature of irreversibility and then suggest that the most reasonable answer has to do with the non-integrability of the dynamic equations due to bifurcations resulting from the non-linearity of driven systems, and that this can be related to Prigogine's proposal of resonances and nonlinearity destroying the integrability.

The most popular explanation of irreversibility, and one that Albert Einstein advocated, is that irreversibility is really only an illusion. The argument goes that the Newtonian equations of motion are reversible and therefore they could not bring any irreversibility into the dynamics. Under this view, Nature, through the time reversible equations of Newton, is deterministic and reversible and therefore, according to the argument, we should, in principle, be able to predict the future of the universe and know its past in exact detail, it is just that we do not have the computational power to determine the coordinates (for example, position and momentum) of such an enormous number of microscopic particles to be able to make any kind of reliable extrapolation to the past or to the future.

However, determinism and reversibility would negate the existence of a "free will" and, in fact, everything occurring at any time in the universe would be completely determined by the universe's initial conditions. The evolution of the cosmos would then just correspond to a great play unfolding as originally scripted through the initial conditions and the reversible Newtonian dynamical laws. There would be no place for true novelty or creation in the universe, and no free will of man.

Even more restrictive on this account is the "recurrence theorem" of Poincare which establishes that, for an isolated mechanical system with reversible dynamics, given sufficient time, the system will return arbitrarily close to its initial condition. However we never see such a regression to initial conditions in Nature. A drop of ink in a glass of water, for example, always disperses homogeneously and we never see the drop of ink reforming again (see figure 3.10). Is it because no truly isolated system exists, except perhaps the universe as a whole, and the universe is just so large in terms of degrees of freedom that the time needed to establish the recurrence result predicted by Poincare's theorem is impossibly

Fig. 3.10. A drop of ink placed into a glass of water will eventually disperse throughout the entire volume of the water and we will never see the drop reforming again, in contradiction to Poincare's recurrence theorem valid for an isolated system and reversible dynamics predicted by Newton's equations of motion. The fundamental reason behind this irreversibility is still being debated in the scientific literature and there is still no general consensus on its resolution. The present author believes that irreversibity is intimately related with the existence of an optimallity principle in Nature for entropy production and has to do with bifurcations at an unstable point in non-linear driven systems (see section 3.3). Image credit: Thesaint, from the stock.xchng.

large (as Einstein believed), or are there problems inherent in the assumptions made in the above arguments which render them invalid?

The statistical approach of Boltzmann to the problem of irreversibility was to presume that the system evolved through macroscopic states of ever greater probability with the probability of a given macrostate being proportional to the number of microstates consistent with that macrostate. Implicit in Boltzmann's view was the *ad hoc* assumption of *a priori* equal probability of the microstates. Boltzmann defined an entropy for the system as,

$$S = k_B \ln W \tag{3.14}$$

where W is the number of microstates consistent with the given macrostate, and k_B is known as Boltzmann's constant. Gibbs later developed a more general description of the entropy for the more realistic case in which not all the microstates have the same probability,

$$S = k_B \sum_i p_i \ln p_i \tag{3.15}$$

where p_i is the probability of microstate i and the sum is over all microstates. Boltzmann suggested that the universe started out in an extraordinarily improbable macrostate in which the conserved quantities were distributed over only a very small number of degrees of freedom (in other words, it began in a macrostate which was highly improbable, of very low entropy, consistent with only very few microstates). What we see today is the causal evolution of the universe towards an ever more probable macrostate, consistent with an ever larger number of corresponding microstates.

Recently, Ilya Prigogine has questioned the fundamental postulate of statistical mechanics, that of nature being found most probably in those macrostates compatible with the greatest number of microstates, with the implicit assumption (Boltzmann's original idea) that each microstate has *a priori* equal probability. Prigogine also insists that irreversibility is not an illusion as Einstein suggested, but due to the fact that interactions in real life are not local as assumed in the traditional statistical mechanics interpretation of thermodynamics, but persistent over both space and time. This allows for *resonances* to form which decay as probabilistic quantum events. Prigogine suggests that the true equations of motion are not reversible because they are probabilistic, just like the decay of an unstable isotope. We cannot predict exactly when an atom of Uranium is going to decay, only the probability of a certain number of decays within a certain time interval.

Although not affecting the formalism of statistical mechanics or thermodynamics, these new ideas on irreversibility, if correct, would have profound implications for our understanding of Nature and, as an unforeseen advantage, may even resolve some of the fundamental paradoxes in the theory of quantum mechanics; such as the role of the observer in bringing irreversibility into the quantum mechanical description of a process[3] (Prigogine, 1996)[292].

In his last book, written shortly before his death, entitled "The End of Certainty: Time, Chaos, and the New Laws of Nature" , published in 1996 [292], Prigogine expounded on these ideas and issues and argues that the apparently irreconcilable notions of the evolution of natural systems towards novel creation through dissipation (irreversibility), or the apparent free will of man, with the

[3] Quantum mechanics, like classical mechanics, is based on a time reversible equation for the dynamics, the Schrödinger equation. Irreversibility is obtained through the colapse of the wavefunction into a particular state, and this collapse into a particular state is determined by the act of observation.

deterministic laws of classical or quantum mechanics, is not somehow the result of our incomplete knowledge of the coordinates or the initial conditions, the positions and momenta of all the $\gg 10^{23}$ particles in a macroscopic system, as assumed in the common interpretation of irreversibility, but rather due to the fact that the fundamental dynamical laws of Nature are in reality not *deterministic* but rather *probabilistic.*

Prigogine argues that thermodynamics was derived from, and has dwelt too long on, ideal systems with local interactions leading to well defined trajectories, and that instead, it is important to consider real systems with non-local interaction where resonances between particles or systems allows the extensive thermodynamic variables (energy, momentum, angular momentum, etc.) to be dispersed over new *non-dynamical degrees of freedom* in a probabilistic manner, thus rendering the system non-intergrable.

However, even if we incorrectly assume integrable systems with point-like particles and interactions, Prigogine emphasizes that the non-linearity inherent in natural systems can lead to trajectories starting with infinitely close initial conditions diverging exponentially with time, and, since no experiment or theory can give infinitely precise initial conditions, the true description of the dynamics in Nature must, in any case, be in the form of probability densities rather than individual trajectories. Irreversibility, therefore, under this view, is not an illusion and cannot be simply derived from an *ad hoc* assumption of an *a priori* equal probability of the microstates, as Boltzmann had suggested. The direction of time is not an illusion due to the "approximations that we introduce in our description of nature", as Einstein insisted, but rather a real constructive force leading to the creation of new dissipative structures such as life and a "free will".

Resonances, in this view, are the means by which Nature conveniently distributes her conserved macroscopic quantities (energy, momentum, angular momentum, charge, etc.) over ever more microscopic degrees of freedom. It appears that Nature creates non-equilibrium dissipative systems which can support these resonances and thereby increase the distribution of the conserved quantities (i.e. increase the entropy production) even though their construction requires going against a free energy gradient that, in systems close to equilibrium, although not forbidden, would have vanishingly small probability. For example, organic pigments are systems that are constructed by Nature against a free energy gradient and these pigments produce resonances when they absorb and dissipate into heat the high energy incoming solar photons. The production of these pigments requires free energy and thus does not occur spontaneously in systems near equilibrium. However, in non-equilibrium situations, in which free energy is available from the environment (for example, from an incident photon flux), such molecules will indeed "spontaneously" emerge under this photon flux since,

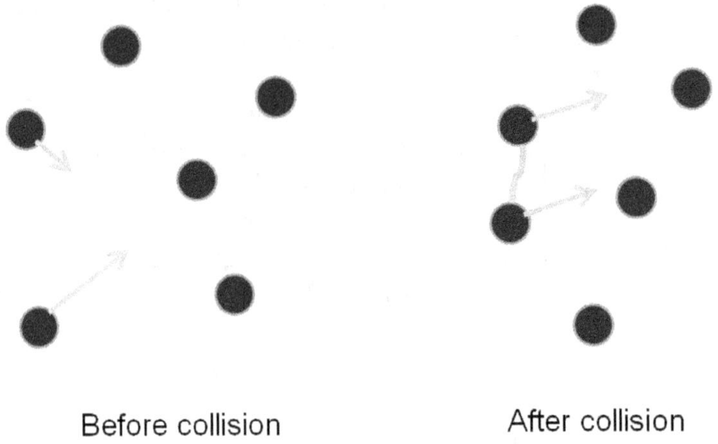

Fig. 3.11. A resonance or correlation is established between two particles as a result of a collision. The resonance decays probabilistically according to quantum mechanical laws and therefore the equations of motion become non-integrable. Only probability densities can be given for the trajectories, and this introduces irreversibility. Adapted from Prigogine (1996)[292].

although the pigments themselves are low entropy systems, they foment greater entropy production by dissipating the same incident high energy solar photon flux that produced them (an autocatalytic system). This will be described in more detail and will be given mathematical form in the following chapters.

The relevance of the latest work of Prigogine has yet to be thoroughly assessed by his peers in the physics and chemistry community at large, and such paradigm changes are notoriously slow to become commonly accepted, even when obviously merited through experiment. My position here is to simply present the reader with what seems to me to be the most reasonable explanation of irreversibility consistent with the physics known today. This new interpretation of irreversibility suggested by Prigogine may be proven to be wrong or in need of reform, but if true would profoundly affect the way scientists view the world. However, its validity or not has no direct impact on the thesis presented here, although it would certainly help to provide a clear understanding and provide a theoretical foundation to the empirical observation and basic premise of this book, that of life as a dynamic dissipative structure self-organized under an external generalized chemical potential.

4. Photon Dissipation and the Origin of Life

Accepting the inescapable fact that life, like any other irreversible process, requires the dissipation of an external potential to originate, persist, proliferate, and evolve, then the first step to be taken in developing a serious theory for the origin of life would be to search for the external potential driving the process of life. From this thermodynamic perspective, it is also clear that it does not make sense to dwell on the living organism itself while ignoring the external potential, this would be tantamount, for example, to ignoring the temperature gradient resulting from the warm ocean surface and the cold upper atmosphere when attempting to understand the origin, persistence, and evolution of a hurricane.

We should begin our search for an external potential which gave rise to the origin of life by considering first the potential that life dissipates today. Starting from our familiar, anthropic, top down perspective of the ecological pyramid, we see that the biological world consists of a few predators that dissipate the chemical potential stored in the proteins, fats and carbohydrates of herbivores, and many herbivores that dissipate the chemical potential stored in the carbohydrates (sugars, starches and cellulose) of plants and cyanobacteria, and we would find an enormous multitude of plants and cyanobacteria that dissipate the chemical potential stored within the molecular energy storehouses of life, such as adenosinetriphosphate (ATP). However, at the true base of this dissipative ecological pyramid, at a microscopic level, we would find an astronomical number of molecular pigment complexes (e.g. chlorophyll, carotenoids, anthocyanins, scytonemins, mycosporines, etc.) that dissipate the free energy in the UV and visible solar photons (that short wavelength light from the Sun incident on Earth's surface).

Given the extraordinary large number of photons incident on Earth's surface (on the order of 10^{22} $m^{-2}s^{-1}$) and of the great number of pigment molecules covering Earth's surface (on the order of 10^{16} m^{-2}; Michaelian, 2015 [218]), and given that most of the highest energy photons arriving at Earth's surface are indeed intercepted and dissipated into heat by these organic pigments, it is logical to conclude that biological evolution has been overwhelmingly concerned with the dissipation of the free energy in the solar photons and it is this dissipation that not only sustains, but, at the same time, in a type of great autocatalytic cycle with positive feedback, is dependent upon, the rest of the biosphere.

The external potential that life dissipates overwhelmingly today is thus the solar photon potential and the dissipation performed by these pigment molecules when in a water environment and exposed to the sun's rays, is the conversion of short wavelength light (UV and visible) into long wavelength light (infrared). A large percentage of the high energy photons (UV and visible) hitting the Earth's surface are converted directly into many more infrared photons by organic pigment molecules in water. For example, one UV-C photon within the wavelength region absorbed by RNA and DNA (230-290 nm) can be converted into between 30 to 40 infrared photons (see Fig. 2.3). This dissipation of the conserved energy is the fundamental thermodynamic function performed by life today (Michaelian, 2011; 2012a)[210, 212]. It is this dissipation of the solar photon potential which drives the dynamic organization of the material of life into organic pigments and their supporting structures (microscopic examples of Prigogine's dissipative structures). Today these structures are formed through complex and indirect cellular biosynthetic pathways, but at the very beginning of life, before the evolution of these complex pathways, they must have been formed directly through the available energy (or, more generally, the low entropy) available in the UV-C component of the solar photon flux of the Archean.

It is more than a curious fact that organic pigments on Earth are absorbing exactly in those wavelength regions where water does not (Stomp et al., 2007)[362] (see figure 4.1). The conventional wisdom addresses this in terms of light niches available for photosynthesis and this sounds plausible until one looks deeper into the process of photosynthesis. Not all pigments donate their energy to photosynthetic reaction centers, most just dissipate their electronic excitation energy directly into heat. In fact, in higher plants and algae, a large number of photosystem II reaction centers do not participate at all in photosynthesis (Chylla and Whitmarsh, 1989)[47]. Photosynthesis accounts for only about 0.1% of free energy usage and is really only truly active at around 700 nm within the red Q_v absorption band of chlorophyll. This means that those photons absorbed at the blue Soret band of chlorophyll must first be dissipated into the red Q_v band energies before the energy can be utilized for photosynthesis. Therefore, even the absorption of photons by chlorophyll for the process of photosynthesis implies a great amount of dissipation.

The wavelengths where pigments absorb, which is just where water is transparent (see figure 4.1), overlap with the wavelengths of the sun's maximum energy output and also with an atmospheric window allowing photons to reach Earth's surface. This extraordinary wavelength overlap of energy source, atmospheric and oceanic windows, and life absorption, is not fortuitous, nor is it

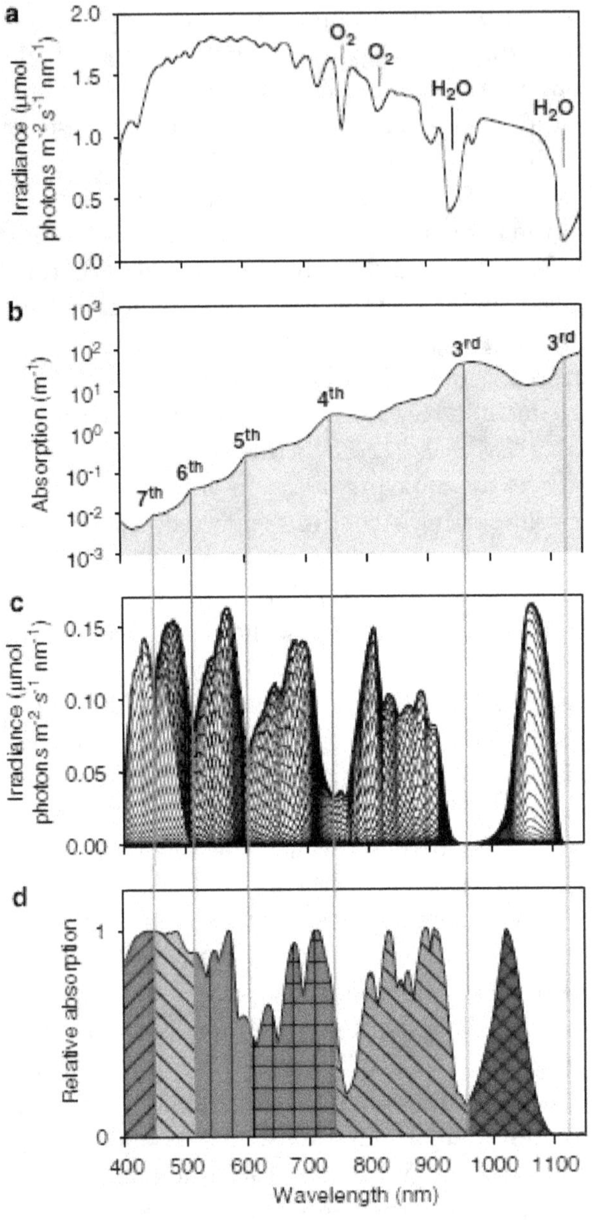

Fig. 4.1. The absorption spectrum of water creates a series of distinct niches in the underwater light spectrum. (a) Light spectrum of the incident solar irradiance at the water surface. Dips in the incident irradiance are caused by absorption of photons by oxygen and water molecules in the atmosphere. (b) Absorption spectrum of pure water, plotted on a log scale. The different harmonics of the stretching and bending vibrations of the water molecule are indicated. (c) Overlay of 100 underwater light spectra at the euphotic depth. The light spectra were obtained using a wide range of different gilvin and tripton concentrations, from the clearest ocean waters to very turbid systems such as microbial mats. (d) Overlay of measured light absorption spectra of 20 phototrophic species, including purple sulfur bacteria, green sulfur bacteria, purple sulfur bacteria, green sulfur bacteria, purple non-sulfur bacteria, cyanobacteria, green algae, red algae, diatoms and chrysophytes. It is obvious from the diagram that pigments in organisms are absorbing in spectrum regions exactly where water does not absorb. Image credit: Stomp et al. (2007)[362]. Reprinted with permission from Macmillan Publishers Ltd.

the result of evolutionary optimization of photosynthesis, but rather the result of a thermodynamic imperative to optimize dissipation of solar photons.

The process of photon dissipation into heat characterizes the thermodynamic work performed by life today more than any other process, including photosynthesis. For example, as stated above, photosynthesis accounts for less than 0.1% utilization (Gates, 1980)[100], while photon dissipation into heat accounts for 99.9% of the utilization of the free energy in sunlight captured by life.

Photon dissipation in pigments couples to, and foments, other dissipative processes such as the water cycle. Absorption and dissipation into heat of sunlight at the leaves of plants increases their temperature by as much as 20 K over that of the ambient air (Gates, 1980)[100]. This leads to an increase of the water vapor pressure inside the cavities of the leaf with respect to that of the colder surrounding air. Water vapor thus diffuses across this gradient of chemical potential from the wet mesophyll cell walls (containing the chloroplasts), through the inter-cellular cavities, and finally through the stoma and into the external atmosphere. There is also a parallel, but less important, circuit for diffusion of water vapor in leaves through the cuticle, providing up to 10% more transpiration (Gates, 1980)[100]. The H_2O chemical potential of the air at the leaf surface itself depends on the ambient relative humidity and temperature, and thus on such factors as the local wind speed and insolation. Diffusion of water vapor into the atmosphere causes a drop in the water potential inside the leaf which provides the force to draw up new water from the root system of the plants.

Evaporation from moist turf (dense cut grass) can reach 80% of that of a natural water surface such as a lake [100](Gates, 1980), while that of a tropical forest can often surpass by 200% that of such a water surface. Single trees in the Amazon Rainforest have been measured to evaporate as much as 1180 l of water per day (Wullschleger et al., 1998)[415]. This is principally due to the much larger surface area for evaporation that a tree offers with all of its leaves as compared to the two-dimensional area projected by the tree onto the Earth's surface. Natural water surfaces, in turn, evaporate approximately 8% more than distilled water surfaces due to the increased UV and visible photon absorption at the surface as a result of phytoplankton and other suspended organic materials, including a large component (up to 10^9/ml) of viral and dissolved DNA, resulting from viral lysing of bacteria (Wommack and Colwell, 2000)[410]. In surface waters with rich ecosystems, neuston (microscopic crustaceans) can increase evaporation 3-fold by stirring the skin layer with their flagella (MacIntyre, 1974)[195].

It is evident from the above that the thermodynamic work of photon dissipation performed by life today has a long history which probably goes back to life's beginnings and has been evolving over time to increase photon dissipation and to couple it to other irreversible processes operating in the biosphere, including the water cycle and ocean and wind currents. However, without the contemporary complex biosynthetic pathways leading to life's contemporary pigments and its universal energy storage molecule, adenosine triphosphate (ATP) (see figure 4.2), life most likely began by utilizing the free energy available in UV-C photons to make and break covalent bonds to form pigments which could absorb and dissipate the solar photons. It is probable then that life began dissipating in this restricted region of the solar spectrum. Photons of too large an energy would lead to ionization of a molecule (the loss of an electron) and thus their destruction, while photons of too low an energy would not be able to break or make covalent bonds to form the pigments. The ideal region of the solar spectrum for starting life would thus be in the UV-C and the high energy part of the UV-B.

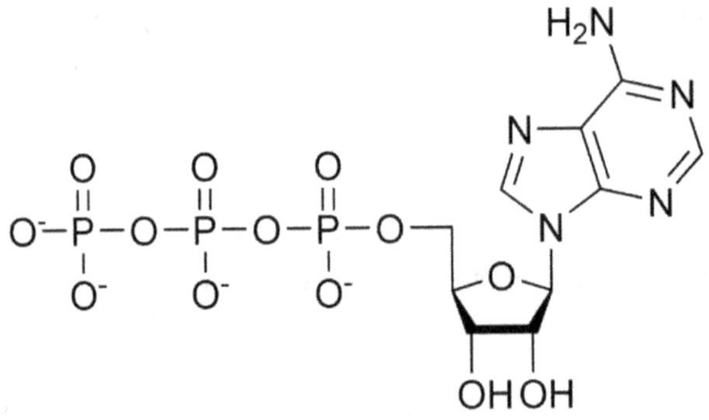

Fig. 4.2. Adenosine triphosphate (ATP). The molecule which today acts as an energy storage device (battery) for most of life's irreversible processes; e.g. muscle contraction, the establishment of electrochemical gradients across membranes, and biosynthetic processes necessary to maintain life, in particular, contemporary pigment production. Hydrolysis, leading to the breaking of any of the phosphoanhydride bonds, releases energy into the local environment, the amount being dependent on the enviornment (particularly the amount of Mg^{+2} ions available to stabelize ATP), but approximately 0.3-0.5 eV for one hydrolysis event leading to adenosine diphosphate (ADP) and 0.6-1.0 eV for two hydorlysis events leading to adenosine monophosphate (AMP). At the beginnings of life, before complex biosynthetic pathways had evolved, pigment production would have had to occur through direct UV-C or UV-B photochemical reactions. A photon in the UV-C range of 230-290 nm would supply directly an energy of between 5.3 and 4.3 eV. Image credit: NEUROtiker, Public Domain.

It is convincingly argued in chapter 16.6 that insects and other animals (including humans) obtain a measure of their thermodynamic relevance to the dissipative biosphere, and therefore a measure of their unambiguous "fitness", in terms of how they aid the organic pigments in improving their photon dissipation by fomenting their proliferation and facilitating their dissemination into initially inhospitable areas by spreading nutrients and seeds through excretion and death (Michaelian, 2009; 2011)[207, 210], and, more recently, with the invention of flowers, through pollination. The programmed death of individual animals can, in fact, be argued to be an intricate component of their fitness function in assuring a continual and spatially homogeneous distribution of photosynthetic life (Werfel, 2015)[397][1]. The dissipation of the chemical potential stored within the organic material of photosynthetic organisms carried out by animal metabolism is only a minor, or secondary, adjunct to the true thermodynamic work performed by animals.

The thermodynamic dissipation theory of the origin of life (Michaelian, 2009; 2011)[207, 210] takes the insight of Boltzmann and the work of Prigogine to their ultimate consequences regarding the origin of life. The theory is based on a simple logical supposition: If the dynamic irreversible process known as "life" is today overwhelmingly driven by dissipating the solar photon flux (UV and visible light) into heat (infrared light), and since the solar photon potential is, and always has been, the most intense external source of free energy (energy available to be distributed over still more microscopic degrees of freedom) prevailing at the surface of Earth, then it is reasonable to assume that life was doing something very similar with solar photons 3.85 thousand million years ago when it is suspected to have first arisen.

Continuity of thermodynamic function would thus argue that life began dissipating sunlight on the Archean ocean surface. However, at the dawn of life the spectral distribution of sunlight arriving at Earth's surface was different from that of today and this will be discussed in detail in the following section.

4.1 The Sun through Time

Through studying nearby spectral type G and luminosity class V (G-V-type) main sequence stars similar to our Sun, of 0.8 to 1.2 solar masses and surface temperatures of between 5300 and 6000 K, which have known rotational periods and well-determined physical properties, including temperatures, luminosities, metal abundance, and ages, Dorren and Guinan (1994)[83] were able to reconstruct the most probable evolution of our Sun's characteristics over time, in particular, of importance here, the evolution of its spectral emission. This "Sun in Time" project has been carried out using various satellite-mounted telescopes including ROSAT, Chandra, Hubble, and EUVE and now has representative photon flux data for our Sun's main sequence lifetime from 130 Ma to 8.5 Ga.

[1] Werfel (2015)[397] considers this continual and homogeneous distribution of photosynthetic life as optimal in terms of available resources for individual organisms, however, it is clear that the most fundamental thermodynamic utility of such a homogeneous distribution of organic pigments is its optimality in terms of photon dissipation. In fact, it is claimed in this book that the most important hallmark of evolution is the homogeneous distribution of pigments over all of Earth's surface.

Over the lifetime of G-type main sequence stars, emitted wavelengths of shorter than 150 nm originate predominantly in the chromosphere and corona (stellar atmosphere) in solar flares resulting from magnetic disturbances. The high rotation rate of a young star gives it an initially large magnetic field, before significant magnetic breaking sets in, and thus intense and more frequent solar flares leading to large fluxes of these very high energy photons. In figure 4.3 the short wavelength flux intensities as a function of the age of a G-type star are given. From the figure it can be seen that our Sun at 500 Ma (at the probable beginnings of life on Earth ~3.85 Ga) would have been 5 to 80 times (depending on wavelength) more intense at these very short wavelengths.

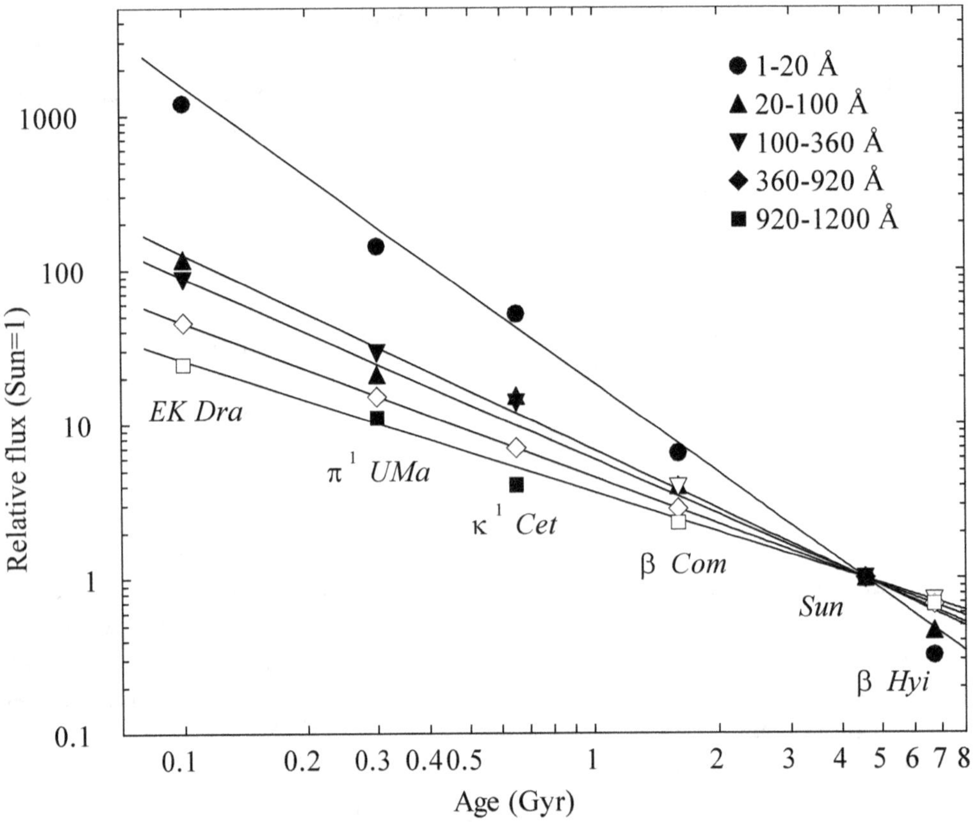

Fig. 4.3. Solar-normalized stellar fluxes (with respect to those of today) vs. stellar age for different wavelength bands for solar-type stars. The names of the proxy stars used of different ages are listed on the graph. Image credit: Ribas et al. (2005)[299]. © AAS. Reproduced with permission.

Wavelengths longer than 150 nm are known to originate on the photosphere and are emitted essentially in a blackbody spectrum (apart from a few strong stellar atmospheric absorption lines) at the star's effective surface temperature (Cnossen et al., 2007)[53]. The effective surface temperature of a star is related to its visible luminosity, L_S, and radius, r, by the Stefan–Boltzmann law,

$$T_{eff} = \left(\frac{L_s}{4\pi\sigma r^2}\right)^{1/4}$$

where σ is the Stefan–Boltzmann constant. The luminosity of a star is an increasing function of its age because hydrogen fusion begins predominantly in the core and gradually proceeds outwards as helium "ash" settles into the core (Karam, 2003)[149]. The temperature of a star is also a monotonically increasing function of its age (Bahcall et al., 2001) and at the origin of life the temperature of our Sun's surface was approximately 1.6% less than today while its radius has increased by more than 11% and it is this increase in radius that is responsible for the greater part of the increase in emitted energy flux. The net result for our Sun is that its effective surface temperature and surface area have been increasing steadily, implying a total integrated energy flux of about 30% less, and an integrated UV flux (176–300 nm) about 15% less, at the origin of life as compared to that of today (Karam, 2003)[149]. However, Compton scattering of the extremely high energy ultraviolet rays from solar flares into the 230 -290 nm window of Earth's Archean atmosphere would have occurred, compensating to some extent the decrease in the UV-C flux.

All of this would imply a photon distribution at Earth's Archean surface having a visible spectrum of black-body nature very similar (with respect to wavelength) to that of today but of ~30% reduced intensity (absorption and scattering in Earth's atmosphere, particularly on water gas, were also more important during the Archean), and an ultraviolet spectrum similar to that arriving today at Earth's upper atmosphere, but with extinction below 230 nm due to CO_2 and N_2 absorption, and between 290 nm and 320 nm due to hydrogen sulfide, H_2S, absorption, and formaldehyde and acetaldehyde absorption[2] (Michaelian and Simeonov, 2015)[217]. Figure 4.4 gives the approximate solar spectrum expected at Earth's surface as a function of time since the present, calculated taking into account solar evolution and atmospheric absorption and scattering using the best available data for atmospheric gasses and their densities over time (see Michaelian and Simeonov, (2015)[217] for details).

The reduced solar output at the origin of life, under atmospheric conditions similar to those of today, would have produced a snowball Earth, a globe completely frozen over from the poles to the equator. However, there is convincing

[2] Both formaldehyde and acetaldehyde are common photochemical reaction products of UV light on volcanic gases such as hydrogen sulfide, carbon dioxide, and water vapor; Sagan, (1973)[309].

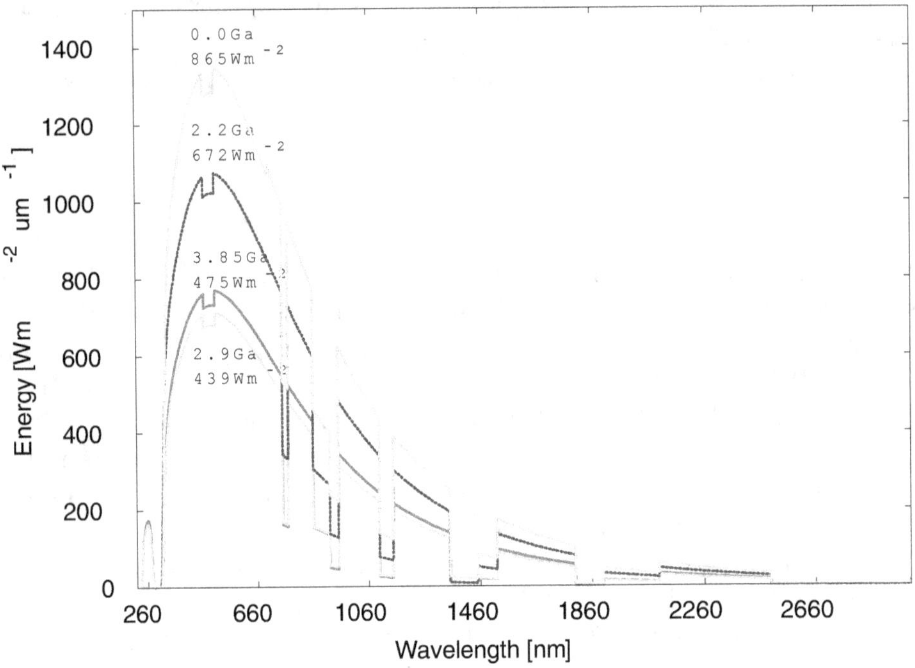

Fig. 4.4. Calculated solar spectra at Earth's surface at particular times since present assuming the standard solar evolutionary model and using the best available data for the atmospheric gasses, their densities, and Earth's surface temperature. The total integrated (over all wavelengths) energy flows midday at the Equator are given below the dates. The integrated energy flux in the UV-C peak centered at 260 nm and calculated betweeen 230 and 290 nm at 3.85 Ga is approximately 5.5 W m^{-2}. An expansion of the short wavelength region is presented in figure 4.7. Image credit: Michaelian and Simeonov (2015)[217]. Creative Commons Attribution 3.0

evidence, in the form of the oxygen isotope content of zircon crystals, for liquid water on the surface of Earth from the Hadean (4.55 - 3.9 Ga) and into the early Archean (3.9-2.9) (Mojzsis et al., 2001)[230]. This contradiction has become known as the "faint young Sun paradox". The resolution of this paradox most likely has to do with the large amount of greenhouse gasses in Earth's early atmosphere, particularly carbon dioxide, CO_2, methane, CH_4, and water vapor H_2O. However, the possibility of a distinct model for the evolution of our Sun should not be completely discarded (Michaelian and Manuel, 2011)[211].

What was most important to the origin of life, however, was the UV-C band between 230 and 290 nm peaking at 260 nm that would have made it to the Earth's surface during the Archean (see figure 4.4). We have estimated that the integrated flux of this light would have been of the order of 5.5 W/m^2 midday at the equator and that it would have been available at the Earth's surface before the beginnings of life and for a duration of about one billion years thereafter (Michaelian and Simeonov, 2015)[217]. This light would have had enough energy to make and break covalent bonds directly, but not enough energy to ionize molecules. Today, this sunlight still arrives at Earth's upper atmosphere but is strongly dissipated by life produced atmospheric oxygen and ozone and therefore no longer reaches the surface.

4.2 Of Pigments and Protectionism

There are, therefore, two very certain premises on which to base a model for the Archean surface solar spectrum; 1) the standard solar model for G-type stars given in the previous section, and 2) with no oxygen in Earth's atmosphere to allow the production of ozone during the Archean, there would probably have been a window of atmospheric transparency that would have allowed ultraviolet light to penetrate to the surface. The eminent scientist, and perhaps greatest science missionary of the 20th century, Carl Sagan was the first to quantify this and he calculated a very intense integrated flux of about 3.4 W/m^2 of ultraviolet light peaking at about 260 nm, in the UV-C, with the spectrum extending from about 240 nm to 290 nm (Sagan, 1973)[309], the short wavelength limit set by absorption of CO_2 and N_2 in the atmosphere and the long wavelength limit set by aldehyde absorption (formed by UV-C light on common volcanic gases such as H_2S, CO_2, and H_2O).

At the time of Sagan's calculation (the early 1970's), ultraviolet light was considered as being fatal to life, given its known potential for destruction of proteins, and Sagan looked for ways in which early life could have been protected from these, what he called, "lethal" rays (Sagan and Chyba, 1997)[311]. This protectionist thinking has unfortunately permeated the origin of life sciences until the present day. For example, the extraordinary large set of organic pigments found in photosynthetic organisms, which cover the region from the UV-C to the red-edge[3], are usually, without careful deliberation, assigned to either a protective role, or to antenna pigments for the photosynthetic apparatus. The argument is based on the common, but mistaken, assumption that plants and cyanobacteria have evolved to optimize the rate of photosynthesis, the fixation of carbon into organic material, under variation of external conditions. The antenna and protective pigments would be relevant in increasing the capture rate of photons for photosynthesis and in protecting the photosynthetic apparatus, respectively, thus obtaining maximum photosynthetic carbon fixation rates. However, Wang et al. (2007)[392] have shown that under variation

[3] The "red-edge" refers to the wavelength at which the absorption spectrum of photosynthetic organisms drops abruptly. This happens at roughly 700 nm for almost all photosynthetic organisms. A thermodynamic explanation of the red-edge, related to optimizing photon dissipation, has been given in (Michaelian, 2015)[218] and is presented briefly in chapter 17.5.2.

of external conditions, plants optimize evapotranspiration, not photosynthesis. In fact, absorbing and dissipating the lethal rays is not the only, nor the best, way to render the rays inoffensive to plants and cyanobacteria, reflection and transmission are much less resource dependent and stressful to the organism.

Other protectionists, advocates of the deep ocean hydrothermal vent hypothesis of the origin of life, argue for the advantage of being shielded from the lethal Archean ultraviolet rays by the overlying water. Still another, more contemporary, example of this protectionist thinking is the suggestion that DNA and RNA evolved their extremely rapid sub picosecond electronic de-excitation characteristics (see chapter 7) in order to afford these molecules protection against photochemical destruction (Mulkidjanian et al., 2003; Serrano-Andres and Merchan, 2009)[238, 333], since photoreactions are much more likely to occur while molecules are in an electronic excited state.

Stability against UV destruction or unfavorable photochemical reactions would certainly have been an advantage in getting life off to a robust start, however, Nature had a simpler way of making the fundamental molecules oblivious to this ultraviolet light, if this were really the intention; by making them either transparent or highly reflective to UV-C light. Dissipation is not a prerequisite for protection, but it is for dynamical structuring of material, so the dissipative perspective on pigment ubiquity makes much more sense than the protectionist perspective. The characteristic strong absorption and rapid dissipation of photons by the fundamental molecules does not seem to be a clever product of natural selection to provide organisms with UV protection, but rather was the fundamental thermodynamic reason for the origin and evolution of life (Michaelian, 2009; 2011)[207, 210]. Today, the structuring of material into pigments and their supporting structures, which together provide for efficient photon dissipation, occurs in response to the externally imposed photon potential, and this is the thermodynamic function of life. The atmospheric window in the Archean between 230 and 290 nm, centered at 260 nm, is exactly where RNA and DNA have maxima in both absorption and dissipation (see figure 4.5).

Cnossen et al. (2007)[53] have carried out detailed simulations of photon absorption and scattering for various hypothetical models of the Earth's early atmosphere. Their models consider different CO_2 concentrations at different pressures and include absorption, Rayleigh scattering, and an estimate of the effects of multiple scattering, besides taking into account the best estimates for the UV intensity expected for a young Sun. They have determined that the ratio of UV-C light reaching the Archean Earth surface compared to that of today would have had a peak at approximately 260 nm and be about 10^{30} times more intense than that of today (see figure 4.5). This exceedingly large ratio is not surprising since today almost all of UV-C light is blocked by the oxygen and ozone in Earth's atmosphere. What is notable, however, is that 260 nm is exactly where RNA and DNA absorbs and dissipates strongly, suggesting that they were UV-C dissipating pigment during the Archean (see figure 4.5).

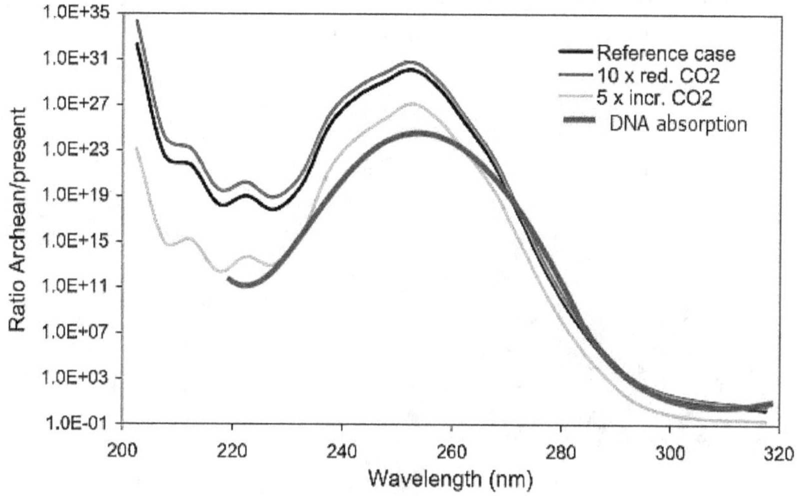

Fig. 4.5. Calculations of Cnossen et al. (2007)[53] of the ratio of UV-C light in the Archean to that of today at Earth's surface for different estimates of the amount of CO_2 in the atmosphere relative to the standard reference case (0.24 bar - lower limit for resolving the faint young son paradox). At the wavelength of 260 nm, where RNA and DNA absorb strongly, the amount of light reaching Earth's surface was 10^{30} times greater during the Archean than what it is today. In blue is the approximate absorption spectrum of short length (25 bp) DNA (the RNA absorption spectrum is very similar). It peaks almost exactly at the wavelength at which the plotted ratio is greatest, which is, in fact, where most UV-C light was reaching the surface. This is a strong indication that DNA and RNA were UV-C pigments during the Archean, and is strong evidence in favor of the thermodynamic dissipation theory for the orgin of life. Image credit: Adapted from Cnossen et al. (2007b)[54] with permssion from the American Geophysical Union.

There is, in fact, clear evidence of anoxygenic photosynthetic bacteria forming stromatolites that thrived between 3.47–3.33 Ga in surface littoral sediments, found in the Baberton greenstone belt of South Africa (Westall et al., 2006)[399]. More recent, but more controversial, evidence indicates that stromatolites may have already formed at 3.7 Ga (Nutman et al., 2016)[253]. This implies that either, for some unknown reason, the UV-C light at Earth's surface was nowhere near as intense as has been repeatedly calculated (Sagan, 1973; Cnossen et al. 2007; Michaelian and Simeonov, 2015)[309, 53, 217], or that the organisms were fully adapted to surviving and proliferating in an intense UV-C environment, which, of course, would have been the case if life had started out dissipating UV-C light (see figure 4.6). Westall et al. (2006)[399] suggest that protectionist and/or RNA or DNA repair mechanisms may have been operating. However, from the thermodynamic dissipation viewpoint, life was efficient at dissipating

UV-C light into harmless heat during the Archean as this is exactly what it was designed to do, being its fundamental thermodynamic reason for existence. In the following section I provide further evidence for the suggestion that not only RNA and DNA, but most of the fundamental molecules, and life in general, were optimized for UV-C dissipation from life's inception and throughout the Archean.

Fig. 4.6. Stromatolites, colonies of photosynthetic bacteria, like the ones shown here from Shark Bay, Australia, were present during the Archean. There is an indication in the fossil record that stromatolites as old as 3.7 Ga may have existed. Careful analysis of anoxygenic stromatolites from $\sim$ 3.47 Ga (see Westall et al., 2006 [399]) has shown that some of these come from a littoral environment, indicating that they must have been thriving under a high flux of UV-C light. Such a result could be understood, and indeed expected, if life started out being driven into existence through dissipation of this wavelength region. Image credit: Paul Harrison (Reading, UK). CC BY-SA 3.0 .

4.3 The Fundamental Molecules of Life are Pigments in the UV-C

An extraordinary fact, and one in favor of the thermodynamic dissipation theory of the origin and evolution of life, is that not only RNA and DNA, but most of the fundamental molecules of life (those found in all three domains of life) have strong UV-C absorption properties in exactly the wavelength region of the predicted Archean atmospheric window (for example, the aromatic amino acids, organic cofactors, vitamins, carotenoids, porphyrins, chlorins, flavins, flavonoids, etc.) and most of these also have strong chemical affinity (attraction) to RNA and DNA (Michaelian and Simeonov, 2015)[217]. The equivalence of the wavelength region of absorption of the fundamental molecules of life with the same wavelength region of the predicted atmospheric window in the Archean (see table 4.1 and Fig. 4.7) and their chemical affinity to RNA and DNA is undoubtedly not coincidental, but should be interpreted as strong evidence for the postulate that incipient life was the non-equilibrium thermodynamic microscopic dissipative structuring of material into pigments to dissipate the prevailing UV-C solar photon potential. The fact that none of the fundamental molecules absorb within the region 285 - 310 nm (see figure 4.7) where the atmospheric aldehydes (formed by UV light on the volcanic gases) strongly absorbed in the Archean (see previous section, figure 4.7, and Sagan, 1973)[309]) is additional evidence in favor of the proposition that the fundamental molecules of life were pigments filling atmospheric and water windows for optimal dissipation of the solar spectrum, just as pigments do today (Fig. 4.1).

Viewed from this thermodynamic perspective, many of RNA's and DNA's salient properties begin to make sense (see chapter 7) and by considering these properties in this new light of dissipation we obtain new insight into the origin and even purpose of life ... yes, from this perspective, life had, and still has, a "purpose" (a reason for being); that of photon dissipation. Photon dissipation was not something previously considered as being fundamental or even relevant to life, but in this new theory it is the process that takes center stage over the entire evolutive history of life. The thermodynamic dissipation theory of the origin and evolution of life suggests that the atoms that make up the bases of RNA and DNA and the other fundamental molecules (see table 4.1) organized "spontaneously" into these molecular pigment structures under the action of the Archean UV-C solar potential in order to dissipate this potential, just as atoms in today's visible pigments in plants and cyanobacteria become organized, albeit through complicated biosynthetic pathways, utilizing the free energy available in visible light, to dissipate the visible spectrum, or, just as the atmospheric material of a hurricane organizes, by utilizing the free energy available in an ocean surface-atmosphere temperature gradient, to more efficiently dissipate this temperature gradient.

Here then is our proposition concerning the origin and evolution of life that takes Boltzmann's insight on the struggle in the animal world for negative entropy and the material organizing principle of Prigogine to their ultimate consequences: *The origin and evolution of life was, and still is, overwhelmingly that*

Molecules common to all 3 domains of life	Absorbance maximum (nm)	Molar extinction coefficient ($cm^{-1}M^{-1}$)
Adenine	261	13400
Guanine	243, 272.75	10700, 13170
Thymine	263.8	7900
Cytosine	266.5	6100
Uracil	258.3	8200
NAD, NADP	260	15000
NADH, NADPH	260, 340	15000, 6200
FMN	260, 375, 445	15000, 10000, 12500,
FAD	260, 375, 450	15000, 9000, 11500
pyridoxal	250, 320	3000, 6000
Phenylalanine	257.5	195
Tyrosine	274.2	1405
Tryptophan	278	5579
Histidine	211	5700
Thiamine	235, 267	11300, 8300
Riboflavin	263, 346, 447	34845, 13751, 13222
Folic acid	256, 283, 368	26900, 25100, 9120
Nicotinamide	262	2780
Pyridoxine	253, 325	3700, 7100
Ubiquinone-10	275	14240
Phytomenadione	248, 261, 270	18900, -, -
Hydroxocobalamin	277.75, 361, 549.75	15478, 27500, 8769
Protoporphyrin (aggregate)		-
Phytoene (Carotenoid)	275	-
Chlorin	250, 431, 669	-, -, -
Scytonemin	253	-
Chlorophyll b	258	-
Phospholipids	230-290	-

Table 4.1. A list of the fundamental molecules of life (those found in all three domains) and their wavelength of maximum absorption along with their extinction coefficients at maximum absorption. The wavelength of the absorption maximum of the molecules are plotted in Fig. 4.7. (Adapted from Michaelian and Simeonov, 2015)[217]. Creative Commons Attribution 3.0

thermodynamically driven process of material organization (microscopic dissipative structuring) into pigments and their dispersal over Earth's surface to absorb and dissipate ever more efficiently and completely the prevailing solar photon spectrum. This process has been evolving to ever greater efficacy and complexity ever since simple aromatic hydrocarbons formed in the protoplanetary disk

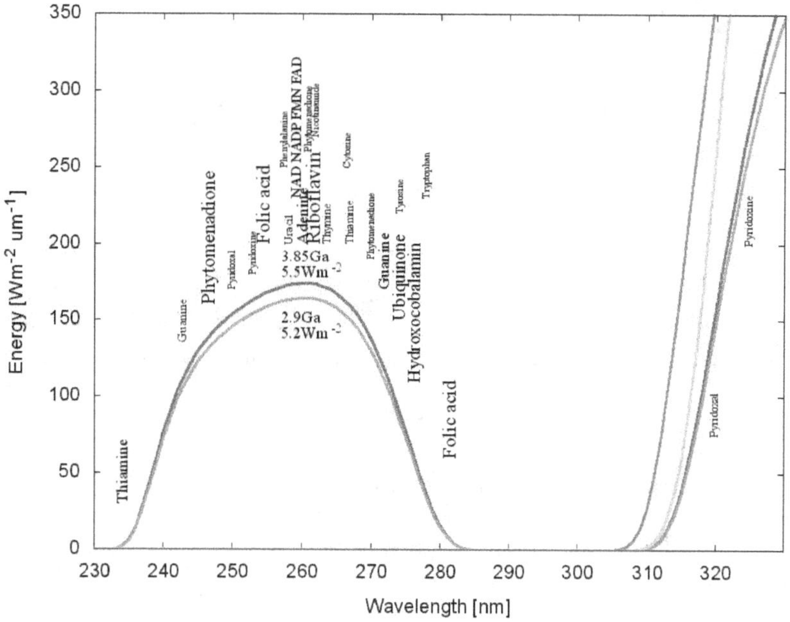

Fig. 4.7. The wavelengths of maximum absorption of many of the fundamental molecules of life (common to all three domains) coincide with the predicted solar spectrum at Earth's surface in the UV-C at the time of the origin of life at 3.85 Ga and until at least 2.9 Ga. The font size of the letter roughly indicates the relative size of the molar extinction coefficient of the indicated pigment (see table 4.1). Image credit: Adapted from Michaelian and Simeonov, (2015)[217]. Creative Commons Attribution 3.0

under UV light even before the formation of Earth itself in the early Hadean (see chapter 5.3), through the formation of UV-C pigments of the Archean, now identified as the fundamental molecules of life, and up until the great assortment of contemporary pigments maintained and proliferated by the biosphere today.

The most abundant contemporary pigments, which almost certainly have a history going back to early life in the Archean since they are also fundamental molecules found in the three domains of life, are the porphyrins and the isoprenoids (e.g. carotenoids)(see figures 4.10 and 4.11). Porphyrins and metalloporphyrins (porphyrins coordinated by a metal ion, of which chlorophyll is an example coordinated by Mg^{+2}) are strong DNA binders[4], and they bind preferentially to the PO_2 group of the DNA backbone and indirectly, through water molecules, to the nitrogen N−7 sites of the guanine bases (Neault and Tajmir-Riahi, 1999)[245]. Chlorophyll b is a strong absorber in the UV-C while chlorophyll a is weaker (see figure 4.8). This is consistent with the suspicion that chlorophyll b is probably a more ancient precursor of chlorophyll a.

[4] For this reason, porphyrins are being used as photosensitizers in photon cancer therapy.

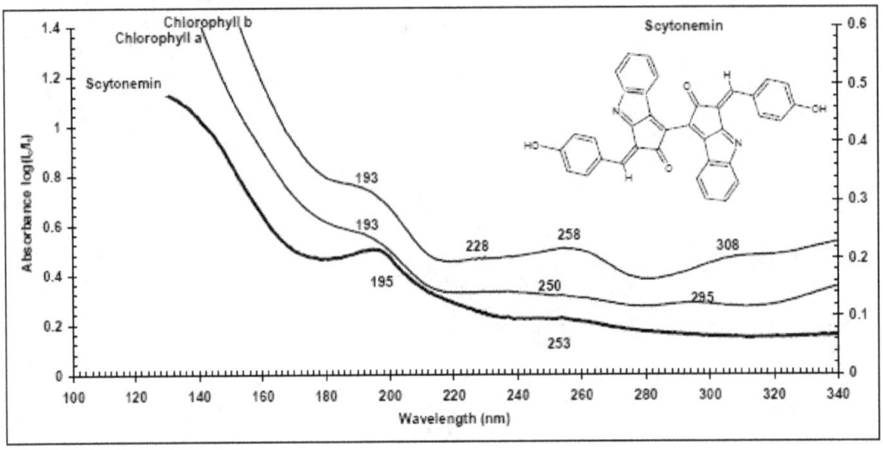

Fig. 4.8. Chlorophyll a, b and scytonemin absorption spectrum from the UV-V to the UV-A regions, taken with synchrotron light. Note that the absorbance scale is logarthimic. Chlorophyll and scytonemin are fundamental molecules, being found in all three domains of life. The fact that they absorb strongly from 230 to 280 nm, corresponding to a window in Earth's atmosphere during the Archean, suggests that they are microscopic dissipative structures. Image credit: Zalar et al. (2007)[427].

Self-assembly of bacteriochlorophyll, BChls c, d, and e, but not BChl a, into supramolecular structures without a protein scaffolding matrix are known to occur in chlorosomes of green photosynthetic bacteria (Jesorka et al., 2012)[139]. It has also been shown that particular porphyrins can self-assemble into stacks forming helical structures within the major groove of DNA (Bouamaied et al., 2008)[23]. When stacked in such arrays, porphyrins and chlorins have significant additional absorption in the UV-C as well as in the visible (see figure 4.9).

Absorption in the visible is attributed to the large number of conjugated[5] double bonds (see figure 4.10 and 4.11), the greater the number, the more closely spaced the electronic excitation energy levels. The de-localization of charge among the p-orbitals in these conjugated molecules stabilizes the molecule and each additional conjugated p-orbital adds to the system stability and pushes its absorption further towards the red (see figure 4.13). It, therefore, must have been a relatively simple matter for Nature to evolve pigments covering ever more of the UV and visible spectrum starting from molecules that absorbed in the UV-C by simply adding conjugated bonds to the molecule. From a porphyrin-, or chlorin-type molecule absorbing and dissipating in the UV-C, little evolutionary effort was therefore expended in bringing the dissipation into the visible

[5] Conjugation in molecules refers to alternating single (sigma) and double (pi) bonds. These bonds form a system of connected p-orbitals that cause a delocalization of the electrons across the entire conjugated portion of the molecule. Charge delocalization causes molecules with extensive conjugation to be more stable than non-conjugated molecules.

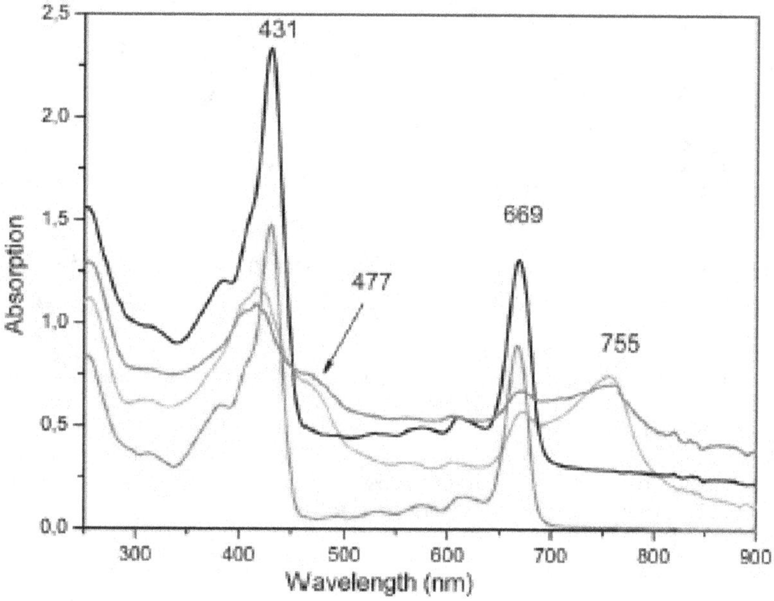

Fig. 4.9. UV-Vis absorption of zinc coordinated Chlorin in various solvents. The lower trace at 250 nm are in dry THF injected into dry n-heptane; the upper trace is Chlorin in dry THF injected into wet n-heptane. Such stacks are suggested to mimic supramolecular stacks of bacteriochlorophyll in chlorosomes of green photosynthetic bacteria. Self-assembly, and therefore absorption, is significantly enhanced when the metal ion is present and the system is in a polar solvent (upper trace at 250 nm). Significant absorption in the UV-C (< 280 nm) is clearly seen. Image credit: Jesorka et al. (2012)[139]. Reprinted with permission from Royal Society of Chemistry.

and simultaneously evolving electron transport for the basis of photosynthesis. This may explain how photosynthesis could have arisen so soon after the origin of life itself [6]. Mulkidjanian and Junge (1997)[237] have, in fact, described evidence based on the sequence alignment between reaction centers and antenna complexes suggesting that a large chlorophyll pigment carrying protein dissipater (what they considered as a UV protector of the primordial cell) was the precursor to the photosynthetic reaction centers of organisms.

Isoprenoids (e.g. carotenoids) are associated extensively today with chlorophyll and with microbial rhodopsin in the only two known energy transducing systems using photons to produce energy for organisms. It is known, however, that the light trapped by the carotenoids, supposedly acting as antenna molecules when associated with rhodopsin, does not participate in the proton pump but is instead dissipated directly into heat. Similarly, when the carotenoids

[6] There is, in fact, some evidence for stromatolite forming anoxygenic photosynthetic cyanobacteria existing at 3.7 Ga (Nutman et al., 2016)[253].

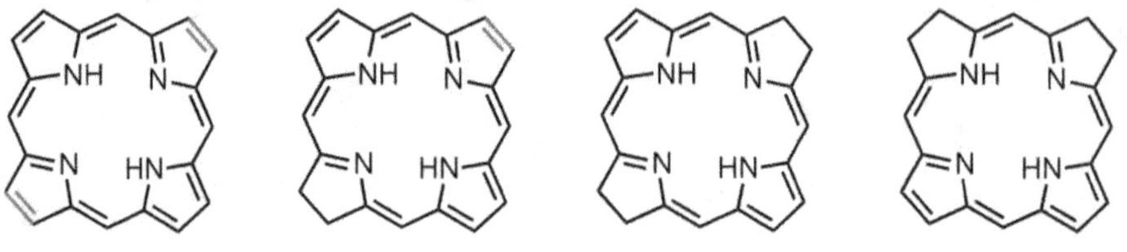

Porphyrin **22 π e-** | **Chlorin** **20 π e-** | **Bacteriochlorin** **18 π e-** | **Isobacteriochlorin** **18 π e-**

Fig. 4.10. The structures of porphyrin, chlorin, bacteriochlorin, and isobacteriochlorin. There are two π electrons (symbolized by π e-) for every double bond in the macrocycle. The greater the number of double bonds, the longer the wavelength of the Soret and Q band absorptions. When these planer structures stack in piles, as found today in the photosynthetic reaction centers of organisms, they have considerable absorption in the UV-C (see figure 4.9). Image credit: Livewire61, Expanded derivative work: H Padleckas, Creative Commons.

are associated with chlorophyll, the light energy captured is not used in the photosynthetic process but instead is also dissipated directly into heat. The carotenoids are considered as protective pigments for the photosynthetic apparatus. Strong absorption in the UV-C of the carotenoids (see figures 4.12 and 4.13) suggests that these pigments were another of the UV-C dissipating pigments of the primordial soup that later formed complexes with porphyrins, chlorins and RNA or DNA. Besides the UV-C chromophore of the carotenoid coupling through resonant energy transfer to the conical intersection of RNA or DNA, the visible chromophore of the carotenoid could also pass the electronic excitation energy to RNA or DNA through vibrational coupling, thus extending efficient photon dissipation into the visible.

Flavins (see figure 4.14) also have strong absorption in the UV-C as well as in the near UV and visible (see figure 4.15). Flavins are pigments found in all three domains of life and must therefore be considered as fundamental molecules of life. Flavins also display strong binding to DNA. Flavins fluoresce, reducing their efficiency for dissipation, for example riboflavin has a fluorescence quantum efficiency of 0.26 (Gordon-Walker et al., 1970)[109] but through binding to RNA or DNA, and employing resonant energy transfer between molecules, this fluorescence would have been eliminated, thereby increasing dissipation. Their role at the beginning of life may have been important in ways other than photon dissipation since flavins reduce pyrimidine dimer formation[7] in DNA by factors

[7] Pyrimidine dimer formation refers to the UV-induced formation of covalently bonded dimers of pyrimidines (thymine or cytosine) when these are located adjacent on the same RNA or DNA strand. This is the most common form of genetic mutation caused by UV light and has been implicated as the principle cause of skin cancers. Such dimer formation would not

β-Carotene

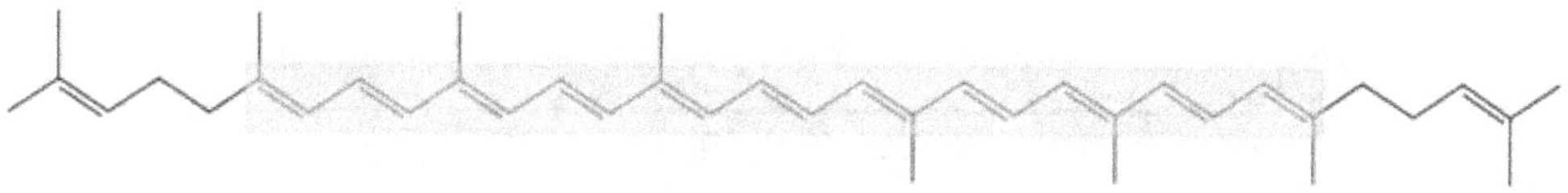

Lycopene

Fig. 4.11. The highlighted parts of the molecules show the conjugated portions. These long chains of alternating single and double bonds add to the delocalization of charge amongst the p-orbitals. With every double bond that is added to a conjugated system, the molecule will absorb light of longer wavelength or, lower energy (see figure 4.13). β-carotene and lycopene have eleven areas of alternating double and single bonds giving them extensive conjugation and causing them to absorb electromagnetic radiation in the visible region of the spectrum. This gives tomatoes and carrots their red color. Image credit: Public Domain.

of up to 5 (Stapleton and Walbot, 1994)[357] and have also been identified as a possible energy conversion molecule at the very beginnings of life, before the advent of the chlorophyll photosynthetic system (Kritsky et al., 2013)[171].

The UV-C wavelengths were best suited to getting life off to an enthusiastic start since there is enough energy in each photon within this wavelength region to allow for the direct breaking and making of covalent bonds, but not enough to cause destructive ionization, thus permitting the photochemical production of these organic pigments out of more simple and common primordial precursor molecules available in the Archean environment, such as hydrogen cyanide (HCN), water (H_2O), hydrogen sulfide (H_2S) carbon dioxide (CO_2) and nitrogen gas (N_2). Under this UV-C flux, and at the high temperatures of the Earth's surface at the time, of around 80 °C at 3.8 Ga (Knauth, 1992; Knauth and Lowe, 2003)[163, 165], falling to 70 ± 15 °C at 3.5–3.2 Ga (Lowe and Tice, 2004)[193], no complicated biosynthetic pathways like those of today would have been needed to make the UV-C pigments (the fundamental molecules of life). Furthermore, of all the wavelength regions reaching Earth's surface during the Archean, the UV-C region had the greatest entropy producing po-

---

have been serious in a scenario like the one we are presenting here for the origin of life (see chapter 12) in which early replication was driven by external environmental factors, and not dependent upon RNA or DNA information content.

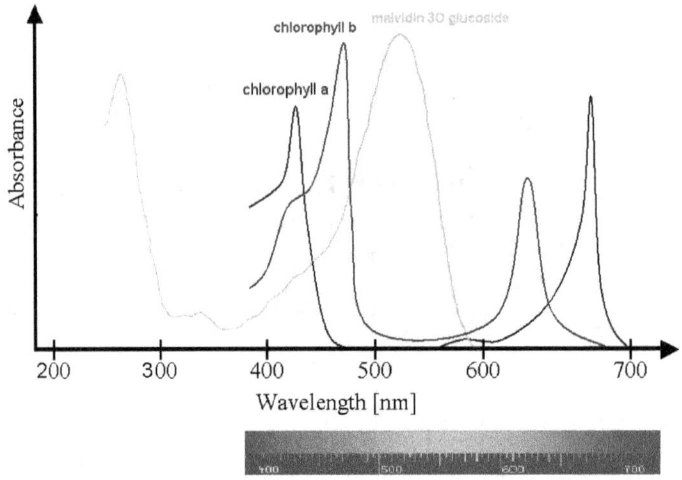

Fig. 4.12. Superposition of spectra of chlorophyll a and b with oenin (malvidin 3O glucoside), a typical anthocyanidin. There is strong absorption of this anthocyanidin in the UV-C region and at the same time in the visible spectrum, mainly in the green, where chlorophylls do not absorb. Image credit: Adapted by NotWith from Aushulz, Creative Commons 3.0.

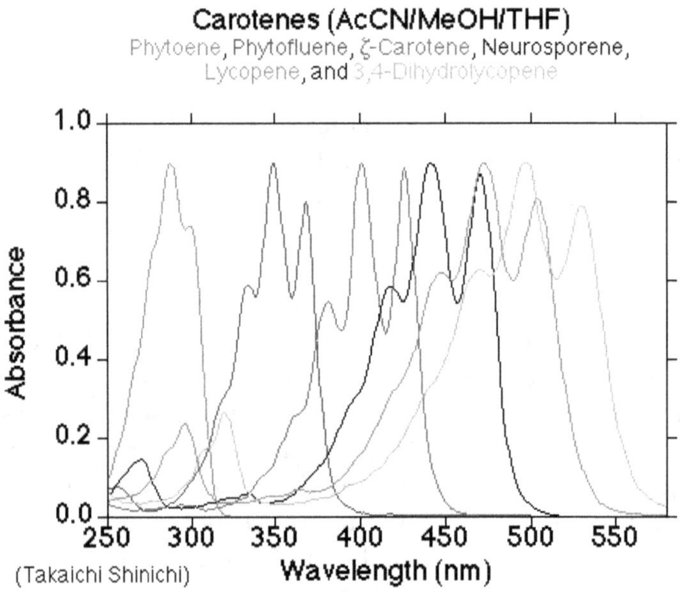

Fig. 4.13. The absorption spectrum of the carotenoids depends on the number of conjugated double bonds in the molecule. The greater the number of double bonds, the more red-shifted is the absorption spectrum (see also figure 4.11). Image credit: Shinichi, (2016)[339], Lipidbank.jp.

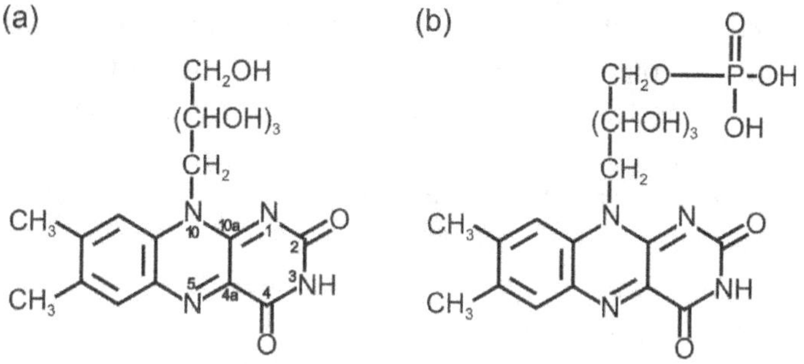

Fig. 4.14. Molecular structure of riboflavin, RBF (a), and flavin mononucleotide, FMN (b). Image credit: Roslund et al. (2011)[303]. Reproduced with permission of the AIP.

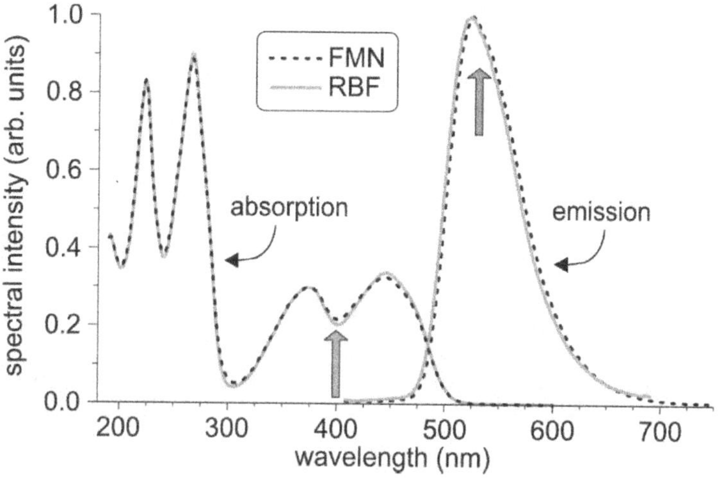

Fig. 4.15. Linear absorption and fluorescence spectra of riboflavin, RBF (solid) and flavin mononucleotide, FMN (dashed) in aqueous media at 298 K. Arrows at 400 and 530 nm indicate the UV pump wavelength giving rise to the fluorescence emission, respectively. Image credit: Roslund et al. (2011)[303]. Reproduced with permission of the AIP.

tential per photon, i.e. since each photon has high energy, this energy can be distributed over many infrared photons and this produces a large amount of entropy per incident photon dissipated (see figure 2.3). The nucleic acid bases and their polymerization into short segments of RNA and DNA, were then, most likely, particular pigments of a rich soup of organic pigments (Michaelian and Simeonov, 2015)[217] that were dissipating UV-C sunlight on the Archean, and perhaps even earlier, on the Hadean ocean surface.

However, if the nucleic acids and the other fundamental pigments, like the aromatic amino acids, cofactors, porphyrins, chlorins, carotenoids, flavins and others, have large positive Gibb's free energies at standard pressure and temperature and are thus difficult to make in the laboratory through thermally induced chemistry, how could a rich soup of these pigments come to have been floating on the Archean ocean surface? In fact, the situation becomes even more mysterious since we are now aware that complex organic pigments are common in interstellar space where conditions (of extreme cold and low material density) appear to be even more adverse to their production (see chapter 5). As mentioned in chapter 3.4 on the microscopic structuring of material through dissipation, and in chapter 5 on the cosmic ubiquity of organic pigments, non-equilibrium thermodynamic chemistry provides the solution to this paradox and we present an explanation of pigment proliferation through autocatalytic photochemical reactions in chapter 6 after first presenting the evidence for the cosmic ubiquity of organic pigments.

5. The Ubiquity of Organic Pigments

Greater than 99% of biotic cellular mass is made up of the elements hydrogen, oxygen, carbon, and nitrogen. These four elements are, respectively, the first, third, fourth, and fifth most abundant elements in the universe, making up more than 70% by mass of the baryonic matter in the universe. Organic chemistry and life on Earth is therefore based upon the most "available" elements in the universe although, as we shall see, this is certainly not the only, nor the most important, reason for their use by life.

The realization that organic molecules are ubiquitous and distributed throughout the cosmos has a long and difficult history, replete with controversy and overt editorial censorship. Until the fourth decade of the 20th century, the possibility of finding complex organic molecular structures in space was not seriously contemplated because of the widely held belief that the extremely low material density in space (between 1000 and 10000 particles per cubic centimeter) and the cold temperatures (10 to 100 K) were not conducive to any appreciable chemical synthesis.

The pioneers of the courageous work to demonstrate the contrary are the astrophysicist Fred Hoyle and astrobiologist Chandra Wickramasinghe (Fig. 5.1). Hoyle's and Wickramasinghe's explanation of the visible and UV absorption spectrum and the infrared emission spectrum of the gas and dust clouds in space evolved from the conventional wisdom of attributing these to interstellar graphite grains in 1962, to organic polymers and aromatic organics in 1974, and finally to the truly startling suggestion of bacteria in space in 1979 (see figure 5.2). In their own estimation (Wickramasinghe, 2010a)[403], the price they paid for this last heretical proposition was that much of their original work on graphite grains and organic molecules in space has been ignored by the astrobiology community and is infrequently cited in the subsequent literature, even though this same community has always had to grudgingly "catch up" with the duo.

Another prominent galactic extinction feature which Hoyle and Wickramasinghe have addressed within their *bacteria in space* paradigm is what are referred to as the diffuse interstellar absorption bands observed in the extinction of light from background stars passing through interstellar clouds. Based

Fig. 5.1. Astrophysicist Fred Hoyle and astrobiologist Chandra Wickramasinghe, pioneers of the idea that life on Earth may have been seeded by life spread throughout the cosmos. Image credit: Fair use (Fred Hoyle) and Creative Commons Attribution-Share Alike 3.0 (Chandra Wickramasinghe).

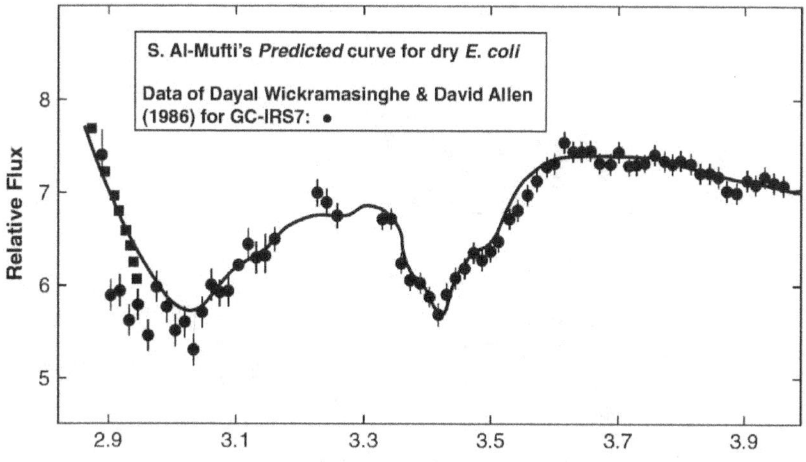

Fig. 5.2. The uncanny fit of the infrared emission spectrum of the galactic center infrared source GC-IRS7 (data) compared with the emission spectrum of dry *E. coli* bacteria (line). The features seen here are assumed to be due to various CH stretching modes (see figure 5.3). Image credit: Wickramasinghe and Allen (1986)[402].

on the average extinction data for several thousand stars in the optical spectral region studied by Nandy (1964, 1965) [242][243], later confirmed by Schild (1977)[322], a strikingly invariant and linear behavior of the logarithm of the extinction versus the inverse wavelength was found with significant slope changes at $\lambda^{-1} = 2.4\ \mu\text{m}^{-1}(\lambda = 416.7\ \text{nm})$ and a further change in slope at $\lambda^{-1} = 3.125\ \mu\text{m}^{-1}(\lambda = 320\ \text{nm})$ (see figure 5.4). At shorter wavelengths the extinction curves cease to be invariant and have different strengths at the $\lambda^{-1} = 4.56\ \mu\text{m}^{-1}(\lambda = 217.5\ \text{nm})$ peak feature (see figure 5.4).

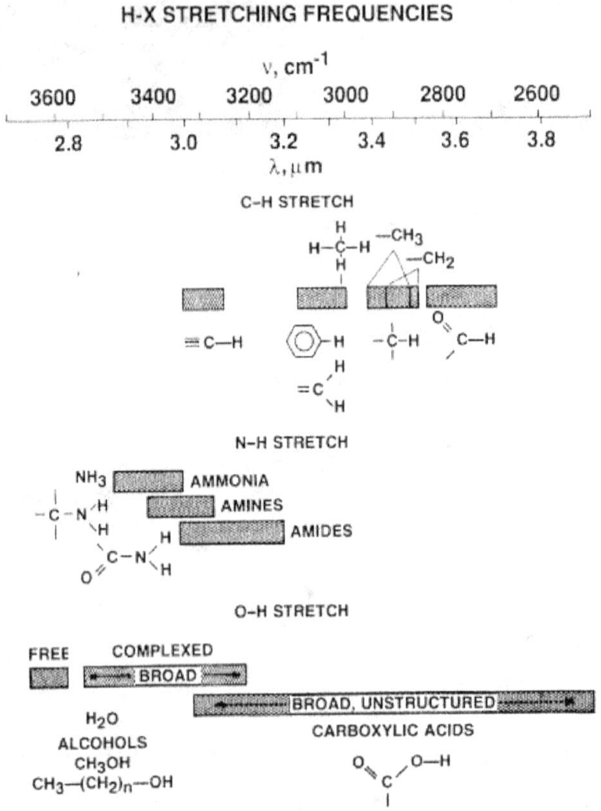

Fig. 5.3. Characteristic vibrational frequencies in the CH, NH, and OH stretch regions according to the particular type of molecule. The unlabeled CH stretching domains (from higher to lower frequencies) correspond to acetylenic, aromatic and olefinic, aliphatic, and aldehydic CH stretches. Image credit: Pendleton and Allamandola (2002)[275]. Reproduced with permission from the authors.

Attempts have been made to fit this extinction curve with Mie scattering theory on solid particle silicates and ices with indexes of refraction in the range $1.33 - 1.66$ but the fitting requires fine-tuning of both the mean size and the size distribution to a level of precision that cannot be justified (Wickramasinghe, 1973)[401]. Hoyle and Wickramasinghe assumed instead hollow bacterial grains (spore shells) with lower indices of refraction $n = 1.167$ for which the average size and size distributions can be significantly relaxed and have found a good fit using Mie scattering theory (see figure 5.5) making some credible assumptions about the size distributions of fragments from cosmic ray, charge induced, destruction of bacterial spores (Wickramasinghe, 2010a)[403]. Wickramasinghe confides, however, that the extinction may someday be better fitted with molecular absorption rather than Mie dispersion, or perhaps with a judicial combination of the two. The extinction peak at $\lambda^{-1} = 4.56\ \mu m^{-1} (\lambda = 217.5$ nm) can be fitted well with a general assortment of biological aromatic molecules (Fig. 5.4).

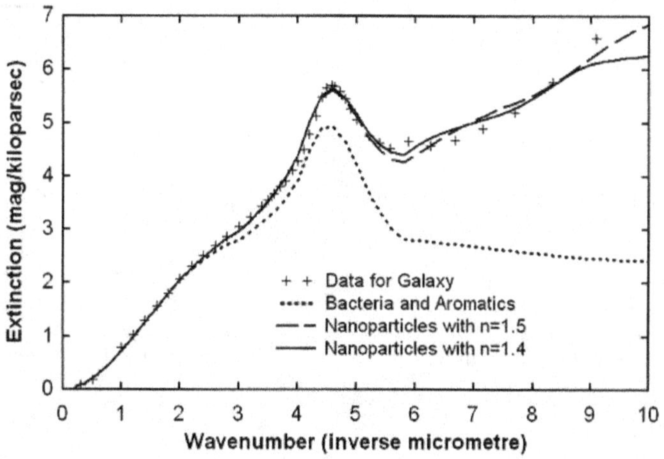

Fig. 5.4. Agreement between interstellar extinction (plus signs) and biological models. Chandra Wickramasinghe found that mixtures of hollow bacterial grains with biological aromatic molecules and nanobacteria provide excellent fits to the astronomical data. The 217.5 nm (4.56 μm^{-1}) peak in the extinction is caused by biological aromatic molecules (see Wickramasinghe (2010)[403] for details). Image credit: Wickramasinghe (2010b)[404].

More recent analysis of higher resolution interstellar spectra, comparing the interstellar near-infrared emission spectral features (between 3 and 4 μm) to those of the E. coli bacteria, has led Pendleton and Allamandola (2002)[275] to conclude that the interstellar 3.4 μm band does not have a biological origin (see figure 5.6). On the other hand, these authors suggest that the striking similarities between the 3.4 μm interstellar band and the corresponding aliphatic CH stretch in the organic extracts of the Murchison and Orgueil carbonaceous meteorites (see figure 5.7) means that some of the interstellar material originally incorporated into the solar nebula may have survived relatively unaltered in primitive solar system bodies.

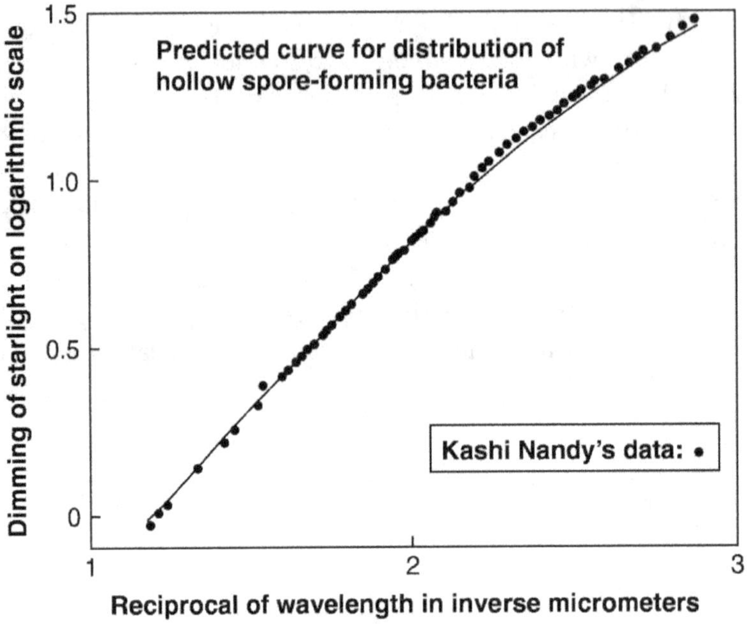

Fig. 5.5. Dimming of magnitude Δm for starlight passing through interstellar gas and dust in the galaxy versus inverse wavelength. Points represent the visual extinction data normalized to $\Delta m = 0.409$ at $1/\lambda$=1.62 μm^{-1} and $\Delta m = 0.726$ at $1/\lambda = 1.94$ μm^{-1} obtained by Nandy (1964)[242]. The curve is the calculated extinction for a size distribution of freeze-dried spore-forming bacteria. The calculation uses the classical Mie scattering theory and assumes hollow bacterial grains comprised of organic material with refractive index $n = 1.4$ and with 60% vacuum cavity caused by the removal of free water under space conditions. Image credit: Wickramasinghe (2010a)[403].

From these spectroscopic studies, Pendleton and Allamandola (2002)[275] conclude that the most reasonable explanation of the material comprising the diffuse interstellar medium is largely hydrocarbon in nature and containing little O or N, with the carbon distributed between the aliphatic and aromatic forms (see figure 5.8).

Our explanation of the similarity between the absorption and emission spectra of freeze dried bacteria with those of the diffuse interstellar medium is different from that of Hoyle and Wickramasinghe. While it seems unlikely that there exist either living or freeze-dried bacteria in space, we do, however, suggest an underlying mechanism of molecular formation based on non-equilibrium thermodynamic principles. Given the well established existence of primordial molecules, such as HCN, H_2O, CO and OH, around star forming regions (see table 5.2), molecular pigments in the UV-C should spontaneously form in the interstellar medium as microscopic dissipative structuring in response to the impressed UV-C photon potential of local stars. The proliferation leading to

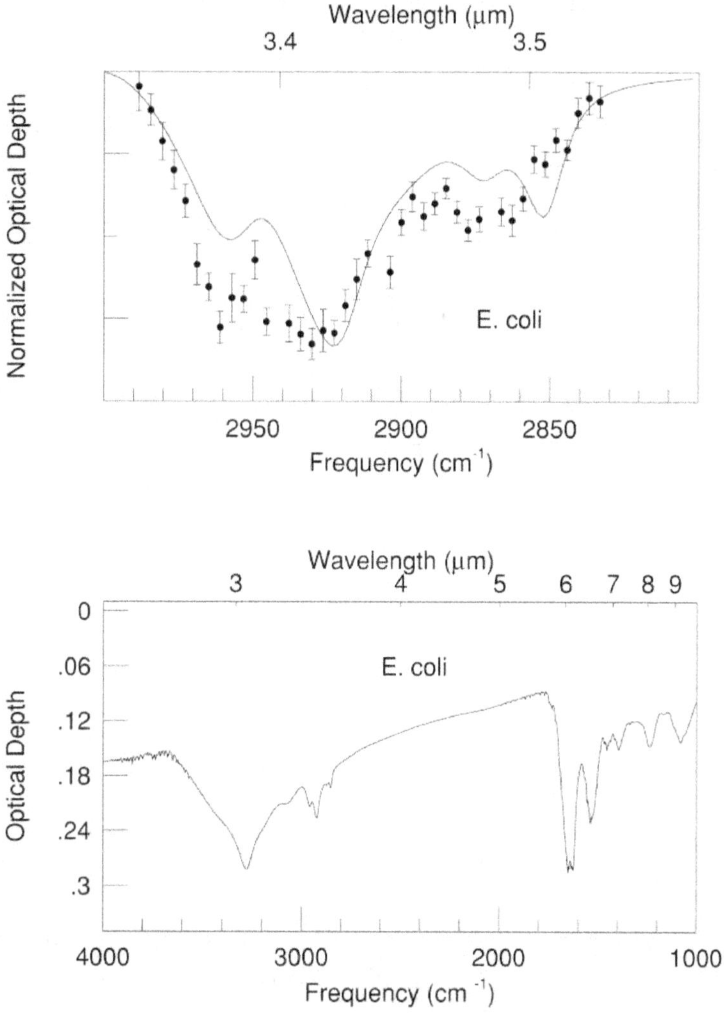

Fig. 5.6. (a) Comparison of the CH stretch of a freeze-dried sample of Escherichia coli W1485 to the diffuse interstellar medium 3.4 μm emisssion band (b) The complete 3600 − 1000 cm^{-1} (2.78−10 μm) spectrum of Escherichia coli W1485. See Pendleton and Allamandola (2002)[275] and references therein for details concerning the data and spectra. Image credit: Pendleton and Allamandola (2002)[275]. Reproduced with permission from the authors.

the ubiquity of these pigment molecules can be explained in purely physical-chemical terms through non-linear, non-equilibrium thermodynamics, without requiring any complex biological input (see chapter 6). The pigments in any existent bacteria on Earth are likewise formed through photon dissipation, albeit now through complex biosynthetic routes, permitting longer wavelength photons to be used which have insufficient energy to make or break covalent bonds directly. However, the principle of evolutionary efficacy, that of adapting

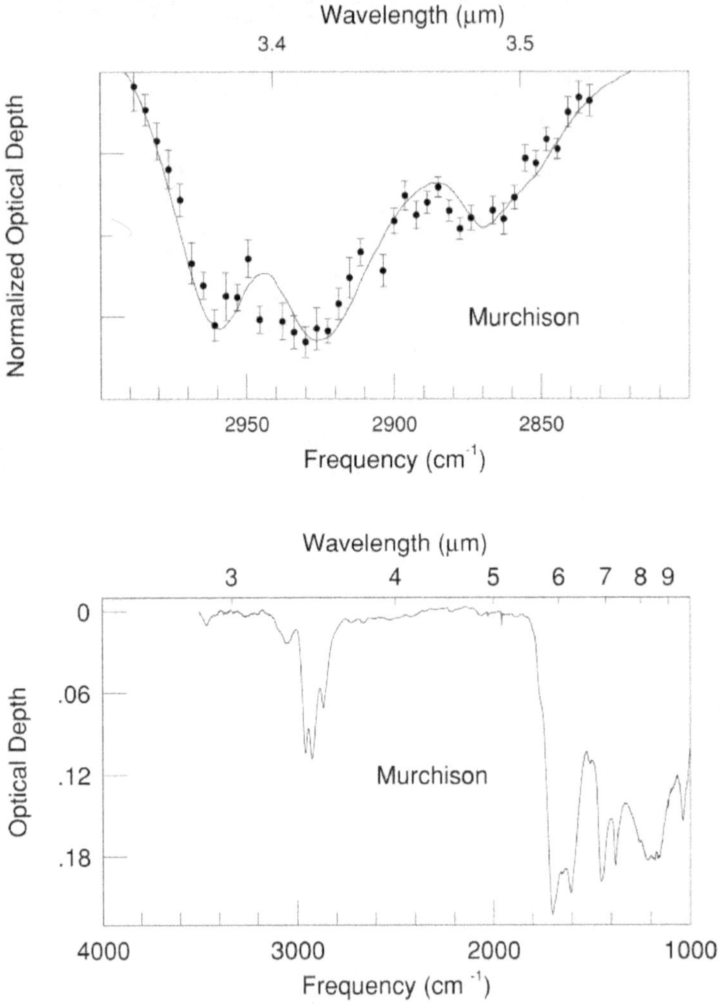

Fig. 5.7. Comparison of the CH stretch of the 600 °C organic extract sublimate from the Murchison meteorite to the diffuse interstellar medium 3.4 μm band. (b) The complete 3600–1000 cm^{-1} (2.78–10 μm) spectrum of this Murchison extract. See Pendleton and Allamandola (2002)[275] and references therein for details concerning the data and spectra. Image credit: Pendleton and Allamandola (2002)[275]. Reproduced with permission from the authors.

existing structures to new roles (or more correctly, thermodynamic efficacy – that of building new solutions from those existing as the system goes through an instability at a bifurcation point as a constraint is changed – see chapter 3.3), may imply that many of these contemporary pigments of life retain the same UV-C absorbing and dissipating properties of their primordial predecessors formed during the Archean. It is suggested here that it is these retained primordial UV-C absorption and dissipation properties of many of the contemporary

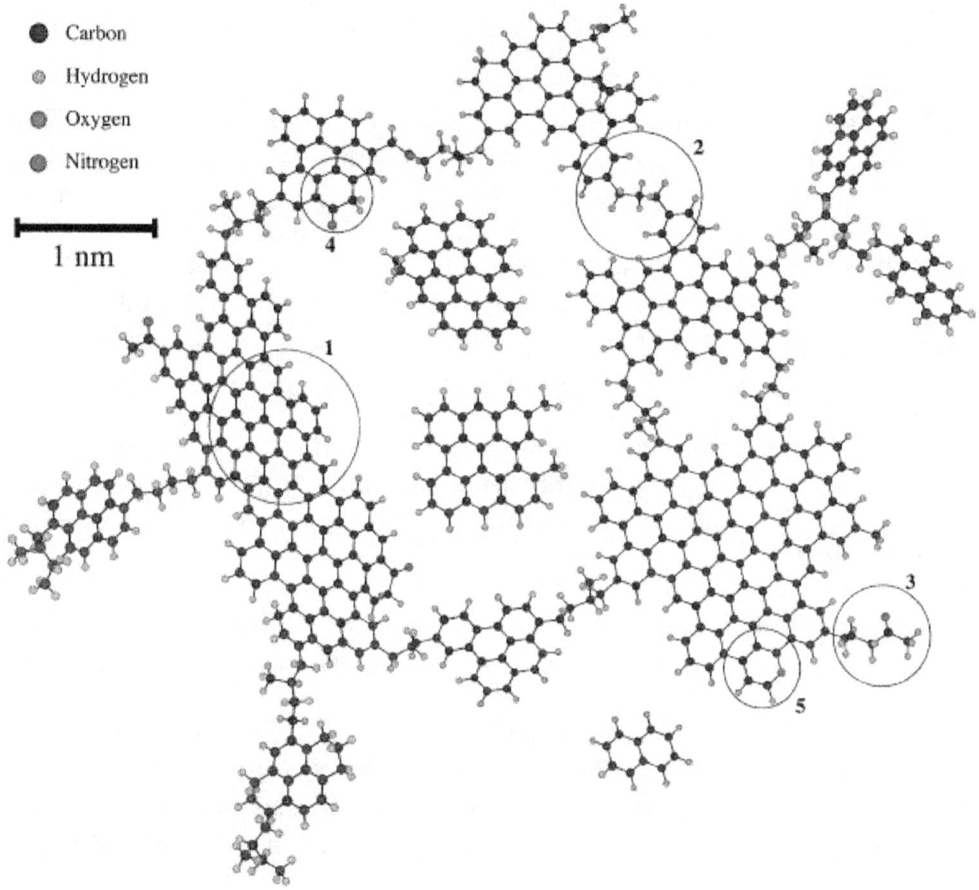

Fig. 5.8. The basic structural and molecular character of carbonaceous, interstellar dust in the diffuse interstellar medium. The molecular details have been deduced from the spectroscopic constraints as determined by Pendleton and Allamandola (2002)[275]. The specific geometries of the aromatic plates and aliphatic components simply represent what is likely. The structure is somewhat splayed out to reveal the molecular structural details; the authors of this image envision the actual structure to be somewhat more closed in. For the interconnected species (not the free-floating entities), the relative numbers of aromatic and aliphatic carbon-hydrogen bonds, as well as their subclassification within type, are all consistent with the observed spectrum. The latter includes the aliphatic - CH_3 to CH_2 ratio, and the relative numbers of aromatic solo, duo, trio, and quartet hydrogens deduced from the interstellar IR emission bands. The approximate volume of this fragment is on the order of 10^{-19} cm^3. Thus, a typical 0.1 μm DISM carbonaceous dust grain would contain approximately 104 of these fragments. The encircled regions are expanded showing the different structural units. See Pendleton and Allamandola (2002)[275] and references therein for details concerning the data and spectra. These structures are reffered to as MAONs (Mixed Aromatic/Aliphatic Organic Nanoparticles – see section 5.2.1). Image credit: Pendleton and Allamandola (2002)[275]. Reproduced with permission from the authors.

pigment molecules in freeze dried bacteria that gives the reasonably good fit to the diffuse interstellar emission and absorption spectra originally considered by Hoyle and Wickramasinghe (see figure 5.2).

The peak at 217.5 nm (4.56 μm^{-1}– see figure 5.4) in the interstellar absorption spectra is interesting since at this wavelength photons still do not have enough energy to ionize some of the polyaromatic hydrocarbons (wavelengths shorter than 203 nm are required - Gudipati and Allamandola, 2004 [118]) which would destroy them, but do have enough energy to facilitate the production of the UV-C pigments from the primordial molecules which form naturally in the protostellar halo (see table 5.2). At wavelengths shorter than ~220 nm RNA and DNA extinction begins to rise again (from its peak at 260 nm) due to ionization (see figure 12.3). At wavelengths longer than 320 nm, there is not enough energy in a single photon to re-configure the covalent bonds of the precursor primordial molecules necessary to form the pigments and thus the production efficiency drops off rapidly. Although significantly less probable, two photon processes could be effective in directly producing pigments at these longer wavelengths.

In the rest of this chapter we take a closer look at the evidence for organic material in interstellar clouds, in condensing gas star forming regions, in the protoplanetary disks around young stars, and in the atmosphere and on the surface of planets, asteroids, and comets. We show that an important correlation exists between the spectroscopic absorption footprint of a particular molecular pigment and the stellar spectrum of available light from the neighboring star, which supports our proposition that these pigments are microscopic dissipative structures formed in response to the impressed UV photon potential of the local star.

5.1 Stellar Organics

The stars in our particular region of the galaxy are made from the remnants of previous generations of stars. The cloud of gas and dust that formed our solar system thus contained a significant amount of the heavier elements produced within these previous stars and during supernova explosions, besides the primordial hydrogen and helium that formed rapidly after the Big Bang. Cameron (1973)[36] has performed an analysis of the concentrations of the different organic elements in our Sun (table 5.1) obtained by analyzing solar wind implantations in moon rocks. The vast majority of stars in the cosmos, being also recent generation stars, are similarly made up of a significant amount of these heavier elements required for forming organic molecules.

It is notable that many of the primordial molecules, such as HCN, H_2O, CO, OH, needed as *materia prima* for the production of the fundamental molecules of life, are found near star forming regions in our cosmos. In the proceedings of

Element	Atomic Number	Relative Abundances (relative to 10^6 Si atoms)
H	1	3.18 x 10^{10}
He	2	2.21 x 10^9
C	6	1.18 x 10^7
N	7	3.74 x 10^6
O	8	2.15 x 10^7
Ne	10	3.44 x 10^6
Mg	12	1.06 x 10^6
Si	14	1.00 x 10^6
S	16	5.00 x 10^5
Ar	18	1.17 x 10^5
Fe	26	8.30 x 10^5

Table 5.1. Relative concentrations of elements in our sun obtained from studying ion implantations in Moon rocks. After Cameron (1973)[36].

the "Fourth International Conference on the Origin of Life and the First Meeting of the International Society for the Study of the Origin of Life" held in 1973, David Buhl published an interesting paper entitled "Galactic Clouds of Organic Molecules" (Buhl, 1973) [30] where he collected all the evidence available up to that year for organic molecules in space.

Particularly interesting was the study presented of the molecular cloud of the Orion Nebula (see figure 5.9). Buhl provided evidence for molecular formation around the central core of a condensing star system. Molecules tended to form at particular concentric regions corresponding to particular temperatures and hydrogen gas densities (see table 5.2). For example, in the case of water, it was found that H_2O emissions were coming from a relatively small region of a few astronomical units (1 astronomical unit, AU, is the distance between the Earth and Sun, $\sim 1.5 \times 10^8$ Km) where the temperature reached 600 K and the H_2 density was approximately 10^{10} molecules per cubic centimeter. From the emission lines of the H_2O molecules in this shell it was determined that water was providing a cooling for the gas as it collapsed to form the star and that the energies involved were quite large 10^{30} ergs s^{-1}, corresponding to about 0.1% of the gravitational energy involved in the collapse of a star of size one solar mass.

A very important primordial molecule for life, hydrogen cyanide[1], HCN, was found to be forming at a larger distances from the core where temperatures were cooler, around 4 K, and H_2 densities were lower, $\sim 10^6$ molecules per cubic centimeter. Another interesting primordial molecule found, which is now known

[1] Oró (1961)[261] and Oró and Kimballl (1962)[262] have shown that UV-C light on a mixture of water and hydrogen cyanide can produce the nucleic acid bases (the nucleobases).

Fig. 5.9. The Orion Nebula is an active star forming region in which many of the primordial molecules necessary for producing the fundamental molecules of life, have been detected. Morris et al. (2016)[236] have demonstrated that the predominant mechanism of CH^+formation in this nebula is through photochemical reactions involving the far UV spectrum. This picture was taken by the Hubble Space Telescope. Image credit: NASA, ESA, M. Robberto (Space Telescope Science Institute/ESA) and the Hubble Space Telescope Orion Treasury Project Team, Public Domain.

to be a building block of sugars and polysaccharides and other important biomolecules, was formaldehyde, H_2CO.

From these simple primordial molecules, and UV light from a star, many fundamental molecules of life can be formed, as indeed has been discovered in the intervening years. Molecules of formic acid HCOOH and methanomide H_2CHN, which together can form the simplest amino acid glycine NH_2CH_2COOH, to ethanol C_2H_5OH have been found in space.

Not only have both the primordial and the fundamental molecules of life been found in space, but there is increasingly more evidence being accumulated for the

Molecule	Temperature (K)	H_2 density (molecules cm^{-3})
H_2CO (6 cm)	10	10^3
CO	6	10^4
HCN	4	10^6
H_2CO (2 mm)	7	10^7
OH	100	10^8
H_2O	600	10^{10}

Table 5.2. Temperature and density corresponding to a particular distance from a star at which the production of a particular primordial molecule is most notable. There is little doubt that UV light from the local star plays an important role in the formation of these molecules since the temperatures in some cases are too low for thermally-induced chemistry. After Buhl (1973) [30].

occurrence of large amounts of more complex organic material throughout the cosmos, particularly in the form of polycyclic aromatic hydrocarbons (PAH's) containing 50 or more atoms. Perhaps up to one half of all the carbon in the universe is now thought to be locked up in these aromatic hydrocarbons. These molecules can be found on the surface of Earth and Mars, in the clouds of Venus, in the atmospheres of the larger planets and on many of their satellites, on asteroids, meteorites, comets, the atmospheres of red giant stars, interstellar nebula, and in the spiral arms of galaxies.

Many of the environments where PAH's are found are of low temperature and pressure, implying that the Gibb's free energy for the formation of these complex molecules in such environments should be positive and large, suggesting that their origin and ubiquity could only be attributed to non-equilibrium thermodynamic photochemical processes. In fact, even in the case of the formation of the simple hydrogen molecule, H_2, in regions of low density, the interaction of a third particle is generally required to achieve energy and momentum conservation. This led to a paradox concerning the formation and ubiquity of these molecules until it was realized that photons must be involved.

PAH's are often associated with sites of strong UV-C light emission by young stars in star forming regions of the cosmos. This strong association of UV light with complex organic molecules we attribute to microscopic material structuring through photon dissipation in the ultraviolet (see chapter 6). This is a non-linear, non-equilibrium thermodynamic and autocatalytic process leading to the conversion of primordial molecules into complex organic pigments such as the fundamental molecules of life (see table 4.1) and the PAH's or MAONs (see figure 5.8) which can then participate in the dissipation of the incident light, resulting in greater entropy production. A relation exists between the wavelength of maximum absorption of these organic pigments and the local stellar photon spectrum, and this provides a guide as to determining which polycyclic aromatic hydrocarbons are most probable in a given stellar environment, a postulate which appears to have been verified on Earth (see figure 4.7). This is discussed in more detail in chapter 5.4.

A important advance in astrochemistry which allowed the discovery of even more complex organic compounds in space came with the development of astronomical infrared (IR) spectroscopy. It significantly extended spectroscopic observing abilities, allowing the detection of various stretching and bending modes of molecules with near and mid-IR observations from space-based telescopes. IR spectroscopy is particularly useful in identifying the structures of organic molecules since different functional groups have very distinctive spectroscopic signatures in this wavelength region (Chang, 2000)[40].

Using data from the Heterodyne Instrument for the Far Infrared (HIFI) which flew aboard the Herschel Space Observatory, Morris et al. (2016)[236] have recently studied the formation of CH^+ molecule (a key species in interstellar chemistry) around the BN/KL source of the nearby Orion Nebula, thought to be the remanent of a collision of two stars about 500 - 1000 year ago. The Orion Nebula is known as a dense molecular cloud and a star forming region (see figure 5.9). Based on accurately measured and velocity-resolved CH^+ rotational transitions, they conclude that, contrary to previous works which attributed the formation of this molecule to shocked gas outflow, their data suggest that the dominant formation mechanism is photochemical production in the far UV.

5.2 Galactic Organics

The literature investigations presented in this and the following two sections were performed together with my collaborator Aleksandar Simeonov at the Ss. Cyril and Methodius University in Macedonia and provide evidence that organic pigments are formed in space as microscopic dissipative structures to dissipate the local stellar photon potential, including the same spectral region of the photon potential that produced them (Michaelian and Simeonov, 2016)[220].

In 1937, Theodore Dunham and Walter Sydney Adams gave the first evidence of molecular absorption lines found in the spectra of background O, B and A type stars (Dunham and Adams, 1937)[88]. The same year Pol Swings and Leon Rosenfeld identified one of the lines at 430.03 nm as an absorption band of the methylidyne radical, CH, making it the first molecule, and, at the same time, the first organic compound, to be discovered in space (Swings and Rosenfeld, 1937)[368]. These results were later corroborated by McKellar (1940)[202] and then by Douglas and Herzberg (1941)[85] with the additional identification of the CH^+ cation in interstellar space.

Twenty six years after these pioneering discoveries, the OH radical was discovered in interstellar medium by radio spectroscopy (Weinreb et al., 1963)[396].

This was followed by the microwave detection of ammonia, CH_4 (Cheung et al., 1968)[44], water (Cheung et al., 1969)[45], formaldehyde (Snyder et al., 1969)[348] - the first large organic molecule to be discovered in space, methanol (Ball et al., 1970)[11], formic acid (Zuckerman et al., 1971)[433], and formamide (Rubin et al., 1971)[305]. The development of millimeter-wave receivers in 1970 led to the even more sensitive method of detection of rotational transitions of molecules such as carbon monoxide, hydrogen cyanide (Snyder and Buhl, 1971)[349], methyl cyanide (Solomon et al., 1971)[350], methylamine (Kaifu et al., 1974)[146], and cyanamide (Turner et al., 1975)[379]. The further development of sensitive radio receivers at higher frequencies has led to the identification of more than 240 organic molecules, recognized today as common constituents of the interstellar medium.

A great puzzle remained however, that of explaining the mechanism for the formation of molecules in space. Two body interactions in space are very rare at these temperatures and densities, and anyway would not facilitate atomic bonding since a third body is normally required to remove energy from the system (Tennyson, 2003)[374]. Three body interactions, however, are almost totally excluded in such rarefied interstellar gases. It was then suggested that organic molecules could have formed on silicon grains but this led to a similarly difficult question of how silicon grains could have formed in space. The puzzle became even more intriguing when it was realized that the formation of stars requires the prior formation of molecular hydrogen.

It was finally realized that photons could play an important role in molecular formation. A high energy photon (>10.2 eV ~ 121 nm) can excite a hydrogen atom to the 2s state which then leads, through electron capture, to a weakly bound H^- ion, which subsequently could interact with an ionized hydrogen H^+ to form a H_2 molecule. Ionization of a hydrogen molecule by an even higher energy photon (the ionization potential of the hydrogen molecule is greater that than of atomic hydrogen) leads, on collision with another H_2 molecule, to the efficient formation of H_3^+;

$$H_2^+ + H_2 \longrightarrow H_3^+ + H.$$

The H_3^+ ion is an efficient catalyzer of many chemical formation reactions in space, especially those involving oxygen (see figure 5.10). These reactions would, of course, become important after first or second generation stars shed into space the heavier elements, O, C, formed through nucleosynthesis in their interiors.

It is now known that large regions of the galaxy M82 are very rich in organics and have a high ratio of [CN]/[HCN] ~5. Galaxies with these high ratios have been associated with photodissociation regions (PDR). Such galaxies are bathed in intense ultraviolet light (Fuente et al. 2005)[97]. It is assumed that photon disassociation processes are taking place although the details remain unknown. In these regions the precursors to the amino acid glycine, formic acid and metanoamine, have been detected in high densities (Hoyle and Wickramasinghe, 1978)[135] and the chemistry appears to be photon dominated.

5.2.1 Unidentified Infrared Emission (UIE) Bands

In 1973 IR spectrophotometry through an atmospheric windows using the 1.5 m Mount Lemmon Telescope, discovered broad, resolved infrared emission features centered at 8.6 and 11.3 μm from two carbon-rich circumstellar nebula (Gillett et al., 1973)[105], which could not be reconciled with a simple population of pure graphitic particles, believed at the time to be common in those types of nebula. Subsequent observations however, revealed that these were just two members of a ubiquitous and prominent family of emission bands whose other components include strong features at 3.3, 6.2, 7.7 and 12.7 μm as well as a complex array of minor bands, plateaus, and underlying continua (Russell et al., 1977; 1978)[306, 307]. Since the carrier of the bands remained unidentified for almost a decade, they were dubbed the Unidentified Infrared Emission (UIE) bands. Observers were initially confounded since these features are much broader than simple atomic lines. The fact that their intensity showed direct correlation with carbon abundance within the source, however, naturally implied a carbon-based carrier.

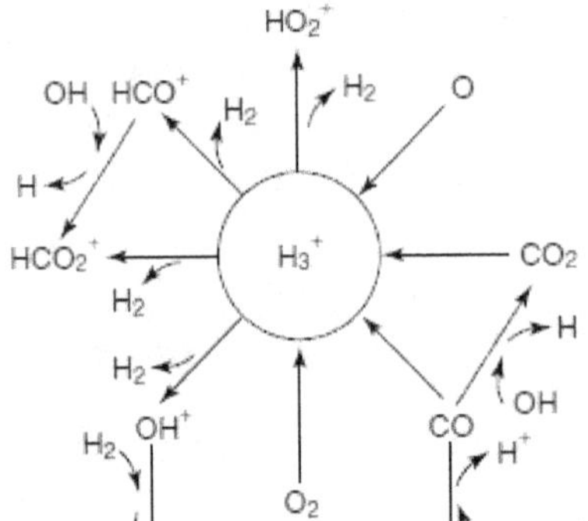

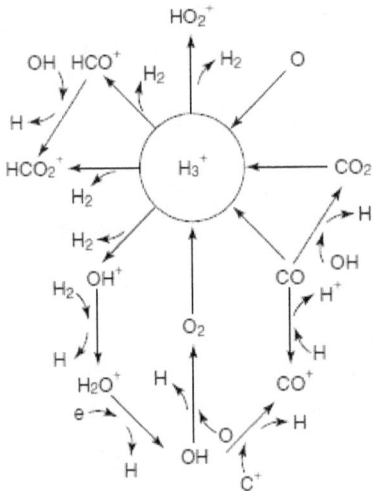

Fig. 5.10. Photochemical reaction pathways in space starting from the UV-C photon excitation of a hydrogen atom into the 2s state finally leading to the catalyst H_3^+ (see text) which is particularly important for producing molecules containing oxygen. Image credit: Miller and Tennyson (1992)[226]. Reproduced with permission from the Royal Society of Chemistry.

Knacke (1977) [162] and Duley and Williams (1979) [87] were the first to suggest that these features are likely due to aromatic organic compounds, although their ideas were at first ignored by the astronomical community. Several years later their suggestion gained acceptance mostly due to the fact that the UIE bands peak at wavelengths corresponding to the stretching and bending modes of various CH and CC bonds in aromatic hydrocarbons, confirmed after the launch of the ISO (Infrared Space Observatory) mission in which the features were observed with better spectral resolution over a broader spectral region, and in a larger sample of sources. Since then it has become quite certain that the UIE features at 3.3, 6.2, 7.7, 8.6, and 11.3 are due to aromatic organic compounds, hence they became known as the Aromatic Infrared Bands (AIB). These emission bands are now known to originate from within different astronomical environments, representing all stages of the life cycle of matter in galaxies. They have been detected in astronomical objects in our own galaxy such as; diffuse interstellar media, emission nebula - HII regions and planetary nebula, reflection nebula and protoplanetary nebula, interstellar cirrus clouds etc.; as well as in external starburst galaxies, such as M82 and NGC1068. In fact, it has been estimated that about 20-30% of the infrared radiation of our galaxy is being emitted in these bands (Snow and Witt, 1995) [347], and up to 20% of the total radiation of distant starburst galaxies (Smith et al., 2007) [346], indicating that their carrier must be an extremely common and abundant cosmic material. Solar System objects, such as interplanetary dust particles (IDPs), carbonaceous meteorites, Martian rocks (Kwok, 2009)[177], comets (Keller et al., 2006)[155], Titan's upper atmosphere (Lopez and Puertas, 2013) [190] are also known to contain these molecules.

A wide variety of hydrocarbon and carbonaceous materials comprising aromatic units have since been proposed as possible carriers of the AIB bands including; hydrogenated amorphous carbon (HAC), quenched carbonaceous composites (QCC), polycyclic aromatic hydrocarbons (PAHs), soot and carbon nanoparticles, fullerenes and fullerene-like particles, coal and kerogen, petroleum fractions etc., the PAH hypothesis being by far the most popular explanation of the UIE phenomenon, hence the UIE bands are also often referred to in the literature simply as the PAH bands.

Polycyclic aromatic hydrocarbons (PAHs) are benzene rings of sp^2-hybridized C-atoms linked to each other in a plane, with H atoms or other functional groups saturating the outer bonds of peripheral C-atoms. According to theoretical calculations, the infrared bands are attributed to the vibrational relaxation of large PAH molecules, containing around 50 carbon atoms on average, which are stochastically heated to high temperatures ($\sim$1000 K) by the absorption of individual high energy UV photons (Puget and Leger, 1989;Tielens et al., 2008)[293, 375]. In the low density environment of the interstellar medium where collisional de-excitation is improbable, these UV-pumped, gas-phase PAH molecules undergo spontaneous de-excitation via infrared fluorescence which gives them the AIB features. This absorption of diffuse UV light in the interstellar medium contributes significantly to the so-called interstellar extinction or reddening, which is the decrease in light intensity and the red-shift in the dominant observable wavelengths of light from a star, caused by the absorption and scattering of photons by the material in the interstellar medium.

Laboratory measurements show that neutral PAHs are primarily excited by UV-photons, with a sharp cutoff in the UV and little absorption in the visible; while PAH ions and clusters are better at absorbing visible wavelengths (Uchida et al., 1998; Li and Draine, 2002)[380, 188].

Although the PAH hypothesis has gained much popularity in the astronomical community, it is not clear whether PAHs can adequately explain the entire UIE phenomenon. The scientific literature has mainly focused on the 3.3, 6.2, 7.7, 8.6, and 11.3 μm aromatic features, but it should be acknowledged that the UIE bands in astronomical spectra are accompanied by aliphatic bands at 3.4, 6.9, and 7.3 μm, which arise from symmetric and asymmetric C–H stretching and bending modes of methyl and methylene groups, and also by unidentified emission features at 15.8, 16.4, 17.4, 17.8, and 18.9 μm. Most importantly, the UIE band features are superimposed upon broad and strong emission plateau features at 6–9 and 10–15 μm that are attributed to bending modes emitted by a mixture of alkane and alkene side groups attached to aromatic rings (Kwok et al., 2001)[175].

IR spectroscopic observations show that the UIE features in reflection nebula are as strong as those in emission nebula, although the former have low temperature central stars with a small portion of UV radiation. Also, UIE features in different reflection nebula, heated by central stars of very different temperatures, exhibit remarkably similar profiles (Uchida et al., 2000)[381], which all might suggest that their excitement is primarily by visible and not UV photons, as the PAH model suggests. In order to avoid the problem of neutral PAHs having low absorption in the visible, the new theory relies on ionized PAHs as possible sources of visible absorption, however PAHs have high ionization potential and it is extremely unlikely that they can become extensively ionized in reflection nebula where the UV background radiation is small and density is relatively high.

Another drawback of the PAH theory is that the UIE features are not sharp as would be expected in the case of molecular emission, with their widths being much broader than can be explained by Doppler effects. Also, no specific PAH molecule has been detected by astronomical spectroscopy yet through its rotational, vibrational or electronic transitions (Salama et al., 2011; Gredel et al., 2011)[318, 116].

To investigate the true nature of the UIE carriers (Kwok and Zhang, 2011; 2013)[179][180] made an analysis of archival spectroscopic observations of the UIE bands at wavelengths of 3–20 μm and showed that the data are most consistent not with pure, free-flying PAH molecules, but with amorphous organic solids that have a mixed aromatic–aliphatic structure. They called this new model MAON (Mixed Aromatic/Aliphatic Organic Nanoparticle). The proposed MAON emission carriers are quite different from the pure hydrocarbon, planar, gas-phase relatively small molecular (~50 C-atoms) PAHs, where the structure, no matter how large, is regular with repeatable patterns. MAONs are amorphous solids of hundreds or thousands of C-atoms and impurity elements such as O, N, and S, with a disorganized, three-dimensional molecular structure comprised of aromatic and aliphatic units, each with variable sizes and random orientations (see figure 5.8).

Since the late 1980's, Renaud Papoular has argued that the observed spectral properties of the UIE bands most closely resemble those of coal and kerogen (Papoular et al., 1989; Papoular 2001)[267, 268], which are actually amorphous, organic, polymeric materials with randomly oriented clusters of aromatic rings connected by long aliphatic chains, with O, N, S functional groups and heterocycles in their structure. This complex structure is very similar to the hypothesized MAON carriers. Both have a mixed sp^2–sp^3 chemistry which can give rise to the discrete aromatic and aliphatic emission features, as well as the broad plateau features (Guillois et al., 1996)[119].

Kwok and Zhang [179] proposed a chemical structure for the MAON carriers very similar to the Insoluble Organic Matter (IOM) found in the Murchison meteorite and identified recently in laboratory analysis by Derenne and Robert (2010) [75]. The structure is very complex and branched, containing polyaromatic units, aliphatic chains of different sizes and heterocyclic rings with N and S atoms, such as the pyrrole, thiophene and carbazole rings. Oxygen is also frequently found in the structure, mostly in the form of hydroxyl, ether or ester functional groups.

More than 70% of the organic matter in carbonaceous chondrite meteorites, such as the Murchison, Murray, Tagish Lake, and Orgueil meteorites, is in the form of insoluble macromolecular solids usually referred to as Insoluble Organic Matter (IOM) (Cronin et al., 1987)[63]. The structure of IOM in all these meteorites shows remarkable similarity, while isotopic ratios point to their interstellar origin. Interestingly, the chemical structure of IOM mostly resembles that of kerogen on Earth (Kerridge, 1999)[156].

The MAON model also theorizes that the coal/kerogen-like UIE carriers are nanometers in size so that their heat capacities are smaller or comparable to the energy of the stellar photons that excite them, which are predominantly visible photons. Upon absorption of a single stellar photon, these nanoparticles are stochastically heated to high temperatures, emitting the UIE emission bands by vibrational relaxation to the ground state.

It is interesting that in the UV-poor environment of protoplanetary (or preplanetary) nebula, the aliphatic chain of the UIE emitters constitutes a significantly larger portion of the entire macromolecular structure when compared to the UV-intense regions of planetary nebula where the aromatic moiety is predominant (Kwok et al. 2001; Kwok 2007)[175, 176]. For example, in the PPN (protoplanetary nebula) phase of stellar evolution, the 3.4 and 6.9 μm aliphatic emission features are as strong as the 3.3, 6.2, 7.7, and 11.3 μm aromatic emission features, but as the star evolves to the PN (planetary nebula) stage (a process of less than few thousand years), the aromatic features become predominant (Kwok, 2007)[176]. This phenomenon has been explained as the result of photochemistry, where the onset of UV radiation modifies the aliphatic side groups through cyclizations and isomerizations, and transforms them into ring structures.

In the following chapter, it is demonstrated how, under a UV-C light potential from a nearby star, non-equilibrium thermodynamic principles based on UV/VIS photon dissipation would foment the formation, aromatization, and proliferation of these organic compounds which absorb and dissipate in the UV-C. Under the same non-linear, non-equilibrium thermodynamic analysis, aliphatic chains absorbing in the near ultraviolet and visible would be the most common under a predominance of near ultraviolet and visible light from a star.

I end this section with an extraordinary quote by Chandra Wickramasinghe on work he performed with Fred Hoyle in the 1970's and 1980's, from his book entitled "A Journey with Fred Hoyle"(Wickramasinghe, 2013)[405];

> "A discovery of a 3.28 μm emission feature in the diffuse radiation emitted by the Galaxy confirmed that aromatic molecules of some kind were exceedingly common on a galactic scale. We argued that the infrared emission, not only at 3.28 μm, but over discrete set of wavelengths – 3.28, 6.2, 7.7, 8.6, 11.3 μm – must arise from the ultraviolet absorption of starlight by the same molecular system that degrades this energy into the infrared. We had shown much earlier that the 2175 $\mathring{A}$ extinction of starlight may be due to biological organic molecules, and it seems natural then to connect the two phenomena. Thus we developed a unified theory for the infrared emission and ultraviolet extinction by the same ensemble of aromatic molecules."

Wickramasinghe and Hoyle, therefore, very early on, characterized very well the fact that the same aromatic organic molecules were responsible for the UV absorption and infrared degraded emission spectra. What we are emphasizing here, however, is the thermodynamic significance of this discovery; that the "exceedingly common on a galactic scale" nature of these aromatic molecules is directly related to their ability to "degrade this energy into the infrared". This relation will be made clear in the following two chapters.

5.2.2 Diffuse Interstellar Absorption Bands (DIBs) and the Extended Red Emission (ERE)

In addition to the UIE bands, other unsolved spectroscopic features commonly seen around stars, in the interstellar medium, and in external galaxies have usually been ascribed to organic carriers whose spectroscopic properties are not known. Among these are the Diffuse Interstellar absorption Bands (DIBs), the 217 nm feature, the Extended Red Emission (ERE), and the 21 and 30 micron emission features. As carriers of these relatively common galactic features, a great variety of possible candidates have been proposed, including PAHs, carbon chains, fullerenes and nanotubes with variations in their hydrogenation and ionization states, carbon anions, polycrystalline graphite, HAC, QCC, crystalline silicon nanoparticles, titanium carbide nanoclusters, silicon carbide grains, nanodiamonds, etc. (Kwok (2012) [178] and references therein). Organic compounds are always favored in this regard, because only carbon has the capacity to produce the rich chemistry necessary to create the large variety of molecular structures to account for these features.

Interestingly, the Extended Red Emission (ERE) feature, which is a photoluminescence process powered by UV photons with a peak emission wavelength around 650-800 nm, has, among other things, been attributed to biological pigments like chlorophyll, because of their absorbance in the UV/blue and fluorescence in the red (Hoyle and Wickramasinghe, 1999)[136].

It may be that in regions of the cosmos, dissipative self-organization of material under stellar photon fluxes has evolved to the point where the emitted spectrum is so far towards the infrared such that the emission lines have become many and too weak to be detectable. It may thus be possible that some of this dissipative structured material is contributing to the baryonic dark matter. Dark matter has been determined to make up some 27% of the mass-energy density of the universe and baryonic dark matter makes up approximately 10% of this dark matter density if the universe is at the critical density for closure, and if the standard model for big bang nucleosynthesis is correct. Since the known baryonic matter makes up less than 5% of the mass-energy density of the universe (NASA, 2016)[244], the amount of dissipative material structuring into this type of baryonic dark matter could exceed 50% of all observable matter in the universe. Other explanations for the invisible baryonic dark matter such as super massive black holes or MACHOS have difficulty in accounting for all of the missing mass. That some of the baryonic dark matter may be in the form of interstellar pigments is a possibility that we are actively exploring (Michaelian, 2016)[221].

5.3 Protoplanetary Organics

Using the Atacama Large Millimeter/submillimeter Array (ALMA) in the Chilean desert (see figure 5.11), astronomers have recently detected the presence of complex organic molecules in a protoplanetary disc surrounding a young star, MWC 480 at 445 light-years from Earth (Oberg et al., 2015)[254] (see figure 5.12). So far, no planets have yet been found in this disc. The colder outer regions of the rotating disc of gas and dust contains large amounts of hydrogen cyanide (HCN) and methyl cyanide (CH_3CN). This region is considered to be similar to our Kuiper Belt region of planetesimals and comets beyond the orbit of Neptune. Cyanides, and most particularly methyl cyanide, are important because they contain carbon–nitrogen bonds, which are essential for the formation of nucleic acids (Sanchez et al., 1966)[319] and amino acids, the building blocks of RNA or DNA and proteins respectively.

Fig. 5.11. ALMA, an arrary of 66 radio telescopes in the Chilean Atacama desert used to detect organic molecules in space. Photo credit: ALMA Collaboration, Creative Commons.

What appears to be the most important of this discovery is that the concentrations of these molecules in the protoplanetary disc are much greater than those found in interstellar gas and dust clouds, indicating that these regions close to stars are efficient nurseries for producing the primordial molecules, the precursors of life on Earth. It is found that the aliphatic hydrocarbons form at a stage when the central star is still cool and not emitting much UV-C light. These compounds will be gradually transformed into aromatic hydrocarbons as the star begins to emit in the UV-C. This is consistent with our suggestion of pigment proliferation due to autocatalytic dissipation of UV-C light (see chapter 6). The high concentrations of the primordial molecules close to a young star undoubtedly has to do with the high intensity of UV-B or UV-C light and the high density of organic elements (H, C, N, O, etc.) in the disc.

Fig. 5.12. ALMA (see figure 5.11) image of the protoplanetary disc around HL Tauri, a star at 445 light years from Earth in the constellation Taurus. The aliphatic hydrocarbons form at a stage when the central star is still cool and not emitting much UV-C light. These compounds will be gradually transformed into aromatic hydrocarbons as the star begins to emit in the UV. The rings of material which will eventually condense into planets or asteroids can be clearly seen. Image credit: ALMA (ESO/NAOJ/NRAO), NSF. Public Domain.

Many abiogenic organic molecules have been found in carbonaceous chondrite meteorites like the Murchison and Allende (see figures 5.13 and 5.14) (Stoks and Schwartz, 1979; 1981)[359, 360] including all the nucleic acid bases, except cytosine (see figure 5.15). These meteorites are dated at 4.566 Ga, the time of formation of the planets, and are thought to have been produced through the condensation of protoplanetary material that was not able to form a planet (see figure 5.12). The similarity of the concentrations of aliphatic to aromatic hydrocarbons suggests that the Murchison meteorite formed in the protoplanetary disc when UV-C light was still not a strong component of the emitted spectrum of the early sun. As the central star heats up and begins to emit more light in the UV-C the aliphatic hydrocarbons are converted into aromatic hydrocarbons (see section 5.2.1 and references Kwok et al. (2001) and Kwok (2007)[175, 176]).

These abiogenically produced organic molecules can be distinguished from their biological counterparts because of their greater diversity in structure, a decrease in abundance with increasing carbon number (i.e. molecule size), the presence of many compounds (such as 80 different amino acids) not used by terrestrial organisms and highly unusual stable isotope ratios. This suggests that these molecules were synthesized under conditions that can only occur outside the Earth. However, eight amino acids commonly used on Earth by life were also found in the Murchison meteorite; glycine, alanine, aspartic acid, glutamic acid, valine, leucine, isoleucine and proline. Other amino acids with a very restricted usage in life were also found, such as b-alanine, a-aminoisobutyric acid (AIB) and sarcosine (Sephton and Botta, 2005)[332].

5.4 Planetary Organics

Planet Earth was for a long time considered to be the only body in the solar system that contains appreciable amount of organic matter, and this was attributed to the activities of life (see figure 16.6). The traditional view of the chemical composition of other solar system bodies (planets, satellites, asteroids, comets, and dust particles) was that they are entirely made up of metals, minerals, ices and gases, with either no, or only traces of, organics. However, with the development of more sophisticated technologies such as infrared astronomy and sending spacecraft to explore solar system bodies with spectroscopic observation and direct sample collection, organic substances are increasingly being recognized as a major product of solar system chemistry.

By comparing the ratio of red to green of solar UV–visible induced fluorescence spectra obtained from the surface of Mars using the Hubble space telescope with the same ratio obtained on fossil samples of organic material of Earth, Pershin (1998; 2000)[278, 279] has claimed detection of ancient pigments from cyanobacteria-like organisms in sediments from former hospitable habitats on the surface of Mars covering a large area in the western Utopia Planitia region of Mars. These detected pigments are reported to be of porphyrin in nature, similar to the chlorophyll molecule.

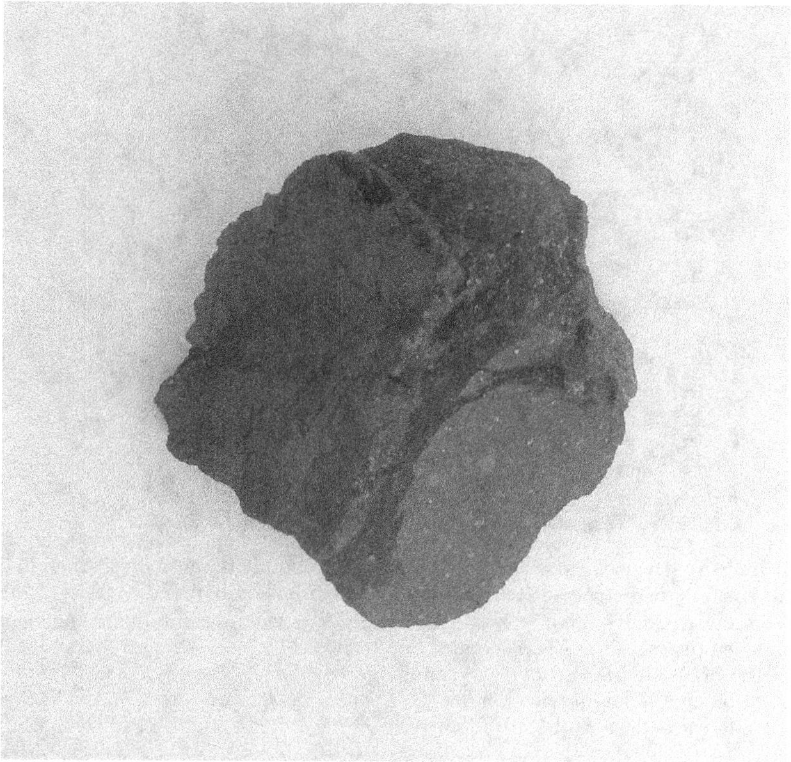

Fig. 5.13. Exterior aspect of a piece of the Allende meteorite which fell over Chihuahuaa, a nothern state of Mexico, near the town of Allende. The meteorite is a carbonacious condrite, known to have formed at the time of the origin of the solar system and containing many organic molecules common to life (see figure 5.15). Image credit: Unknown, Public Domain.

One of the first hints of the presence of organic matter on the surface of asteroids came from their deep red colors and very low albedos (0.01-0.15). Such intense colors and extremely low albedos are difficult to explain by inorganic minerals and ices alone but can easily be explained by organic pigments with a complex kerogen-like polymeric structure (Gradie and Veverka, 1980)[113]. Roush and Cruikshank (2004)[304] have noted that terrestrial dark pigmented organic polymers like tar sands, asphaltite, anthraxolite, and kerite, which all have complex aromatic/aliphatic molecular structure, have very low albedos and deep red colors, reminiscent of asteroids.

Saturn's largest moon Titan (see figure 5.16), the only known natural satellite to have a dense atmosphere, has attracted much attention since the Cassini–Huygens unmanned spacecraft arrived in 2000 and revealed that Titan's atmosphere contains haze-like solid particles that are most probably the result of the condensation of organics (see Fig. 5.16). The observations support the hypothesis that methane and nitrogen molecules excited by UV photons re-

Fig. 5.14. Interior aspect of the Allende carbonaceous chondrite meteorite. Allende has small, spherical to subspherical structures called chondrules (all chondrites have these). Allende also contains whitish, irregularly-shaped patches called CAIs (calcium-aluminum inclusions), composed of high-temperature Ca-Al-Ti silicates and oxides. The blackish, fine-grained, carbon-rich matrix consists of Fe-olivine and poorly graphitized carbon. A few tiny specks of metallic iron-nickel alloy also occur. The olivine chondrules in Allende rocks date to 4.560 billion years. The CAIs in Allende rocks date to 4.568 billion years. Image credit: James St. John, CC BY 2.0

act to form polymeric hydrogenated carbon-nitride compounds, called tholins, that give the distinctive thick layer of orange-brown haze in Titan's lower stratosphere (Waite, 2007; Nguyen et al., 2007)[390, 247].

Cassini RADAR observations found that these organic nanoparticles condense on surface sand grains that are blown by winds into longitudinal dark-colored dunes on the surface of Titan. Most interestingly, some of these organic nanoparticles end up dissolved in the numerous lakes and rivers of liquid methane and ethane in Titan's polar regions, which exhibit active liquid-gas phase cycling (see Fig. 5.17), similar to the water cycle on Earth, although at a much lower temperature (-179.5 °C) (Atreyaa et al., 2006) [8]. It is probable that Titan-based organic molecules are acting as catalysts to the methane cycle analogously to the way Earth-based pigments act as catalysts to the terrestrial water cycle by dissipating the incident solar photons into heat which is then utilized as the heat of evaporation (Michaelian, 2012b)[213].

Polyaromatic hydrocarbons (PAHs) have also been detected in large quantities in Titan's upper atmosphere (Lopez and Puertas, 2013) [190].

The Cassini spacecraft has also found evidence of mixed aromatic/aliphatic hydrocarbons on other Saturnian satellites: Iapetus, Phoebe and Hyperion (Cruikshank et al., 2008)[67]. Iapetus is the third largest satellite of Saturn,

Compounds	Abundances %	µg g⁻¹ (ppm)	Reference
Macromolecular material	1.45		(Chang *et al.* 1978)
Carbon dioxide		106	(Yuen *et al.* 1984)
Carbon monoxide		0.06	(Yuen *et al.* 1984)
Methane		0.14	(Yuen *et al.* 1984)
Hydrocarbons:			
aliphatic		12–35	(Kvenvolden *et al.* 1970)
aromatic		15–28	(Pering & Ponnamperuma 1971)
Acids:			
monocarboxylic		332	(Lawless & Yuen 1979; Yuen *et al.* 1984)
dicarboxylic		25.7	(Lawless *et al.* 1974)
α-hydroxycarboxylic		14.6	(Peltzer *et al.* 1984)
Amino acids		60	(Cronin *et al.* 1988)
Alcohols		11	(Jungclaus *et al.* 1976b)
Aldehydes		11	(Jungclaus *et al.* 1976b)
Ketones		16	(Jungclaus *et al.* 1976b)
Sugar-related compounds (polyols)		~24	(Cooper *et al.* 2001)
Ammonia		19	(Pizzarello *et al.* 1994)
Amines		8	(Jungclaus *et al.* 1976a)
Urea		25	(Hayatsu *et al.* 1975)
Basic N-heterocycles (pyridines, quinolines)		0.05–0.5	(Stoks & Schwartz 1982)
Pyrimidines (uracil and thymine)		0.06	(Stoks & Schwartz 1979)
Purines		1.2	(Stoks & Schwartz 1981b)
Benzothiophenes		0.3	(Shimoyama & Katsumata 2001)
Sulphonic acids		67	(Cooper *et al.* 1997)
Phosphonic acids		1.5	(Cooper *et al.* 1992)

Fig. 5.15. Organic material and their abundances found inside the Murchison carbonaceous condrite meteorite which fell near the town of Murchison, Australia in 1969. This material could only have been formed through UV-C photochemical interaction with the primordial gases in the protoplanetary disc. Image credit: Sephton and Botta (2005)[332]. Reproduced with permission from Cambridge University Press.

locked in synchronous rotation about the planet, with the leading hemisphere and sides of dark reddish-brown color and extremely low albedo (0.03–0.05), and most of the trailing hemisphere and poles of bright color and high albedo (0.5–0.6). This difference in coloring and albedo between the two hemispheres is striking. Temperatures on the dark region's surface reach about 130 K (-143 °C) at the equator, while the bright surfaces reach only about 100 K (-173 °C) due to less sunlight absorption and therefore less dissipation into heat. The dark material is believed to consists primarily of lag from the sublimation of ice from the warmer areas of Iapetus's surface, and Earth-based observations have shown it to be carbonaceous, most likely in the form of HCN polymers (Spencer and Denk, 2010; Buratti et al., 2005)[354, 32].

Trans-Neptunian objects (TNO) are a group of minor bodies that orbit the Sun in the outer solar system beyond the planet Neptune. Photometric observations in the visible have found the colors of some of these objects to be intensely

Fig. 5.16. This composite image shows an infrared view of Saturn's largest moon Titan from NASA's Cassini spacecraft, acquired during the mission's 10,000 Km flyby on Nov. 13, 2015. The spacecraft's visual and infrared mapping spectrometer (VIMS) instrument made these observations, in which blue represents wavelengths centered at 1.3 microns, green represents 2.0 microns, and red represents 5.0 microns. A view at visible wavelengths (centered around 0.5 microns) would show only Titan's hazy atmosphere. The near-infrared wavelengths in this image allow Cassini's vision to penetrate the haze and reveal the moon's surface. The scene features the parallel, dark, dune-filled regions named Fensal (to the north) and Aztlan (to the south), which form the shape of a sideways letter "H". The sand grains comprising the dunes are thought to be covered with dark colored organic nanoparticles. Image credit: NASA/JPL/University of Arizona/University of Idaho. Public Domain.

red (Jewitt and Luu, 2001)[140], which can be an indication of the presence of organic material on their surface.

Pluto was visited for the first time on July 14, 2015 by the probe New Horizons. It was found to have a surface color varying between charcoal black, dark orange and white (see figure 5.18). Its tenuous atmosphere was found to consist of nitrogen (N_2), methane (CH_4), and carbon monoxide (CO), which are in equilibrium with ices of the same substances on Pluto's surface. According to the measurements by New Horizons, the surface pressure is about 1 Pa (10 μbar). The white colored regions are probably ices of nitrogen, methane and carbon monoxide. The orange color region near the north pole has been associated with

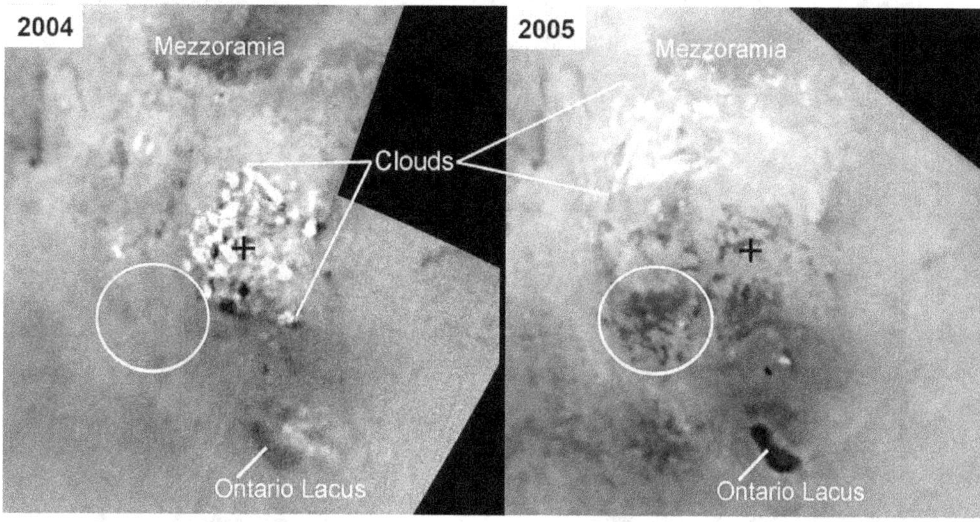

Fig. 5.17. These mosaics of the south pole of Saturn's moon Titan, made from images taken almost one year apart, show changes in dark areas that may be lakes filled by seasonal rains of liquid hydrocarbons. The deepest lakes have been determined to have a depth of about 200 m, with the average being about 50 m. The image on the left was acquired July 3, 2004. That on the right was taken June 6, 2005. In the 2005 image, new dark areas are visible and have been circled in the labeled version. The very bright features are clouds in the lower atmosphere (the troposphere). Titan's clouds behave similarly to those on Earth, changing rapidly on timescales of hours and appearing in different places from day to day. During the year that elapsed between these two observations, clouds were frequently observed at Titan's south pole by observers on Earth and by Cassini's imaging science subsystem. It is likely that rain from a large storm created the new dark areas that were observed in June 2005. Some features, such as Ontario Lacus, show differences in brightness between the two observations that are the result of differences in illumination. These mosaics use images taken in infrared light at a wavelength of 938 nanometers. The images have been oriented with the south pole in the center (black cross) and the 0 degree meridian toward the top. Image resolutions are several kilometers (several miles). Image credit: NASA/JPL/Space Science Institute. Public Domain.

UV photochemical reaction products on methane. The dark charcoal regions near the southern pole are thought to be a kind of "tar" made of complex hydrocarbons called tholins which form by UV light and cosmic rays interacting with methane and nitrogen in the atmosphere.

Exceptionally rich in organic content and organic diversity are comets. The tail of comet Kohoutek was extensively studied in the UV, visible, infrared and radio wavelength bands. Emission bands attributed to cyanide CN, methyl radical CH_3, the hydroxyl radical OH, and amino group NH_2 were found. From observations of Kohoutek at radio wavelengths, more complex organic molecules such as acetonitrile CH_3CN and hydrogen cyanide HCN were found (Hoyle and Wickramasinghe, 1982)[135]. Dust samples from the coma of comet Wild 2 were collected by the Stardust spacecraft and returned to Earth in 2006. The samples

Fig. 5.18. An image of Pluto obtained by the New Horizons spacecraft on July 14 of 2015. The orange color near the north pole are thought be photochemical products from UV sunlight absorption on methane in the atmosphere. The white colored regions are most likely frozen nitrogen, methane and carbon monoxide, while the dark charcoal color is probably rich in hydrocarbons like thiolins. Image credit: New Horizons collaboration. Public Domain.

were analyzed with various techniques, giving evidence of rich organic content, but mostly in the form of aromatic-aliphatic polymers similar to the IOM of meteorites but with higher N and O content. Of particular significance is that the aromatic compounds are not pure hydrocarbons but have a very high N content, where N is incorporated predominantly in the form of aromatic nitriles (R–C=N); a fact of great astrobiological implications as comets may have contributed to Earth's prebiotic chemical inventory (Clemett et al., 2010; Keller et al., 2006)[51, 155]. Most of all, though, they indicate that nitrogenous aromatic pigment compounds can be readily created in an environment of organic elements subjected to ultraviolet radiation.

The ubiquity of these organic pigment molecules both in interstellar space and on solar system bodies cannot be explained by traditional equilibrium thermodynamics but requires a non-equilibrium explanation involving the dissipation of UV photons. It will be shown in the following chapter that if a non-equilibrium route to these molecules can be found by Nature, and if these molecules are efficient in dissipating the same photon potential that produced them, then we have a an autocatalytic photochemical dissipative process that leads to microscopic self-organization of these pigments and to their proliferation. This non-equilibrium thermodynamic explanation for the ubiquity of these molecules has relevance to the possibility of life, both as we know it, and as we may not know it, throughout the universe. Within this non-equilibrium thermodynamic framework there is unambiguous continuity from the photochemical production and proliferation of pigment molecules to the origin and evolution of life, since, as shown in the following chapters, the evolution of life can be succinctly described as just that process which enhances pigment dissipation efficacy and foments pigment dispersal over the whole surface of Earth where sunlight is available.

6. Microscopic Dissipative Structuring and Autocatalytic Photochemical Proliferation

In an autocatalytic chemical reaction, one of the products of the reaction acts as a catalyst for the reaction itself. Most reactions occurring in living systems today are autocatalytic. An example is the glycolysis cycle which produces ATP from glucose and is auto-catalyzed by the enzyme phosphofructokinase (PFK). It is generally believed that the first chemical reactions of life must also have been autocatalytic. Auto-catalytic reactions, as all chemical reactions, arise to dissipate a chemical potential, but they are extraordinarily efficient at dissipating the chemical potential because of the positive feedback provided by the catalyst product. The rate equations are non-linear and, as we have seen in chapter 3.1 and as we will see in further detail here, this leads to a spontaneous generation of order in both space and time (Prigogine, 1967)[289]. A simple example of a autocatalytic reaction is the following;

$$A + B \rightleftarrows 2B \tag{6.1}$$

here the product of the reaction B acts as a catalyst for the conversion of A into B. The rate equations for the concentrations of A and B are nonlinear.

For an autocatalytic chemical reaction under the non-equilibrium conditions of a constant supply of the reactants and a constant sink of the products (a fixed external chemical affinity for the reaction) it can be shown that the stationary state concentrations[1] of the catalyst product may grow to many orders of magnitude larger than that which would be expected under near equilibrium conditions (Prigogine, 1967)[289]. The greater the catalytic activity, the greater the amount of catalyst formed in the non-equilibrium steady state and the greater the rate of the dissipation of the chemical potential and, therefore, the greater the rate of entropy production (see section 6.3).

Autocatalytic chemical reactions have been understood for some time as being processes necessary to bootstrap life. However, extensive searches for these have so far been unsuccessful (see chapter 1.2.8). Instead of an "autocatalytic

[1] Given a system over which constant external conditions exist, it can be shown that the system will evolve towards a stationary state in which the thermodynamic variables remain constant or fluctuate cyclicly or even chaotically, but around fixed average values. This, in thermodynamic language, is known as an out-of-equilibrium *stationary state*.

chemical reaction" dissipating an external chemical potential, however, here I suggest that the origin of life would have been based on dissipating the prevailing solar photochemical potential due to four principle reasons; 1) as evidenced in the previous chapter, the dissipation of stellar photon potentials in the cosmos has led to an ubiquity of organic compounds, 2) the solar photon potential has always been many orders of magnitude greater in its free energy content available for dissipation than all extinct or extant generalized chemical potentials at Earth's surface together, including deep sea hydrothermal vents, 3) life today is almost completely concerned with dissipating the solar photon potential, and 4) this solar photon source has been relatively constant on the billions of years (10^9 a) time scale.

The advantages of photochemical reactions over normal thermally induced chemical reactions occurring in the electronic ground state are various (König, 2015)[169]:

1. Reactions may occur that are very endothermic since the absorbed photon donates its energy to the molecule. The energy of a photon of 260 nm is 4.77 eV or 110 Kcal/mol.
2. In the electronic excited state antibonding orbitals are occupied and this may allow reactions which are not possible for electronic reasons in the ground state.
3. Photochemical reactions can involve singlet and triplet states. Thermal reactions usually only involve singlet states. Therefore, photochemical reaction intermediates may be formed which are not accessible under thermal conditions.

The optical, electronic and physical properties of RNA and DNA (to be discussed in detail in chapter 7) make them excellent candidates for photochemical and autocatalytic reactions. Autocatalytic photochemical reactions are indeed known and have even been controlled by DNA. For example, Dutta and Mokhir (2011)[89] present the controlled conversion through photochemical oxidation of 2',7'-dichlorofluorescin (pro-P) into 2',7'-dichlorofluorescein (P), which is a fluorescent dye. P itself, when electronically excited through photon absorption, can act as a catalyst (by emitting a photon of the required frequency) for the photochemical conversion of pro-P into P, which provides the autocatalytic nature of this photochemical reaction. Controll is achieved by using complementary oligonucleotides but of different lengths with a photosynthesizer E connected on one oligo and a photoquencher Q connected to the complementary oligo. In this double helix configuration there is no fluorescence of the photosynthesizer E since the electronic excitation energy is dissipated by the quencher Q which is in close proximity. However, by introducing a complimentary oligo of the long oligonucleotide attached to the quencher into the reaction chamber,

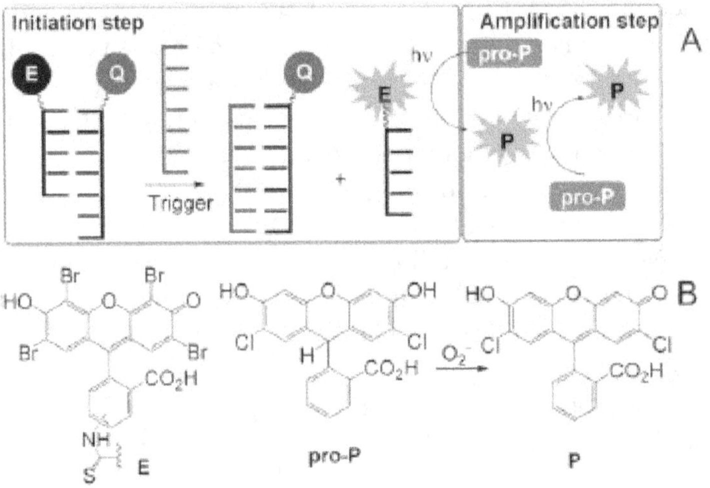

Fig. 6.1. The autocatalytic photochemical conversion of pro-P into P controlled by a particular nucleic acid sequence. The photosynthsizer E is eosin and the quencher Q is known as "black-hole-quencher-3". See text for details. It is easy to imagine that such controlled autocatalytic photochemical reactions involving nucleic acids, but in the UV-C region and with the fundamental molecules of life, were important in the first steps of the origin of a dissipative life. At UV-C wavelengths RNA and DNA could act as their own quencher. Image credit: Dutta and Mokhir (2011)[89]. Reproduced with permission of the Royal Society of Chemistry.

this releases the short oligo with the photosynthesizer attached (because of the greater binding energy of the same length oligos) which can then fluoresce, since the quencher is now nowhere near, and therefore act as light source for the photochemical oxidation of pro-P to P, increasing the overall reaction rate of pro-P to P (see figure 6.1).

Routes have indeed been found to the production of nucleobases using only hydrogen cyanide, HCN, in a water and ammonia solution with inorganic catalysts (Oró, 1961; Oró and Kimball, 1962; Saladino et al. 2005b; 2007b)[261, 262, 313, 317] and to nucleic acids, ribose sugar, and even lipids using HCN, and hydrogen sulfide, H_2S, in water, under UV-C light (Powner et al., 2009; Patel et al., 2015)[288, 269]. However, these experimental results of the group of Sutherland were probably not the reactions prevalent at the beginnings of life since these do not appear to be autocatalytic and therefore would not contribute to the nucleotide proliferation to be described in section 6.3, although these reactions may have contributed modest amounts of fundamental molecules to the prebiotic organic soup. Instead, in the following sections, we will consider in detail the photochemical pathways to the nucleobases first discovered by Ferris and Orgel (1966)[96] and another photochemical pathway to the nucleobase polymerization in the structuring of single strand RNA and DNA;

both reactions, in fact, display all the characteristics of microscopic dissipative structuring and appear to be autocatalytic.

6.1 Dissipative Structuring of the Nucleobases from HCN under UV-C Light

The most abundant carbon-containing 3-atom molecule in the observable cosmos is hydrogen cyanide HCN. Oró (1961)[261] and Oró and Kimball (1962)[262] were the first to discover a thermal reaction pathway from HCN to the purine nucleobase adenine at high temperatures in a water and ammonia solution. Later, Saladino et al. (2005b; 2007b)[313, 317] showed that all the nucleobases except guanine could be formed from formamide H_2NCOH (a common product of HCN and H_2O) in a water environment at high temperatures of between 100 °C and 160 °C using a number of distinct available minerals acting as catalysts. However, Ferris and Orgel (1966)[96] had earlier discovered a low temperature generic photochemical pathway to all the nucleobases except guanine without the necessity of the inorganic catalysts as given in the reaction scheme shown in figure 6.2. Recently, Barks et al. (2010)[12] showed that guanine could be derived from formamide under UV irradiation at lower temperatures.

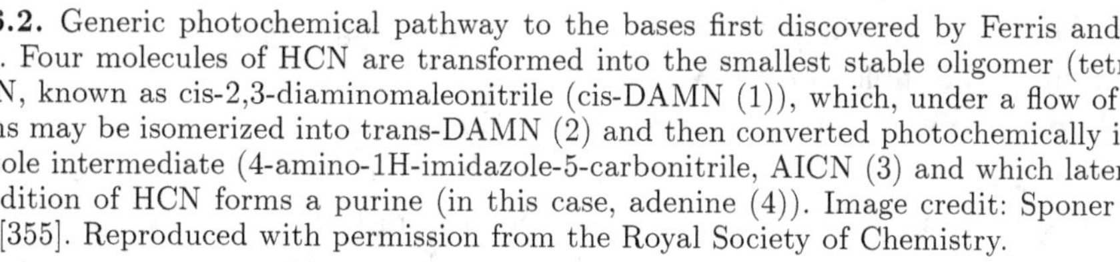

Fig. 6.2. Generic photochemical pathway to the bases first discovered by Ferris and Orgel (1966). Four molecules of HCN are transformed into the smallest stable oligomer (tetramer) of HCN, known as cis-2,3-diaminomaleonitrile (cis-DAMN (1)), which, under a flow of UV-C photons may be isomerized into trans-DAMN (2) and then converted photochemically into an imidazole intermediate (4-amino-1H-imidazole-5-carbonitrile, AICN (3) and which later, with the addition of HCN forms a purine (in this case, adenine (4)). Image credit: Sponer et al., (2016)[355]. Reproduced with permission from the Royal Society of Chemistry.

The formation of the nucleobases at high temperature from formamide employing inorganic catalysts may have been operative well before the origin of life at 3.85 Ga, perhaps even during the Hadean (4.56 - 3.9 Ga) when surface temperatures were very high. However, near the beginning of the Archean at 3.85 Ga Earth's surface temperature is estimated to have cooled to ~80 °C (Knauth, 1992; Knauth and Lowe, 2003)[163, 165], too low for normal thermal chemistry with formamide, and the photochemical pathway to the production of the bases would appear to be the only option. In fact, a much more important reason for concentrating on the photochemical reaction pathways discovered by Ferris and Orgel is that these would provide a route for microscopic dissipative structuring and autocatalytic proliferation, which appear to be the hallmarks of life even today. In the following, we therefore consider only the generic photochemical pathway to the bases depicted in figure 6.2.

Although the details of the actual processes leading from (1) to (3) (see figure 6.2) were previously not sufficiently determined, Boulanger et al. (2013)[24] have recently analyzed in detail the possible mechanisms of the required photochemical and chemical steps using computational chemistry employing density functional theory and time dependent density functional theory to determine the minima and transition states in the electronic ground state and excited state respectively, and employing chemical kinetics. The fast decay of the excess energy due to UV-C photon absorption and internal conversion to vibrational energy, with 2/3 of the excess energy being dissipated within 0.2 ps to the surrounding water molecules, implies that the maximum free energy barrier height for a hot ground state thermal reaction would be approximately 30 Kcal/mol (1.3 eV) (Boulanger et al., 2013)[24]. Any barrier higher than this would have to involve a distinct photon absorption. They find that there is, therefore, only one sequence of steps that is thermodynamically and kinetically compatible with all previous experimental results (see scheme 5 of Boulanger et al. (2013)[24]).

The full reaction requires the photo-excitation of cis-DAMN (1), trans-DAMN (2), and AIAC with UV-C photons of greater than 4.0 eV (see scheme 5 of Boulanger et al. (2013)[24]). Koch and Rodehorst (1974)[166] have found that irradiation with UV-C light leads to photoisomerization of cis-DAMN giving a photostationary state with a large predominance of trans-DAMN (2) isomer over cis-DAMN (1). This isomerization under the UV-C flux is an obvious example of microscopic dissipative structuring involving internal isomeric degrees of freedom of the molecule DAMN.

Perhaps the most important result that Boulanger et al. (2013)[24] find, and a result of great importance to the thesis of this chapter, is that, although energies of greater than 4.0 eV are added to the system through photoexcitation by UV-C photons, most of this energy, rather than being used in the photochemical reactions, is instead simply dissipated to the ground state of the relevant structure and surrounding solvent through internal conversion, occurring in less than 0.2 ps (Boulanger et al., 2013)[24]. In fact, Koch and Rodehorst (1974)[166] have shown that cis-DAMN (1) is excited about 300 times (on average) before a single cyclization event takes place. The formation of the purines from HCN is thus a very UV-C dissipative process and the experimental data presented by Koch and Rodehorst, along with the molecular dynamic simulations of Boulanger et al., indicate that the observed process of purine formation from HCN under a UV-C flux is an example of microscopic dissipative structuring in which molecular isomerizations and finally new structures arise "spontaneously" in order to dissipate the same external generalized chemical potential that was required to produce them.

6.2 Dissipative Structuring of Single Strand DNA

Assuming that the nucleobases had been formed through some particular microscopic dissipative structuring as suggested in the previous section, the following step in the evolution of life towards a greater dissipating system must have been the polymerization of the bases into single strand RNA and DNA. There are both thermodynamic and non-equilibrium thermodynamic reasons for the polymerization of the bases into strands, these being, respectively, the greater stability against hydrolysis and the greater potential for photon dissipation through the provision of a scaffolding to allow antenna type molecules to attach themselves to RNA or DNA and, through resonant energy transfer, use the conical intersections of these acceptor molecules to rapidly dissipate the UV-C induced electronic excitation energy of the donor into heat (see chapter 8.2). In this section I show how microscopic dissipative structuring can lead to the formation of single strand RNA and DNA once the nucleobases and UV-C light were existent in the environment, and show that this process is autocatalytic.

Consider the following simplified reaction scheme for the autocatalytic photochemical production of single strand DNA or RNA polymers of length, say, 10 nucleotides. More details of the reaction in the context of the prevailing Archean environmental conditions, which I have called Ultraviolet and Temperature Assisted Replication (UVTAR), and for which some tentative experimental evidence exists, will be given in chapter 12.

$$\gamma_{260} + PO_3 + N_s \longrightarrow N_t^* \tag{6.2a}$$
$$10N_t^* + ssDNA_{10} + Mg^{2+} \longrightarrow dsDNA_{10} + Mg^{2+} \tag{6.2b}$$
$$\gamma_{260} + dsDNA_{10} \longrightarrow 2ssDNA_{10} \tag{6.2c}$$

In reaction (6.2a), a 260 nm UV-C photon interacts with a phosphate group PO_3 and a nucleoside N_s to produce an activated nucleotide N_t^*. In the second reaction (6.2b), 10 activated nucleotides interact (over night at somewhat colder temperatures and with the aid of Mg^{2+} ions as catalysts for the polymerization) with a single strand DNA segment of 10 nucleotides $ssDNA_{10}$ to produce a double strand DNA, $dsDNA_{10}$. In the final photochemical reaction (6.2c), occurring during the day, a 260 nm UV-C photon interacts with a double strand DNA and, through dissipation into local heat, denatures it into two single strand DNAs (see chapter 12.2). Note that because of the strongly non-equilibrium nature of these reactions involving high energy photons, the backward rate constants would be essentially zero. The overall photochemical reaction can thus be written as;

$$11\gamma_{260} + 10PO_3 + 10N_s + ssDNA_{10} \longrightarrow 2ssDNA_{10} \tag{6.3}$$

which is just an autocatalytic photochemical reaction of the form (6.1) for the production of single strand DNA, with $A = 11\gamma_{260} + 10PO_3 + 10N_s$ and $B = ssDNA_{10}$.

Assuming a relatively constant supply of 260 nm photons, phosphate groups, and nucleosides, it would be expected that the steady state concentrations of single strand DNA ($ssDNA_{10}$) could become many orders of magnitude greater than what would be expected if these reactions were taking place closer to equilibrium (for example with thermally induced chemistry). The distance from equilibrium is determined essentially by the photon potential which is derived from the free energy dissipated during the conversion of the incident beam of photons (UV-C) into the outgoing beam (which would have an almost black-body spectrum at the temperature of the DNA in its water solvent environment). These affinities, representing the generalized force driving the photochemical reactions, can be derived using either an approximate calculation employing photon pressures, or a more exact calculation using an expression derived by Planck for the entropy of an arbitrary beam of photons (see section 6.3.1 and the Appendix respectively for the calculation of affinities for the two different approaches).

Note that in order to achieve a stationary state at which the concentration of single strand DNA polymers are large, a relatively constant and similarly large supply of 260 nm photons, phosphate groups, and nucleosides would be required. The UV-C photon flux from the Sun during the Archean would have been large and constant; with an integrated flux between 230 and 290 nm of about 5.5 Wm^{-2} (see chapter 4.3) there would have been approximately 7×10^{18} photons m^{-2} s^{-1} in the UV-C region arriving at the Earth's surface at midday at the equator. The concentrations of the phosphate groups and the nucleosides would also have to be supported by similar photochemical autocatalytic reactions as postulated for the production of single strand DNA (Eq. (6.3)) but with primordial molecules as the reactants A. A particular example of producing the nucleosides from the primordial molecules of hydrogen cyanide HCN and water H_2O using UV-C light has been given in the previous section 6.1.

Folowing Prigogine's non-equilibrium thermodynamic analysis of autocatalytic chemical reactions (Prigogine, 1967)[289], I have shown (Michaelian, 2010)[208] that, if under the imposition of the solar photon potential and given a constant supply of reactants and a constant sink (dispersal) of the products, a photochemical route can be found to the production of a pigment molecule (such as that given above for the production of the UV-C pigment ssDNA), and if that pigment molecule is efficient at dissipating the same photon potential that was required to produce it, then, a similar situation to that of the autocatalytic chemical reaction would exist, only that a photochemical potential, rather than a chemical potential, would be dissipated. In a similar manner as with the catalyst in the autocatalytic chemical reaction, the concentration of that pigment in the photochemical autocatalytic reaction at the thermodynamic stationary state (when the rates of all dissipative processes are no longer changing, given a constant imposed external photon potential and a constant source of reactants and sink of products) would become many orders of magnitude larger than what would be expected under near equilibrium conditions. A mathematical derivation of this has been given in reference (Michaelian, 2013)[214] and is presented in the following section, 6.3. Understanding the derivation will require a minimum knowledge of the basics of non-equilibrium thermodynamics which can be obtained by studying carefully section 3.1 in chapter 3 or in greater detail by studying Prigogine's book "An Introduction to the Thermodynamics of Irreversible Processes" (Prigogine, 1967)[289]. The derivation shows that the greater the efficacy of the particular pigment in dissipating the solar photon potential, the greater the amount of that pigment in the stationary state that would be expected and the greater the global entropy production of the process in the stationary state.

Above we have given an example of the autocatalytic photochemical production of single strand DNA based on the catalytic template nature of single strands and the activation of nucleotides through UV-C photons and the UV-C induced denaturing of DNA through photon dissipation. However, other types of autocatalytic photochemical reactions are also possible for producing the fundamental molecules. For example, suppose that a UV-C photon is capable of converting a molecule A plus a molecule B into a pigment molecule C through a photochemical reaction. Now, photochemical reactions are like thermal reactions in the sense that higher temperatures imply more rapid displacements in phase space (velocities and space coordinates of the molecules) which is helpful in finding the optimal collisional orientation between A and B for the reaction to take place. Rate constants for both chemical and photochemical reactions are thus normally an increasing function of the local solvent temperature. If the pigments C produced in a photochemical reaction are effective photon dissipaters, converting the photon energy rapidly into heat, their local presence can therefore increase the rate of the photochemical reaction producing the pigment itself. It therefore acts as a catalyst for its own production and we have another kind of autocatalytic photochemical reaction. Not only could the pigment provide local heat, but it may also be able to transfer energy in other ways, such as through resonant energy transfer, through an electron transfer, or fluorescence, etc. (see the remarks at the beginning of this section for a description of what is possible for photochemical reactions). I believe that an example of this is the production of the nucleosides as described in section 6.1 in which a cyclization event is stimulated by the local heat of dissipation. In such autocatalytic reaction schemes, the concentration of pigments C will increase to a point at the steady state in which it is many orders of magnitude over what it would have been if the pigment did not have its catalytic properties, or if the reaction were close to equilibrium. Dispersal of the newly formed pigment catalyst by whatever mechanism, including normal diffusion, will thus imply a continual production of the pigment. This is the reason we offer for the ubiquity of organic pigments in the cosmos (see chapter 5).

It is now possible to envision nested sets of autocatalytic photochemical reactions occurring, each feeding off a more fundamental photochemical reaction process by using its products and the resulting chemical potential generated[2] (see figure 2.4). The overall coupled process generating more entropy than could the individual processes summed (if indeed they could be artificially separated). This hierarchy of autocatalytic dissipative cycles is exactly the scheme employed by life today, such as, for example, in the metabolic Crebs cycle or the carbon fixation in the photosynthetic cycle.

Finally, it is emphasized that it is the dissipation of the solar photon potential by these fundamental molecules of life (pigments in the UV-C) through these types of autocatalytic photochemical reactions mentioned above that was, and still is, driving their proliferation over the sunlit Earth. It is this driven proliferation which we identify at macroscopic scales as the vitality of life. These molecules and their organismal support structures do not have a mysterious "will to survive and proliferate" as the Darwinists are obliged to see it. They are driven to proliferation and evolution through dissipating the external photon potential, and recognizing this fact provides a non-tautological basis for understanding evolution from within a thermodynamic paradigm more encompassing than the Darwinian paradigm. *Natural selection* is not operating as the Darwinists see it, and we should hence forth consider as more fundamental and accurate the non-equilibrium thermodynamic explanation given here and refer to selection as *thermodynamic selection*. How the increase in concentration and entropy production of the pigments are related to their autocatalytic dissipative nature is made explicitly clear in mathematical terms in the following section.

The rest of this chapter may be skipped by those with an adverseness to mathematics without loosing too much continuity of argument, but I would

[2] This is similar to what Kaufmann suggested (see chapter 1.2.8), but with the most fundamental driving reaction being photochemical and involving the fundamental molecules as catalytic agents for their own production through the dissipation of the solar UV-C flux.

strongly recommend it to the brave as the key to understanding the vitality of life and the mechanism of evolution lie within it.

6.3 Autocatalytic Proliferation of Organic Pigments

In this section we consider the production of any organic pigment (e.g. a nucleobase, aromatic amino acid, carotenoid, flavin, scytonemin, mycosporine-like amino acid, etc.) as an autocatalytic photochemical reaction promoting the dissipation of the solar photon flux. It will be shown that if organic pigments are good at the irreversible process of photon dissipation, then the photochemical reaction leading to their production will be thermodynamically selected and then their concentration will increase on the Earth's surface, much over and above that which would be expected if the pigments did not act as catalysts or if the system was close to equilibrium. This proliferation of organic pigments over the surface of Earth to increase the solar photon dissipation, as viewed from our thermodynamic perspective, is precisely what biological evolution is all about. The Earth is densely covered with photon dissipating pigments. They are, in fact, found wherever there exists sunlight, the dissipative medium of water, and nutrients (precursor molecules for making the pigments) (see figure 6.3).

All of evolution, both biotic and coupled abiotic-biotic, can only be understood in these non-equilibrium thermodynamic terms. There is no other plausible physical-chemical reason for the appearance of life and the ensuing evolution of life on Earth other than that of increasing the rate of photon dissipation in the interaction of Earth with its solar environment. There are, of course, other, but almost thermodynamically irrelevant, parasitic life (for example, humans) living off small sources of chemical potential produced as a side product of photon dissipation. However, it is the positive feedback in promoting proliferation through spreading nutrients which animals provide to the photon dissipating plants and cyanobacteria which is much more relevant than their dissipation of these secondary chemical potentials (see chapter 16.6).

The derivation to be given here for proliferation of organic pigments under photochemical reactions parallels that given by Prigogine (1967)[289] for a purely chemical autocatalytic reaction. However, instead of an affinity[3] derived from a chemical potential which depends on the concentrations of the chemical constituents, we use an affinity derived from a photo-chemical potential which can be determined in two distinct ways; first, to be given in the following section, the affinities can be determined approximately from the pressure of a photon

[3] The "affinity" of a chemical reaction is a measure of the force of the reaction which depends on the chemical potentials (changes in the Gibb's free energy of the system with concentration of the reactants and products) and the stociometric coefficients of the reaction. The affinity over the temperature times the rate of the reaction gives the entropy production (see chapter 3.1).

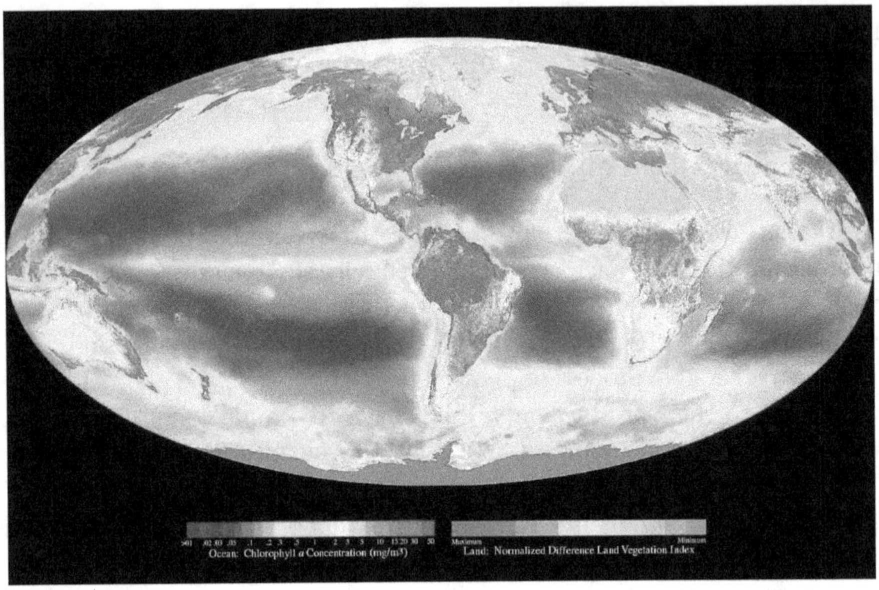

Fig. 6.3. The actual distribution of chlorophyll *a* over Earth's surface. The dark blue areas correspond to ocean waters with very little chlorophyll for lack of nutrients at the surface, and the lighter band around the equator corresponds to high concentrations of chlorophyll where colder ocean water has welled up to the surface carrying nutrients with it for pigment production. Over land, pigments are only lacking where there is insufficient water, such as the Sahara desert. Other pigments have a very similar distribution since they are almost always associated with chlorophyll. This spread of pigments over all of Earth's surface is one of the important hallmarks of evolution and speaks for the thermodynamic function of life, which is photon dissipation. Image credit: SeaWiFS Project, Goddard Space Flight Center and ORBIMAGE, Public Domain.

gas (Herrmann and Würfel, 2005)[130] which goes as the fourth power of the temperature of the gas for an equilibrium distribution of photons (black-body), and secondly and more accurately, given in the Appendix, the affinities can be determined from a formula for the entropy of an arbitrary light beam derived by Planck in the early 1900's (Planck, 1901;1906;1914)[283, 283, 285].

In both derivations, the generalized flows corresponding to these derived generalized forces (affinities) will be the flows of energy which are transformed from the solar spectrum incident on the material (or Earth's surface) to the outgoing emitted spectrum determined by the temperature of the emitting material (or Earth's surface), and the flow of energy going into the production of the pigments. The following subsection can be skipped on first reading by those with an adverseness to mathematics, however, it is strongly recommended to the brave as within it lies the answer to one of the most difficult questions regarding the vitality of life, that of its incessant replication.

6.3.1 Autocatalytic Proliferation of Pigments: Affinities Derived from the Pressure of a Photon Gas

The dissipative process which life is performing and which I will describe in this section in mathematical language is the transformation of a black-body spectrum of a high temperature (that of the surface of the Sun $\sim$ 5760 K) into a black-body spectrum of a lower temperature (that of the Earth's surface $\sim$ 287 K). It is assumed that the rate of energy conversion from the high temperature spectrum to the low one will be linearly proportional to the difference of the photon pressures of the different spectra[4]. This is analogous to assuming that the rate of a chemical reaction is proportional to the difference in the concentrations of the reactant and product (with equilibrium and rate constants set equal to one), or that the flow of material is proportional to the gas pressure differences, or that the flow of electrical current is proportional to the difference in electrical potential. These latter two linear relationships are valid when the mean free paths (between scattering events) of the particles are small with respect to the size of the system, i.e. that the system attains local equilibrium, while for the former chemical reactions (or photon dissipation), it is only required that the reactants and products (or photons) retain a Maxwell-Boltzmann (or Boltzmann) distribution of the velocities (or energies) of the particles (or photons) involved.

From the formalism of classical irreversible thermodynamics (Prigogine, 1967)[289], the entropy production $\mathcal{P}$ is a product of generalized forces, X, times generalized flows, J, (see chapter 3.1), summed over all k irreversible processes operating within the system,

$$\mathcal{P} = \frac{d_i S}{dt} = \sum_k J_k X_k \geq 0. \tag{6.4}$$

For a chemical reaction, the generalized force X is the affinity over the temperature, A/T, where the affinity A for the reaction is equal to the stoichiometric coefficients of the reactants multiplied by the chemical potentials plus the same of the products (the stoichiometric coefficients are negative if on the left side of the equation describing the reaction, being the reactants, and positive if on the right, being the products). The chemical potentials in turn depend on the concentrations of the reactants and products and the free energy differences of the reactant and product states. The generalized flow J for the irreversible process of a chemical reaction is the rate of the reaction. For our photon dissipation reaction to be considered here, the generalized forces are the photon spectrum affinities determined from the photon pressures, and the generalized

[4] Photons carry momentum and thus, for a collection of photons in equilibrium (a black-body spectrum), a unique equilibrium pressure can be defined which is found to go as the fourth power of the temperature. This relation is derived in the Appendix, equation (A.15).

flows are the flows of energy from the incoming short wavelength spectrum into the outgoing long wavelength spectrum, or into pigment production.

The change of the entropy production $d\mathcal{P}$ can be decomposed into two parts, one related to the change of forces and the other to the change of flows

$$d\mathcal{P} = d_X\mathcal{P} + d_J\mathcal{P} = \sum_k J_k dX_k + X_k dJ_k, \tag{6.5}$$

where the sum is over all k irreversible processes occurring within the system.

In the whole domain of the validity of thermodynamics of irreversible processes, and under constant external constraints, the contribution of the time change of the forces to the entropy production is negative or zero,

$$d_X\mathcal{P} \leq 0. \tag{6.6}$$

This is known as the *general (or universal) evolutionary criterion* or the *Glansdorff-Prigogine criterion* since it was first established by Glansdorff and Prigogine (1964)[107] (see also chapter 3.1 and Prigogine (1967)[289]). The general evolutionary criterion is the most general result that has thus far been obtained from Classical Irreversible Thermodynamics.

For systems with constant external constraints, the system will eventually come to a steady state in which case (Prigogine, 1967)[289],

$$d_X\mathcal{P} = 0. \tag{6.7}$$

With this background in classical irreversible thermodynamics, we can now consider our particular irreversible processes consisting of the conversion of energy through different photon spectra. First, the incident flow of energy of the photon spectrum coming from the surface of the Sun $I_S(\lambda)$, which can be approximated by a black-body spectrum for convenience, $I(T_S)$, with T_S=5760 K, is converted, by absorption on primordial molecules followed by either a photochemical reaction leading to a pigment molecule, or by the dissipation directly into heat from an intermediate spectrum of intermediate temperature (also assumed to be black-body), $I(T_I)$ with say T_I= 1000 K corresponding roughly to the vibrational temperature of a molecule of 50 atoms or so after absorbing a single UV-C photon. This spectrum is then converted through interaction of the molecule with its solvent surroundings into the emitted photon spectrum of the Earth's surface $I(T_E)$ with T_E= 287 K.

This process can be equivalently looked at as one in which, at the intermediate temperatures of the vibrationally excited molecule, a photochemical reaction may take place in which some of the free energy of the vibrationally excited molecule is diverted into the making or breaking of covalent bonds leading to the production of organic pigments and an emitted photon spectra of the newly formed pigments which is dependent upon the flow of free energy that went into making the pigments which will be proportional to the concentration C of the pigments. This emitted photon spectrum of the pigments $I(T_P)$ is also assumed to be characterized by a black-body spectrum of temperature T_P.

These newly formed pigments themselves act as catalysts for the overall dissipation process of $I(T_S) \rightarrow I(T_E)$ and for the photochemical reaction process of $I(T_I) \rightarrow I(T_P)$ by, for example, providing local heat of photon dissipation to increase the probability of the photochemical reaction leading to a pigment (as mentioned earlier, photochemical reactions are also temperature dependent), making the dissipation process autocatalytic. A schematic diagram for this autocatalytic photochemical process can be given as follows,

$$\begin{array}{ccc} I(T_S) \rightleftharpoons^1 & I(T_I) & \rightleftharpoons^2 I(T_E) \\ & \Updownarrow^{\;3} & \\ & I(T_P) & \end{array} \tag{6.8}$$

Since the system is far from equilibrium, the backward rate constants for the spectra conversions can be considered as being essentially zero, which is also the case for the photochemical reactions leading to the pigments (reaction 3). Note, however, that if a high energy photon can produce a pigment, a similar photon could also destroy it. There is escape from this predicament if the pigment is fortuitously produced with a conical intersection (see figure 7.5) allowing rapid non-radiative dissipation, thereby neutralizing the destructive potential of the UV-C photons. By the same manner, any pigments not being produced with a conical intersection, but having chemical affinity to a molecule that has one (such as DNA or RNA), so that the pair could operate in the donor-quencher mode, would also be spared from the destructive potential of the UV-C photons. Such a selection process would tend to build up pigments with conical intersections or pigments with chemical affinity to those that have one. Sagan (1973)[309] and Mulkidjanian and co-workers (2003)[238] have emphasized this kind of "selection pressure" as an important force in building the protective dissipative mechanism of the conical intersection in RNA and DNA. However, here we consider protection as being necessary but not representative of the real nature of the conical intersections, which we insist were derived through thermodynamic selection based on the effectiveness of the pigment in dissipating the prevailing solar spectrum.

Since the photon spectra are all considered to be black-body[5], they are completely characterized by one variable which is the black-body temperature T,

[5] As a cautionary note, I remark that using a black-body approximation is wrought with error since the radiation arriving at the Earth's surface from the Sun is very directional and only approximates a black-body in its spectrum characteristics, and the spectrum of the radiation leaving Earth is shaped by radiation windows in Earth's atmosphere (see

or equivalently by the photon pressure P which goes as T^4. Thus, the conversion of the energy through the different spectra and the energy flow into the formation of the pigments can be characterized, in a first approximation, by the pressures corresponding to the assumed black-body spectra. We can therefore write system (6.8) alternatively in terms of pressures,

$$\begin{array}{ccccc} P_S & \overset{1}{\rightleftharpoons} & P_I & \overset{2}{\rightleftharpoons} & P_E \\ & & \Updownarrow 3 & & \\ & & P_P & & \end{array} \tag{6.9}$$

where P_P is the photon "pressure" related to the black-body temperature of the pigments after production which is related to their concentration. It is also assumed that the rate of production of the organic pigments (reaction 3) is linearly related to the photon pressure difference $P_I - P_P$. This pressure difference is essentially related to the free energy available for the formation of the pigment. If the photon pressure P_P associated with the newly formed pigments is low, then more photon energy will flow into the production process, pigments will be produced at a greater rate.

In terms of the affinities and flows of the different processes, we can write Prigogine's general evolutionary criterion, Eq. (6.7), once arriving at the stationary state, in the following form,

$$d_X\mathcal{P} = d\mathcal{P} - d_J\mathcal{P} = d\left(\sum_{\rho=1}^{3} \frac{A_\rho}{T_\rho} v_\rho\right) - \sum_{\rho=1}^{3} \frac{A_\rho}{T_\rho} dv_\rho = 0, \tag{6.10}$$

where A_1 is the affinity for the conversion of energy from the solar photon spectrum into energy of the vibrationally excited intermediate molecule photon spectrum, A_2 is the affinity for the conversion of energy of the excited intermediate molecule photon spectrum into energy of the emitted Earth photon spectrum, and A_3 is the affinity for the photochemical reaction producing the organic pigment.

For the case of equilibrium photon distributions (black-body spectra), the affinities will go as the logarithm of the ratio of the photon pressures (Herrmann and Würfel, 2005)[130],

$$A_1 = kT_I \log \frac{P_S}{P_I} \qquad A_2 = kT_E \log \frac{P_I}{P_E} \qquad A_3 = kT_P \log \frac{P_I}{P_P}. \tag{6.11}$$

figure 2.6). However, the purpose of the calculation here is to show qualitatively how the proliferation of organic pigments over Earth's surface can be explained from a non-linear irreversible thermodynamic analysis of the photon dissipation process. A more accurate derivation of this, devoid of these approximations, will be given in the Appendix using Planck's formula for the entropy of an arbitrary beam of photons.

The v_ρ in Eq. (6.10) are the rates of the corresponding energy conversion (dissipation) processes which we assume are related to the differences of the photon pressures (see discussion at the beginning of this section) attributed to the different black-body spectra. Since the produced organic pigments are assumed to act as catalysts for the conversion of energy from the solar spectrum to the intermediate spectrum (i.e. the local heat of photon dissipation by the pigment can be dissipated in the solvent environment (reaction 2) or can act as a catalyst for the photochemical production of a new pigment (reaction 3)), and since P_p is related to the concentration of the pigments, then the rate of the first dissipation process, $P_S \to P_I$, is multiplied by a factor $(1 + \alpha P_P)$ where α represents the effectiveness of the organic pigment as a catalyst for spectrum conversion from a black body at high temperature into one of low temperature (i.e. $\alpha \to \infty$ for an excellent catalyst, and $\alpha \to 0$ for a completely ineffective catalyst). Therefore, the rates of conversion, assuming all constants of proportionality equal to one for convenience (again, we are only interested in showing qualitatively the non-equilibrium dynamics of pigment proliferation), are given by

$$v_1 = (1 + \alpha P_P)(P_S - P_I) \qquad v_2 = P_I - P_E \qquad v_3 = P_I - P_P \tag{6.12}$$

Note the non-linear relation between the forces, Eq. (6.11), and flows, Eq. (6.12).

Using Eq. (6.10) for the steady state together with Eq. (6.12), taking the Boltzmann constant $k = 1$ for convenience, and observing that the free forces can be characterized in terms of the two free pressures, P_I and P_P (since P_S and P_E are fixed and given by the Sun surface temperature to the fourth power and the Earth surface temperature to the fourth power respectively) gives

$$\begin{aligned}&\frac{\partial}{\partial P_I}\left[(1 + \alpha P_P)(P_S - P_I)\log\frac{P_S}{P_I} + (P_I - P_E)\log\frac{P_I}{P_E} + (P_I - P_P)\log\frac{P_I}{P_P}\right]\\ &+(1 + \alpha P_P)\log\frac{P_S}{P_I} - \log\frac{P_I}{P_E} - \log\frac{P_I}{P_P} = 0\end{aligned} \tag{6.13}$$

$$\begin{aligned}&\frac{\partial}{\partial P_P}\left[(1 + \alpha P_P)(P_S - P_I)\log\frac{P_S}{P_I} + (P_I - P_E)\log\frac{P_I}{P_E} + (P_I - P_P)\log\frac{P_I}{P_P}\right]\\ &-\alpha(P_S - P_I)\log\frac{P_S}{P_I} + \log\frac{P_I}{P_P} = 0\end{aligned} \tag{6.14}$$

At the steady state, we have (see Eq. (6.7)),

$$T d_X \mathcal{P} = \sum_i v_i dA_i = 0$$

which, since the overall affinity $A_1 + A_2 =$ constant, so the variation $\delta A_1 = -\delta A_2$, we have in the steady state that,

$$v_1 = v_2, \quad v_3 = 0, \tag{6.15}$$

which implies that at the steady sate no free energy is being diverted into further pigment production, $v_3 = 0$. This is a very interesting point as it says that in a climax ecosystem (a thermodynamic system which has arrived at a steady state) the concentration of pigments is a constant. Although in a real climax ecosystem pigment production will continue to occur due to pigment degradation or destruction by animals and the environment, the bulk of the free energy arriving from the Sun is not sequestered into photosynthesis (into making more pigments), but rather through dissipation directly into heat. This indeed has been measured in real ecosystems which approximate steady states where it can be verified that only approximately 0.1% of the free energy is sequestered into photosynthesis (Gates, 1980)[100].

Also, at the steady state, solving Eqs. (6.13) and (6.14) gives,

$$\begin{aligned} P_P = P_I &= \frac{1}{2\alpha}[\alpha P_S - 2 + [4 + 4\alpha P_S(1-\gamma) + \alpha^2 P_S^2]^{\frac{1}{2}}] \\ &\rightarrow \frac{1}{2}(P_S + P_E) \quad for \quad \alpha \rightarrow 0 \\ &\rightarrow P_S \quad for \quad \alpha \rightarrow \infty \end{aligned} \tag{6.16}$$

where we have defined $1-\gamma \equiv P_E/P_S$ (γ is, therefore, a measure of the "distance" from equilibrium of the system). Therefore, since P_S is much greater than P_E (the pressures go as the temperature to the fourth power for black-body spectra) equation (6.16) indicates the photon pressure related to the organic pigments P_P, or in other words the pigment concentration, or the free energy which has gone into pigment production, has increased due to its catalytic activity in dissipating the solar photon spectrum into the Earth emitted spectrum and forming new pigments.

The entropy production of the energy conversion processes, including catalytic activity of the organic pigment, is given by a sum of flows times forces,

$$\frac{d_i S}{dt} = \sum_k v_k \frac{A_k}{T_k} = (P_S - P_I)(1 + \alpha P_P) \log \frac{P_S}{P_I} + (P_I - P_E) \log \frac{P_I}{P_E} + (P_I - P_P) \log \frac{P_I}{P_P}. \tag{6.17}$$

Although it will not be demonstrated here, it can also be shown that the entropy production at the stationary state shifts to larger values as a result of the catalytic activity (see Prigogine (1967)[289], for the corresponding case of chemical reactions).

These results give the non-linear irreversible thermodynamic explanation for the proliferation of organic pigments over Earth's surface, and, indeed, over the cosmos. Their concentration can become much greater than that expected based on their Gibb's free energy under near equilibrium situations, due to their catalytic nature in dissipating the solar spectrum, depending on the ratio of $P_S/P_E = (T_S/T_E)^4$ which is a measure of the distance of the system from equilibrium.

Considering now a much more complex system than the idealized system presented above, with many coupled irreversible processes operating, given the above demonstration of how free energy is channeled into the photochemical production of absorbing pigments in such a manner so as to increase the overall rate of entropy production, it is easy to visualize an associated biotic-abiotic coevolution of Earth's physical and chemical characteristics towards an out-of-equilibrium thermodynamic stationary state with ever greater global entropy production for Earth in its interaction with its solar environment. This theme will be taken up again in the chapter on the biosphere (chapter 17).

In referring to purely chemical reactions, Prigogine (1967)[289], in fact, noticed that such a result sheds light on the problem of the occurrence of complicated biological molecules in steady state concentrations which are of orders of magnitude larger than the equilibrium concentrations. In his 1967 book "Thermodynamics of Irreversible Processes" (Prigogine, 1967)[289] Prigogine writes, "Thus, for systems sufficiently far from equilibrium, kinetic factors (like catalytic activity) may compensate for thermodynamic improbability and thus lead to an enormous amplification of the steady state concentrations of the catalyst. Note that this is a strictly non-equilibrium effect. Near equilibrium, catalytic action would not be able to shift in an appreciable way the position of the steady state."

The search for autocatalytic routes to pigment proliferation should concentrate on photochemical reactions in the UV-C region as photons in this region have sufficient energy to make and break covalent bonds, but not enough energy to cause ionization. As listed in chapter 6, photochemical reactions have numerous advantages over normal thermal chemical reactions which occur in the electronic ground state at ambient temperatures. A plethora of routes leading to dissipative structuring are available through photochemical reactions, including resonant energy transfer, electron transfer, proton gradients, fluorescence and phosphorescence. We have given an example of a photochemical autocatalytic route to nucleobase formation and proliferation in section 6.1 based on the Ferris and Orgel photochemical mechanism which appears to be autocatalytic. Other examples of photochemical routes to the nucleotides have been given by Powner et al. (2009) and Patel et al. (2015)[288, 269] and for producing RNA nucleotides from 2-Aminooxazole has been given recently by Szabla et al. (2013)[370], however these do not appear to be autocatalytic and therefore probably not directly related to the origin of life. Many other routes to the fundamental molecules of life will undoubtedly be discovered and some of these will be found to be autocatalytic as prebiotic chemists come to realize the importance of the non-equilibrium thermodynamic principle of dissipative molecular structuring and proliferation, particularly under an imposed UV-C photon potential.

Examples of present day autocatalytic photochemical reactions tied to solar photon dissipation are those of UV light-induced mycosporine production in plants (Sinha et al., 2002)[345] and scytonemin (see figure 4.8) production in cyanobacteria (Mushir et al. 2014)[241]. It has been shown that scytonemin production in present day cyanobacteria can be stimulated by exposure to only UV-C light at 254 nm (Dillon and Castenholtz, 1999)[81], surely a relic of an epoch 2.9 billion years ago when cyanobacteria were constantly exposed to this light. The pigment scytonemin has, in fact, been recovered from 3.446 Ga ancient stromatolites (see figure 4.6) which are anoxygenic photosynthetic bacterial mats (Edwards et al., 2007)[91], indicating that the pigment had been around since very near to the beginnings of life. The pigments mycosporine and scytonemin can thus be considered as catalysts that promote the dissipation of UV light into heat. The biologists, however, see this UV induced mycosporine production as an evolved protectionist response of a plant or cyanobacteria to its harsh environment. We have presented arguments and evidence to the contrary in chapter 4.2 (see also chapter 19.13), and instead here suggest that the production of mycosporine or scytonemin under UV light, or of chlorophyll under visible light, has nothing to do with a hypothetical "vital" force inducing a metaphysical "will to protect itself" or "will to survive", but rather with the non-linear irreversible thermodynamic imperative of increasing the entropy production founded on the well known and well characterized non-equilibrium thermodynamic principles derived from the symmetries of nature.

The implicit and unjustified assumption of the "will to survive" in the Darwinian theory can now be replaced with an explicit and physically founded "will to produce entropy" (colloquially speaking) in the thermodynamic dissipation theory of the origin and evolution of life. However, we should no longer speak about individual selection (although this still retains an approximate meaning in a different context) but rather about hierarchal selection, from the fundamental molecules (RNA or DNA and the other UV-C pigments) up to the biosphere involving both biotic and abiotic components. Random variations may spontaneously occur at any level, including the individual or lower pigment levels, and these are "selected" (a better word would be "proliferated") based on how good a catalytic agent they are in fomenting the global entropy production given the local and global constraints (sunlight, nutrients, water, etc.). What is optimized is not individual "fitness", which, in fact, cannot even be unambiguously defined, but rather "global entropy production" which can, and has been, accurately measured (see chapter 2). This thermodynamical fitness function of global entropy production has, in fact, been proposed as a useful indicator of ecosystem health (Michaelian, 2015[218] - see also chapter 17.5).

The autocatalytic photochemical dissipation reaction leading to pigment proliferation described above can also be looked at equivalently as two coupled irreversible processes; 1) the photochemical reactions leading to the pigment, and 2) the photon dissipation performed by the pigment. The two process are coupled by, for example, the heat of dissipation increasing the rate of the photochemical reaction leading to the pigment, and together these two processes increase the global rate of photon dissipation and therefore will be thermodynamically selected. Lars Onsager had already shown that two irreversible processes will couple, respecting, of course, symmetry constraints, the conservation laws and physical constants of Nature, if, and only if, the coupling leads to greater entropy production than the separate processes (Onsager, 1931)[256].

As a result of these autocatalytic photochemical reactions, a thick soup of organic pigments is exactly what would be expected at the surface of the oceans during the Archean when UV-C light was intense, independently of the small equilibrium probabilities due to the large Gibb's free energy required for their production. As Prigogine stated, in non-equilibrium situations, kinetic factors like catalysis can be more important than thermodynamic probabilities. In fact, if the above arguments are correct, a thick layer of organic pigments would therefore also be expected on any cosmic body containing a mix of primordial molecules such as H_2O, CO_2, HCN, H_2S, etc. and exposed to UV-C light while being shielded somehow from vacuum UV (which causes ionization and therefore molecular degradation). One would, in fact, expect a great ubiquity of these organic pigments throughout the cosmos and the evidence for this has been given in chapter 5 (see also Michaelian and Simeonov (2016)[220] and figure 5.16).

Part III

Origin of Life's Thermodynamic Characteristics

7. Salient Characteristics of RNA and DNA

When in 1953 James Watson and Francis Crick discovered the double strand helical structure of RNA and DNA they immediately recognized how structural characteristics were related to function; how the linear sequence of the four distinct bases provided a mechanism for information storage, and how the structure lends itself to duplication by employing a single strand as a template for the construction of a new complementary strand (see figures 7.1 and 7.2). Since those early days, many other characteristics of RNA and DNA uniquely suited to the functions of reproduction and evolution have been enumerated; these include, being the molecules involved in the translation of the stored genetic information into functional proteins (transfer RNA, tRNA), being molecules with catalytic activity for forming peptide bonds between amino acids to synthesize enzymes and proteins (ribosomal RNA, rRNA - see figure 3.8), being suited to high fidelity reproduction while at the same time being susceptible to mutation, a prerequisite for evolution through thermodynamic selection. So related are these characteristics of RNA and DNA to the mechanisms of reproduction and evolution in living organisms that these molecules are universally considered as being at the very foundations of life.

RNA and DNA are, in fact, two of a long list of "fundamental" organic molecules, those which can be found in all three of the great domains of life; the archea, bacteria, and the eukaryote. A common characteristic that all these fundamental molecules share is their ability to absorb and dissipate in the UV-C region. This property has been discussed in relation to their proliferation in chapter 6 where we have shown how their catalytic nature in dissipating the imposed solar spectrum guarantees their proliferation under non-equilibrium thermodynamic principles.

However, RNA and DNA also have a number of other salient but less known characteristics which, until recently, have only been considered, at best, as providing auxiliary functions to reproduction and evolution, and, at worst, as being mere curiosities or accidents. These salient characteristics include;

i) the strikingly large absorption coefficient for light in the UV-C wavelength region and the extremely rapid dissipation of the electronically excited singlet state to the ground state which happens on sub-picosecond ($<10^{-12}$ s)

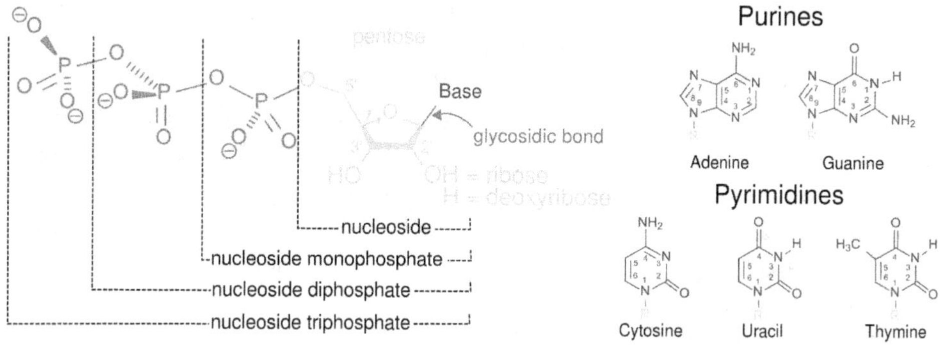

Fig. 7.1. The nucleobases of which RNA and DNA are made. These structures are conjugated aromatics which absorb strongly in the UV–C. They may be considered as the primordial pigments of the Archean era. The vertexes in the diagram at which no element is indicated refer to carbon atoms. Standard numbering of the atoms is given. The "R" refers to the location of the attachment of the Ribose sugar shown on the left part of the diagram with phosphate groups attached. Together, the nucleobases, with ribose and the phosphate groups are known as "nucleotides". Image credit: Boris, Public Domain and Lhunter2099 CC Share Alike 4.0.

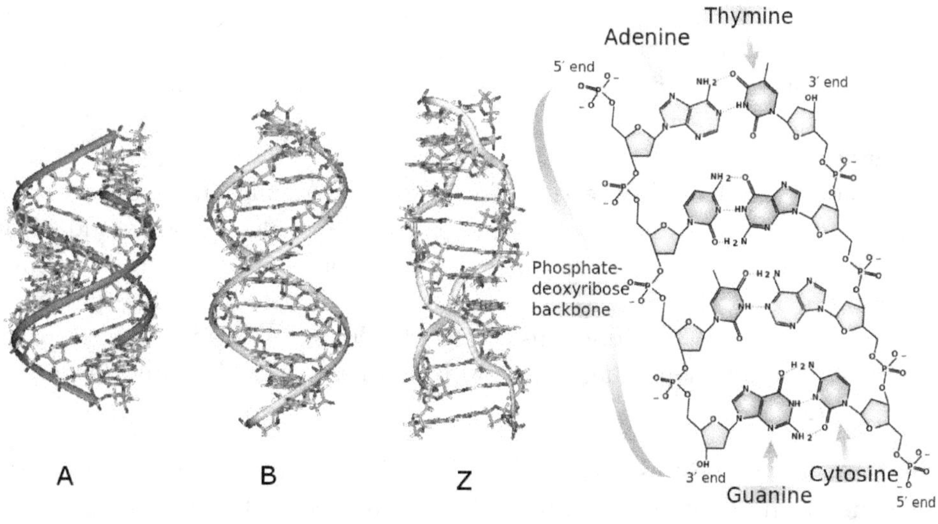

Fig. 7.2. There are three different (depending on the environment) stable helical structures of DNA labled as A, B, and Z. The linear sequence of four distinct nucleotides (adenine, guanine, cytosine and thymine) is used to code information in the molecule. The bases are connected to a sugar (ribose) and phosphate backbone. The hydrogen bonds between the nucleotides guanine and cytosine and between adenine and thymine break during denaturing leaving two separate single strands that can act as templates for the construction of complementary strands. These characteristics suggest that RNA and DNA were molecules at the very foundations of the origin of life. However, these molecules have other salient but little known characteristics, such as extremely rapid non-radiative excited state decay times which suggest that they arose as UV-C photon dissipating structures. Image credit: Adapted from Thorwald and Madeleine Price Ball, Creative Commons.

time scales through a conical intersection (see Figs. 7.4 and 7.5), and similarly rapid vibrational cooling of the hot molecule to the temperature of its water solvent environment (see Fig. 7.7),

ii) approximately 30-40% higher efficiency in UV-C absorption and dissipation for single strand RNA and DNA as compared to double strand helix, an effect known as "hyperchromism" (see figure 7.3),

iii) Watson-Crick base-pairing which provides rapid vibrational de-excitation, two orders of magnitude more rapid than any other of the possible base pairing, even though these other pairings are more energetically favored (Serrano-Andres and Merchán, 2009)[333],

iv) a much broader UV-C absorption spectrum for Watson-Crick base-pairings than other pairings (Abo-Riziq et al., 2005)[1],

v) an exclusive right-handed chirality, referred to as *homochirality*,

vi) the recently discovered enzymeless reversible denaturing of DNA under UV-C light (Michaelian and Santillán, 2014)[216] (see chapter 12.2),

vii) a sharp denaturing curve in temperature with a melting temperature close to the prevailing temperature of Earth's surface at the origin of life (see Fig. 7.6), the sharpness decreases with increasing heterogeneity of the base sequence (Marmur and Doty, 1959)[198],

viii) extraordinary stereochemical fitting of the side chain of amino acids into the cavity left by removal of the second nucleobase from the respective codon (Hendry et al., 1981)[128],

ix) a strong chemical affinity (attraction) to some particular amino-acids (Yarus et al., 2009)[418] and to other fundamental molecules (those found in all three domains of life) such as enzymatic cofactors and protoporphyrin (precursors of chlorophyll and heme); and even a strong chemical affinity to some enzymes common to all domains of life and thought to be involved in primitive energy metabolism and nucleotide synthesis (Krypides and Ouzounis, 1995)[172].

These salient properties of RNA and DNA, although having been investigated independently in great detail, have not hitherto been integrated into a more encompassing framework for the origin and evolution of life. It is an important goal of this book to show that the thermodynamic dissipation theory of the origin of life (Michaelian, 2009; 2010; 2011)[207, 208, 210], which proposes that life got started by dissipating the prevailing Archean UV-C solar spectrum, provides a consistent framework for explaining all of RNA's and DNA's most salient features.

It is often argued that UV light would have been dangerous to RNA and DNA since it has been shown that UV radiation of all wavelengths, UV-A (390-320 nm), UV-B (320-280 nm) and UV-C (280-220 nm) can cause damage to DNA and RNA leading to mutation in today's organisms and thereby affecting

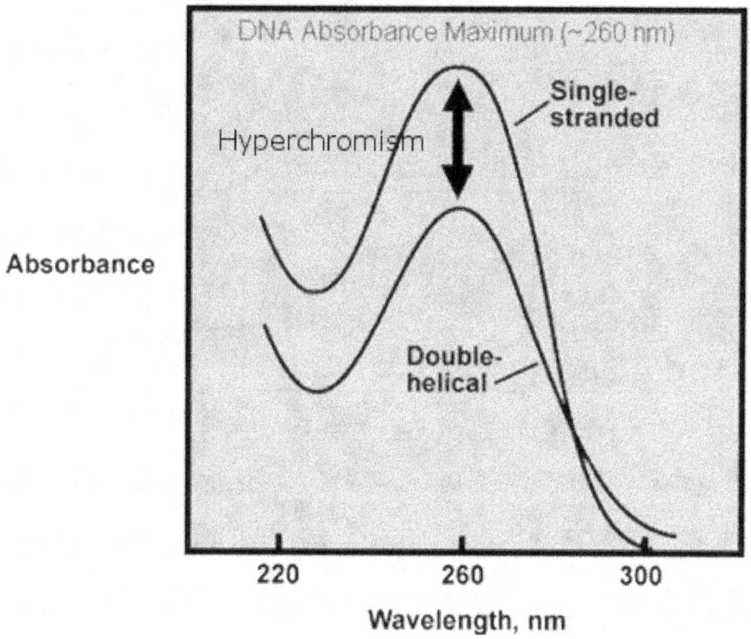

Fig. 7.3. DNA and RNA absorb strongly at 260 nm in the UV-C. Single strand DNA and RNA absorbs from 25% to 40% more UV-C light than the native double strand helix, a salient characteristic known as hyperchromism. This effect is due to a relaxation of the tight stacking of the bases on denaturing and thus a reduction in the shadowing or screening of the bases.

the fidelity of reproduction. In particular, this radiation causes the formation of cyclobutane pyrimidine dimers (CDPs)(Cadet, 1992)[35] which consist of covalent bonding between successive pyrimidine bases on the DNA strand, the most common being that formed between consecutive thymines. CDPs makes up about 50 to 80% of the UV photoproducts on DNA, with the rest being mainly pyrimidine(6,4)pyrimidone (Mitchell and Nairn, 1989)[227]. However, the quantum efficiency for cyclobutane dimer formation at its maximum production wavelength of 280 nm is low, about 10^{-4}, and the photoreversion to the native non-dimer state has been observed at a somewhat shorter wavelength of UV-C light, at 239 nm (Setlow and Carrier, 1963; Garcès, 1982)[334, 98]. Therefore, under the Archean UV-C solar spectrum which included both these wavelengths at similar intensities (see figure 4.7), an equilibrium would have been established between dimer formation and dimer reversion (Setlow ad Carrier, 1963)[334].

Most importantly, however, although the probability for dimer formation is relatively constant at temperatures below its melting temperature, when in single strand formation, it has been shown that the probability for dimer formation is a linear and strongly decreasing function of temperature due to the decreasing

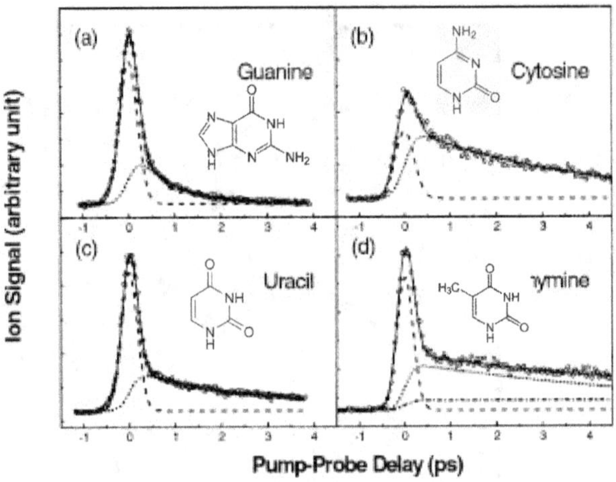

Fig. 7.4. Ultra-fast decay to the ground state of the electronic singlet excited state $S_1$of the nucleobases (induced with 260 nm UV-C pump light). The graph for adenine is not shown but is similar to that of guanine. There is also a very rare base called hypoxanthine which is found only in tRNA which also has sub-picosecond excited state decay time (Guo et al. 2013)[120]. There appears to be two time constants contributing to the signals. The ultra-fast subpicosecond component is known to be due to direct energy conversion through a conical intersection. The overall ground state recovery rate of a few picoseconds may be due to vibrational cooling to the solvent. Conical intersections (see figure 7.5) are ubiquous in pigment molecules and are not accidents but rather the result of dissipative micro-structuring obeying Attard's "second law of non-equilibrium thermodynamics" which says that Nature organizes into structures and processes which provide the greatest number of phase space paths between configurations (see chapter 3.1.1). Image credit: Kang et al. (2002)[148]. Reprinted with permission from the American Chemical Society, copyright (2002).

probability for base $\pi-$stacking (Norlund et al., 1993)[250]. Therefore, at the high surface temperatures existing during the Archean, and at least during the afternoon when it is postulated that RNA or DNA was in single stranded form, photon-induced dimerization was probably not an issue.

Although dimer formation may have been an issue during the early morning hours, it is furthermore relevant that many of the fundamental molecules provide DNA with protection against dimer formation and other photodamage. For example, the flavonoid pigments reduces dimer formation by factors of up to 5 (Stapleton and Walbot, 1994)[357]. Flavonoids are among the fundamental molecules of life (see table 4.1) and absorbs in the UV-C (see figure 4.15). It has also been known for some time that tryptophan can act as a protective agent to DNA damage under high UV light fluxes (Arcaya et al., 1971)[5]. In particular, it was shown that the amino acid tryptophan, which is a residue of *Escherichia coli* photolyase (PL), Trp-277, alone can act as a rudimentary photolyase and can break pyrimidine dimers with a high quantum efficiency of 0.56 using 280 nm wavelength light (Kim et al., 1992)[159] (see figure 7.8), just that wavelength

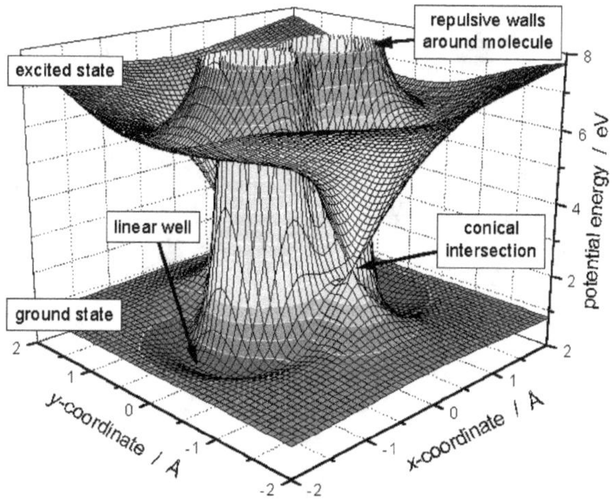

Fig. 7.5. Schematic representation of a conical intersection which permits the rapid dissipation of the electronic excitation energy into heat due to the formation of an energy degeneracy of the lower energy vibrational states superimposed on the excited electronic state with the vibrational states of a somewhat coordinate deformed ground state structure (see Fig. 7.7). Conical intersections are prominent in many pigments. It is suggested here that conical intersections are microscopic dissipative structures. Image credit: Canuel et al. (2005)[38]. Reprinted with permission form AIP Publishing LLC.

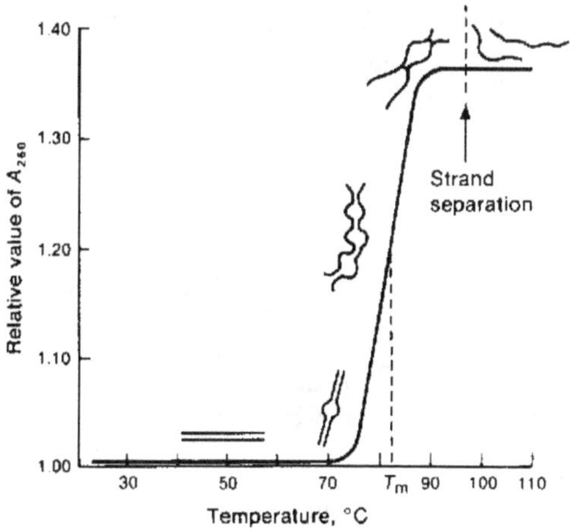

Fig. 7.6. DNA denaturing as monitored by measuring the hyperchromicity of the absorption at 260 nm. DNA has a well defined denaturing temperature similar to the temperature of Earth's surface at the time of the origin of life (~80 - 85 °C). This fact provides an intriguing insight into the origin of life which will be discussed in chapter 12 on enzymeless reproduction of RNA and DNA at life's origin. Image credit: Unknown. Public Domain.

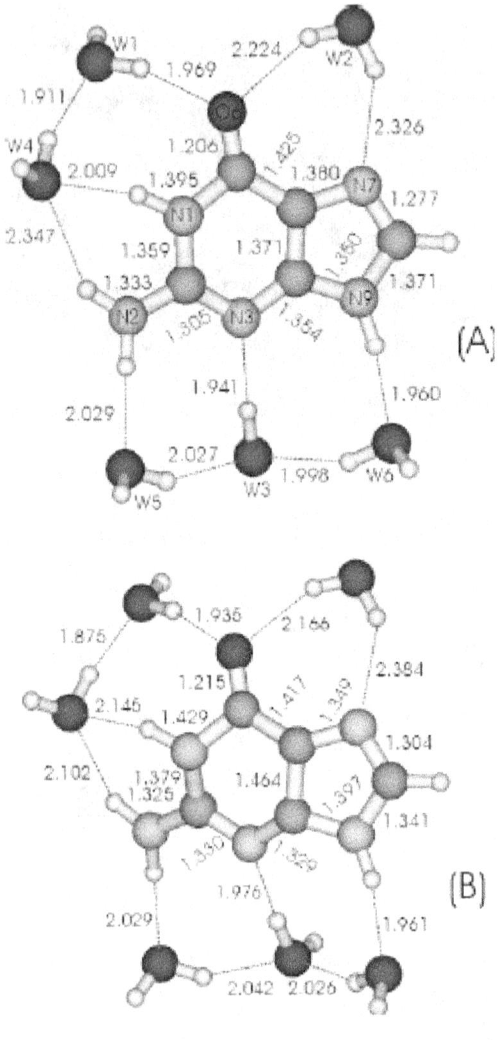

Fig. 7.7. Ground and lowest singlet $\pi\pi^*$ excited-state structures of the hexahydrated guanine: (A) ground-state structure and bond distances; (B) excited-state structure and bond distances. Distances are in Å. The vibrational energy of the bases after de-excitation through a conical intersection is rapidly passed to water molecules in its environment through the high frequency normal modes of the water molecules which interact with the base through hydrogen bonds. For the case of guanine in an environment of 27 water molecules (not shown) similar calculations of the S_1 excited state show that the lowest energy configuration corresponds to a deformed 6 member ring configuration of guanine and this may facilitate the de-excitation through a conical intersection (see Shukla and Leszczynski, 2009)[341]. Image credit: Shukla and Leszczynski, (2005[340]). Reprinted with permission from the American Chemical Society, copyright (2005).

of maximum dimer formation. Tryptophan is another fundamental molecule of life (see table 4.1) and also has chemical affinity to RNA and DNA.

Finally, it is emphasized that at the origin of life, fidelity of reproduction was not important to RNA's and DNA's primordial function of photon dissipation and neither to its primordial replication mechanism (see chapter 12). Fidelity was only required later in the evolution of life as biosynthetic pathways became increasingly more complex to deal with; i) the difficulty of replicating in ever colder oceans, ii) evolving dissipation at other solar photon wavelengths, and iii) the dispersal of pigments into new Earth environments (see Chapters 12 and 13).

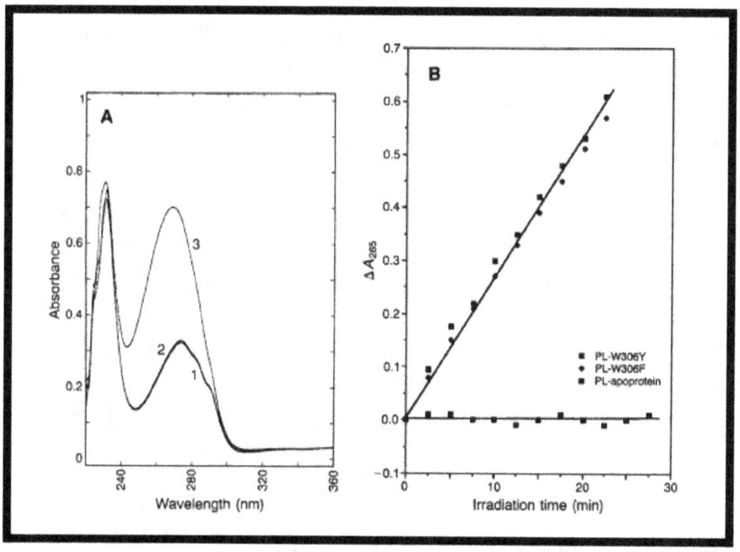

Fig. 7.8. Tryptophan photorepair of thymine dimers produced in UV-irradiated $(dT)_{15}$ by irradiation at 280 nm with a photolyase enzyme PL-(W306F) containing the cofactor non-photoreducible FADH°. (A) Absorption spectra of the PL-(W306F) (3 μM FADH) containing UV-irradiated $(dT)_{15}$ (100 μM dimer) in phosphate buffer (50 mM NaCl/10 mM potassium phosphate, pH 7.0). Curves: 1, before irradiation; 2, after a 20-min irradiation at 366 nm (absorption spectrum was superimposed onto curve 1); 3, after an additional 20-min irradiation at 280 nm. (B) Absorbance increase at 265 nm due to dimer repair is plotted vs. irradiation time at 280 nm under conditions described above. Image credit: Kim et al. (1992)[159].

8. Polymerization of the Nucleotides into RNA and DNA

8.1 UV-C Activation of the Nucleotides

To form single strand RNA and DNA segments out of nucleotides, it is necessary that the nucleotides polymerize, i.e. form phosphodiester bonds between a phosphate group and the 3' and 5' carbon atoms of two adjacent sugars to link a backbone for the chain (see figures 7.1 and 7.2). However, isolated nucleotides will not spontaneously polymerize in this manner since the Gibb's free energy of formation is large and positive. In living systems today, the nucleotides are first chemically "activated" through a sequence of chemical reactions which involve attaching two phosphoryl groups to a nucleotide (see figure 8.1 and figure 4.2 for an example of the phosphorylation of the adenine nucleotide).

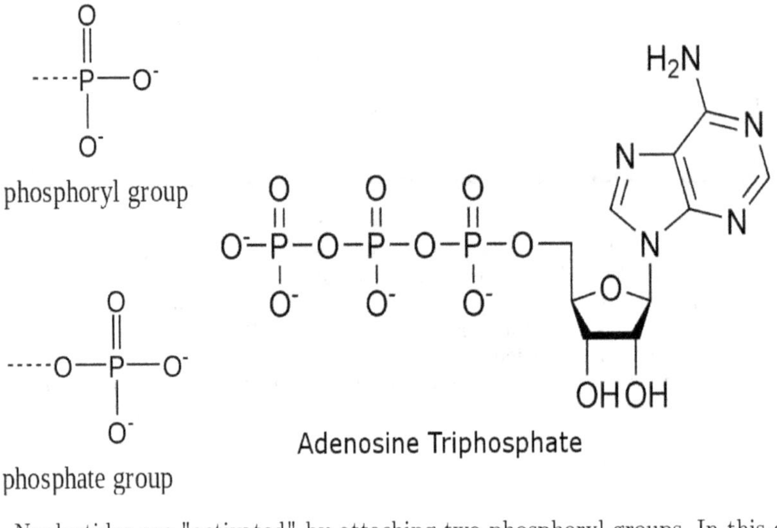

Fig. 8.1. Nucleotides are "activated" by attaching two phosphoryl groups. In this case forming adenosine triphosphate. Without activation, the nucleotides will not polymerize (i.e. join together to form strands). In living systems today, activation is accomplished through chemical reactions in complex biosynthetic pahtways. At the origin of life, the activation of nucleotides could have been performed by UV-C photons interacting with phosphate (Szostak, 2012)[371]. Image credit: Strater. Creative Commons.

Once activated in this manner, the nucleotides can use the free energy obtained by cleaving off a phosphoryl group through hydrolysis, to form the phosphodiester bond between consecutive nucleotides. At the beginning of life, however, it is unlikely that the complex chemical pathway leading to activation of the nucleotides existed and it is necessary to look for an alternative environmental factor that could have been responsible for the activation.

In experiments (Szostak, 2012)[371], it has been shown that ultraviolet light appears once again to be the necessary ingredient, able to activate nucleotides in a solvent environment rich in phosphate. Under very general conditions in a water solvent environment, the activated nucleotides will stack one above the other due to hydrophobic and van der Waals forces, allowing their sugar and phosphoryl groups to come into close proximity (Raszka and Kaplan, 1972)[298]. Hydrolysis of the two phosphoryl groups will provide enough energy ($\sim$1 eV) to form the phosphodiester bond required.

An intriguing idea is that UV-C light interacting with ADP or AMP in a phosphate environment formed not only the first activated nucleotides but also the general energy units of ATP for use in a large number of processes requiring free energy in early life. Experimental evidence for ultraviolet-induced, nonenzymatic formation of ATP from ADP, AMP and adenosine and phosphate groups has been obtained by Ponnamperuma et al. (1963)[286]. Other experiments have also shown that abiogenic flavin conjugated with a polyamino acid matrix (glutamic acid, lysine, and glycine), forms a photophosphorylation pigment that photocatalyzes the phosphorylation of ADP to form ATP, and could have been present in the prebiotic environment (Kritsky et al., 2013 and references therein)[171]. Activation of the other nucleotides could have occurred similarly during daylight hours under the UV-C component of the Archean sunlight.

Other experimental data obtained using chemical activation of the nucleotides indicates that long random sequence linear DNA oligonucleotides can be template synthesized from a random pool of short oligonucleotides with high fidelity (with fidelity increasing with temperature) (James and Ellington, 1997)[138]. In our proposed ultraviolet and temperature assisted replication (UVTAR) mechanism (see chapter 12) for the early non-enzymatic replication of life, UV-C light activation of the nucleotides during daylight hours would substitute for today's chemical activation through complex biosynthetic pathways.

8.2 Why Polymerization?

As emphasized in chapter 3, any evolution or even formation of a more complicated molecule that will not occur spontaneously under equilibrium conditions because of free energy constraints, must, therefore, require non-equilibrium conditions and as such must contribute to a greater dissipation of some imposed generalized chemical potential. Could the polymerization of nucleotides into RNA or DNA at the beginnings of life have contributed to an increase in the dissipation of the imposed solar photon potential and therefore polymerization of the bases be thermodynamically justified and thus explained? In the following I describe how polymerization does, in fact, lead to, not only more stable, but also more complex structures with greater photon dissipation efficacy.

Comparison of the relative stabilities of the polymer and their monomer precursors, with respect to the key chemical bonds (3-phosphoesteric, 5-phosphoesteric, and -glycosidic), for the nucleotides in RNA and in DNA has been made by Saladino et al. (2005; 2006a)[312, 314]. For RNA, it was found that for water temperatures between 60 and 90 °C, the polymer was more stable than the individual monomers (Saladino et al., 2006a, 2007)[314, 316]. Free phosphates in the solution (Saladino et al., 2006b)[315] and a particular pH window (pH 5-6, but depending somewhat on oligo sequence) (Ciciriello et al., 2008)[48] further stabilized the polymers over the monomers. Ocean surface temperatures and pH values at the origin of life are predicted to have been within this range where the polymer stability is favored over the monomer.

However, stability alone cannot explain polymerization since the Gibb's free energies are higher in the polymer state. There must be an explanation from non-equilibrium thermodynamics related to dissipation. There are two time constants in the extremely rapid non-radiative decay of the photon induced excited states of single strand RNA and DNA. The fast one is sub picosecond, 200 fs (one femtosecond, 1fs, equals 10^{-15} seconds) and the second time constant is of the order of 100 ps (one picosecond, 1 ps, equals 10^{-12} seconds). The fast component is similar to the ultrafast 200 fs decay of the individual bases which occurs through a conical intersection (see Fig. 7.5), while the slow component has been assigned to charge migration along stacked bases (Bucher et al., 2004)[28] which can occur before charge recombination and final decay through a conical intersection. However, the long time constant, which appears in single strand RNA or DNA but not in the individual bases, is very dependent on base stacking which can be easily disrupted in single strand RNA and DNA by either high temperatures or by a low pH (Su, 2012)[364]. Therefore, because of the high surface temperatures existing at the time of the origin of life (~80 °C, Knauth, 1992; Knauth and Lowe, 2003[163, 165]), and most probably a lower ocean surface pH (more acidic than today) due to an atmosphere of high CO_2 concentration[1], no important penalty with respect to photon dissipation efficiency would have had to be paid for polymerization of the bases into single strand RNA or DNA.

[1] Ocean water acidity is increased (pH decreased) by a high concentration of CO_2 in the atmosphere through the chemical reaction $CO_2 + H_2O + CO_3^{2-} \rightarrow 2HCO_3^-$ giving carbonic acid. Since the beginning of the Industrial Revolution, the pH of surface ocean waters has decreased by 0.1 pH units. Since the pH scale, like the Richter scale, is logarithmic, this change represents approximately a 30 percent increase in acidity.

But, what could have been gained in dissipation that would have promoted polymerization thermodynamically?

It is well known that many organic molecules can act as acceptor quencher molecules of the electronic excitation energy of donor molecules if the acceptor and donor molecules are in close proximity (Vekshin, 2005)[386]. There are a number of process giving rise to this energy migration from one molecule to another, including; *electronic excitation energy transfer* (EET), *Förster resonant energy transfer* (FRET) and electron-hole transfer (Dexter) (see figures 8.2 and 8.3).

For the case of EET, if one of the molecules is excited electronically by an incident photon, and if in the process of vibrational energy dissipation (internal conversion) in the excited electronic state of the "donor" pigment, states are reached that are in resonance with certain strongly vibrating states of the "acceptor" then a resonant energy transfer can take place. Usually the energy transfer is "down hill", that is from the molecule that absorbs at a smaller wavelength to one which absorbs at a longer wavelength, but this is not always the case, it can depend strongly on the particular overlap of the vibrational states superimposed on the electronic excited states of the two molecules. An example of an up hill energy transfer is that found in the photosynthetic system of the bacterium *Rhodopseudomonas viridis* where the acceptor reaction center bacteriochlorophyll b has its first singlet excited state at much shorter wavelengths (980 nm) than the large number of donor bacteriochlorophyll b molecules that do most of the absorbing (1,050 nm).

In the case of FRET, for energy transfer to proceed, an overlap is required of the donor emission (fluorescence) spectra with the acceptor absorption spectra, so this process is down hill. The energy transfer is through non-radiative dipole–dipole coupling and the efficiency of this energy transfer is inversely proportional to the sixth power of the distance between donor and acceptor. The observation of FRET can therefore be used to establish the close proximity between molecules. The case of Dexter energy transfer is similar to that of FRET only than and electron-hole pair is transferred from one molecule to another

As mentioned above, RNA and DNA are superb quenchers, dissipating their electronic excitation energy within sub-picosecond times. They could, therefore, have acted as acceptors to quench the UV-C induced excitation of other fundamental molecules of life having a large UV-C absorption cross section (see Fig. 4.7) but slower non-radiative decay times or large quantum yields for radiative decay such as fluorescence, both of which reduce the efficiency of dissipation. For example, the aromatic amino acid tryptophan absorbs strongly in the UV-C but has a long non-radiative decay time of nanoseconds and a probability of getting trapped in an even longer lived triplet spin state of about 20% (leading to phosphorescence) compared to that for RNA and DNA of about 0.01% (Bensasson, 1983; Cadet and Vigni, 1990)[18, 34]. Intercalations of tryptophan with DNA involving the stacking of the tryptophyl ring with the bases is known from monomer studies to lead to the total quenching of both the tryptophan and base fluorescence (Montenay-Garestier and Hèlene, 1968; 1971)[232, 233]. In double strand DNA, the quenching persists on intercalation and the double strand opens somewhat at the site of the tryptophan (Rajeswari et al., 1987)[295].

The donor–acceptor complex of tryptophan with RNA or DNA would have thus constituted a greater dissipating system than the sum of its individual molecular components and would therefore have been thermodynamically selected through a *dissipation-replication* mechanism to be presented in the following chapter (see also Fig. 8.4). Certain linear segments of native RNA and DNA of a particular base sequence do indeed have chemical affinity to tryptophan and to other UV-C dissipating molecules. In fact, codons and anti-codons (contiguous triplets of base pairs, each coding for a particular amino acid) of RNA and DNA have strong chemical affinity to their respective aromatic amino acids (Yarus et al., 2009)[418].

Polymerization of the bases would thus have provided a scaffolding for the attachment of other fundamental UV-C dissipating molecules, and this, through the donor-acceptor EET or FRET or Dexter mechanism, would have led to a greater dissipating complex than the sum of its parts (see Figs. 8.2 and 8.3) and so be thermodynamically selected. Besides leading to greater dissipation, if the affinity of the donor chromophore molecule to the RNA or DNA acceptor was selective (for example, to only particular codons) then this dissipation-replication mechanism would also have led to information accumulation in RNA and DNA and to a thermodynamic advantage, in terms of dissipation, of evolving a faithful replication mechanism (see following chapters).

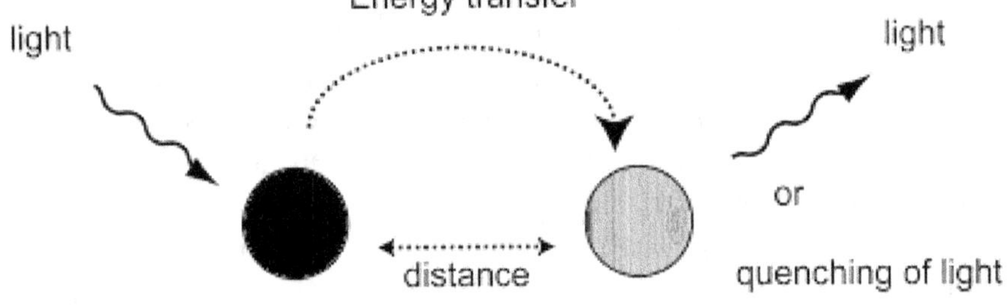

Fig. 8.2. Molecules can exchange electronic excitation energy depending on the distance betwen them, for example through electronic excitation energy transfer (EET) or Förster resonance energy transfer (FRET). Since RNA and DNA are exceptional dissipators of electronic excitation energy, the thermodynamic efficacy of the photon dissipating complex of donor plus acceptor would be greatly enhanced with respect to the individual molecules acting separately. This is sufficient non-equilibrium thermodynamic reason for an association of the two molecules under a UV-C photon potential.

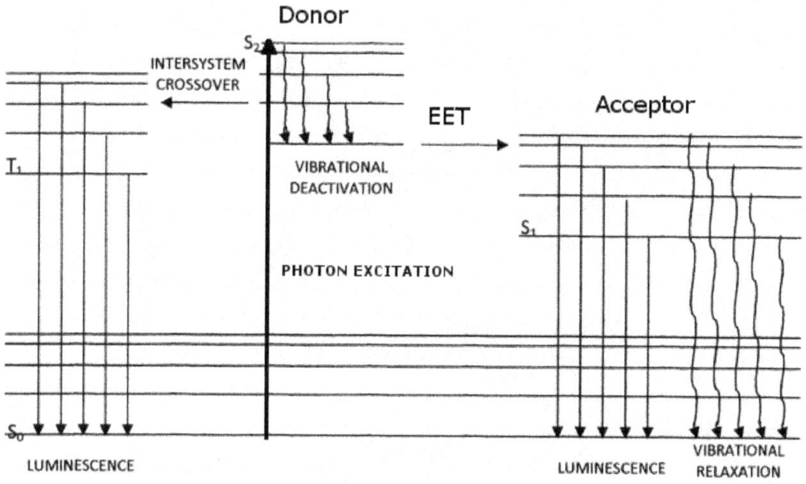

Fig. 8.3. Schematic diagram of electronic excitation energy transfer (EET) between a donor and acceptor molecule. The wavy arrows represent non-radiative decay modes. If the acceptor molecule (in our case RNA or DNA) has an ultra-fast non-radiative decay mode (vibrational relaxation), greater thermodynamic efficacy in dissipation is attained through the formation of the bimolecular complex. Image credit: Adapted from Dr. R. A. Wheatley. Creative Commons.

Finally, there exists yet another non-equilibrium thermodynamic reason for the polymerization of the nucleotides. It has been shown that the separate nucleotide purine bases, particularly Guanine, can get trapped in triplet excited states from which, in water at pH 7, they emit phosphorescence at approximately 400 nm with a quantum efficiency for GMP (guanine mono phosphate) of 0.02 and a time constant of roughly 0.2 s (see figure 8.5). However, in double strand native DNA the phosphorescence is now quenched 10 fold giving a quantum yield of 0.002 and appears to come from thymine residues with the peak shifted to 435 nm (Rahn et al., 1966)[294]. Therefore, even if the possibility did not exist for DNA/RNA acting as a template for resonant energy transfer from UV-C antenna molecules, there is a dissipative (i.e. non-equilibrium thermodynamic) reason for favoring the polymerized state over free floating nucleotides.

8.3 RNA and DNA as Quencher Acceptor Molecules

As has been explained in chapter 4, the fundamental molecules of life are absorbers in the UV-C region and these have chemical affinity to RNA and DNA. Their electronic excitation energy can be transferred to RNA and DNA through resonant energy transfer thereby reducing the excited state life-times of these pigment molecules by various orders of magnitude and thus giving the com-

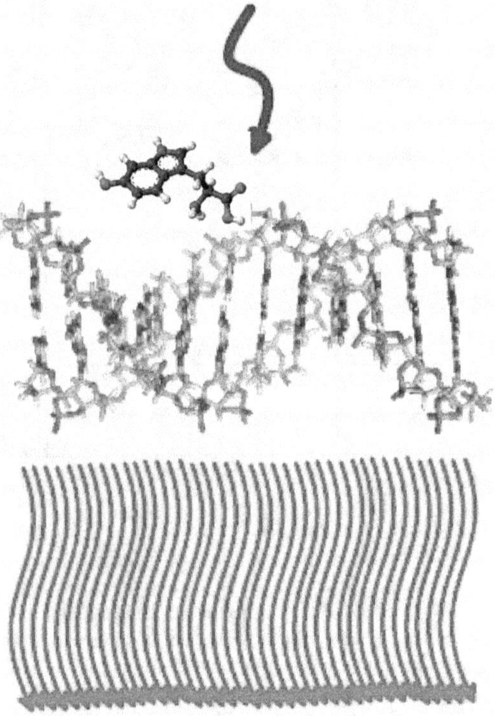

Fig. 8.4. Tryptophan donor attached to a segment of DNA. The complex acts as a greater photon dissipating complex than the sum of its parts because the donor molecule (tryptophan) with a slow de-excitation time of nanoseconds takes advantage of the rapid de-excitation time (picoseconds) of the aceptor molecule (DNA) through resonant energy transfer. The complex, amino acid + DNA, is a greater dissipater than the sum of its parts and is thus thermodynamically selected. Tryptophan can also intercalate between consecutive base pairs without dissrupting the structure of DNA.

plex as a whole, RNA or DNA+fundamental molecule, much greater dissipation efficiency than the individual components acting separately.

Förster energy transfer can take place whenever there exists some overlap between the absorption spectra or between the fluorescence spectrum of the donor and the absorption spectra of the acceptor. However, it is also possible for the acceptor (in this case RNA or DNA) to dissipate partially some of the energy of a donor molecule even if the excitation energy of the donor lies at a significantly larger wavelength than the absorption wavelength of the acceptor and even without the necessity of chemical affinity but, instead, through diffusional collision. Two absorbing molecules, one electronically excited, can also come together and form excimers (for molecules of the same type) or form exciplexs (for molecules of different types). This allows for the fractional energy transfer

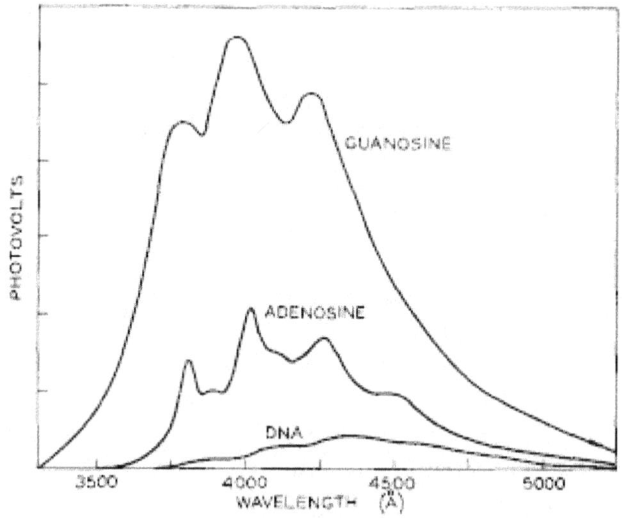

Fig. 8.5. Comparison of native DNA phosphorescence with that of adenosine and guanosine, the two nucleosides with strongest phosphorescence in neutral solution. As can be seen from the graph, polymerization of the nucleosides into DNA reduces phosphorescence 10-fold, providing another dissipative reason for polymerization. Image credit: Rahn et al. (1966)[294]. Reproduced with permission from AIP Publishing LLC.

in these systems. According to Vekshin (2005, p.101)[386] the mechanism for dissipation in such systems can be divided into the following stages:

1. Photoexcitation of the donor molecule.

2. Vibronic relaxation in this molecule.

3. Diffusional collision of donor and acceptor molecules during the lifetime of the donor.

4. Electronic-vibrational interaction between the lower level of the electronically excited state (S_1) of the donor and the vibrational levels of the ground state of the acceptor. These are essentially new levels in the donor produced by the formation of the excimer or exciplex.

5. Some portion of energy is transferred to the vibrionic levels of the ground state of the acceptor.

6. Emission of photons with an energy corresponding to the decay from the intermediate levels, giving rise to a structureless excimer or exciplex band.

7. Transfer of vibrational energy from the acceptor molecule to the solvent or the emission of IR quanta from the acceptor.

There are, therefore, a number of distinct ways other than Förster resonant energy transfer by which electronic excitation energy can be transferred from one molecule to another, which, if the accepter molecule was a quencher, would increase the overall dissipation rate.

9. RNA+DNA+Fundamental Molecules World

As we have emphasized above, the optical, electronic, and dissipative properties of RNA and DNA are similar and therefore, from the perspective of the thermodynamic dissipation theory of the origin of life, in which proliferation was driven by the physical environmental conditions (particularly the dissipation of the solar UV-C flux) there is no reason to presume that and RNA World preceded a DNA World. In fact, since all UV-C pigments must have been proliferated through the same autocatalytic non-equilibrium mechanism, it is much more reasonable to assume that at the very beginnings of life there existed a RNA+DNA+Fundamental Molecules World, consisting of a thick soup of these UV-C pigments, many in physical association. Whatever associations among these pigments which fomented photon dissipation would have been thermodynamically selected and by this means evolved towards ever greater entropy production efficacy.

When the sequence lengths of the RNA or DNA are relatively short, as they must have been at the beginning of life (say less than 50 base pairs long), there exists many kinds of stable associations among the different kinds of strands, each complex with a different set of physical parameters for the complex. For example, the strands can form duplexes, triplexes and even cuadruplexes (four strands intertwined in a single helical structure). Furthermore, they form either homo structures, such as DD and RR (duplex DNA and RNA respectively) or hybrid, hetero duplex structures such DR (one strand of DNA with a complimentary strand of RNA). The optical and electronic properties of these hetero structures still remain to be delineated in detail, but there is reason to presume that they are similar, but distinct, to those of the homo structures. For example, Barone et al. (2000)[13] have shown that the hetero structures (DR) have optical and physical properties usually intermediate between those of the homo structures DD and RR. However, there are interesting differences. For example, for some sequences the hetero structure DR is even more thermodynamically stable than its homo structure analogue DD of the same sequence, while RR of that sequence is the most strongly bound (27 bp duplex RNA having a high melting temperature of 64.7 °C). This is the case for duplexes or hybrids with a purine-rich RNA strand, but the situation is destabilizing if the RNA strand is pyrimidine rich. This stability difference is due to the presence of a 2'-OH group in the ribose moiety in RNA and of a C-5 methyl group in thymine moiety of DNA (Barone et al., 2000)[13].

The DD structures tend to take on the B-form of DNA (see figure 7.2) while the RR and DR structures tend to take on the A-form of DNA. For this reason, and those related with the stronger binding energy (due to greater overlap of the bases in the A-form), RR is found to have the highest rigidity followed by DR, with DD having the lowest (for purine-rich RNA strands). It is also noteworthy that in going from the B-form to the A-form of greater overlap of the adjacent bases, the rigidity of the helix structure also increases and this leads to strong inhibition of the photoreactivity of DNA under UV light (Becker and Wang, 1989)[17].

For short length nucleic acid, there is also the possibility for parallel duplex formation (3´-5´:3´-5´) as well the normal anti-parallel (3´-5´:5´-3´) duplex structure. Parallel duplexes have "reverse Watson-Crick pairing" and are usually less thermodynamically stable than the nominal anti-parallel duplexes. However, at least for short segments (12 bp) and DNA duplexes, the stability is a strong function of pH and in going from 7.0 to 5.0 Szabat et al. (2015)[369] have observed inversion from the anti-parallel to the parallel motif, as determined from changes in the circular dichroism spectrum, in particular, going from a positive to a negative peak in the range 210-220 nm.

The other fundamental molecules of life, most of these pigments in the UV-C (see figure 4.7), would also have had much relevance to the beginning of life considering the utility of these in fomenting dissipation; as antenna molecules, intercalating molecules (for example the amino acid tryptophan or one of the flavonoid pigments) for reducing RNA or DNA UV damage and for increasing by orders of magnitude the DNA extension rate (see chapter 12). Other amphipathic fundamental molecules (having both polar and non-polar ends) would have had utility in preventing RNA or DNA sedimentation.

The possibility of Nature discovering useful combinations of physical, optical, and electronic properties for favoring dissipation through the particular autocatalytic photochemical reactions which drove the beginnings of life would thus seem to be much greater for a RNA+DNA+Fundamental Molecules World than it would for a simple RNA World. A more detailed critique of the RNA World from the perspective of the thermodynamic dissipation theory is given in chapter 19.4.

10. The Ocean Surface as the Cradle of Life

In contradistinction to theories that suggest an origin of life at a hydrothermal vent located at the bottom of the oceans and argue for the advantage of being shielded by water from the prevailing incident UV-C light, the thermodynamic dissipation theory of the origin and evolution of life sees the absorption of UV-C light and its dissipation as being absolutely essential to the origin and proliferation of the fundamental molecules of life through autocatalytic photochemical reactions, and to the origin and incipient evolution of life itself through the formation of ever more complex RNA or DNA-pigment complexes and the coupling of these with other irreversible processes, leading to ever greater photon dissipative efficacy. From this perspective, the most appropriate location for the origin of life would have been at the ocean surface, where the maximum illumination of UV-C light would occur, entering the window in Earth's primitive atmosphere centered at 260 nm where RNA and DNA absorb and dissipate strongly (see figures 4.7 and 10.1). At these wavelengths, pure water is relatively transparent, about as transparent as it is for green light of 550 nm (see Fig. 10.2).

The ocean surface would also have been an optimal location for the efficient coupling of the heat generated through photon dissipation by UV/VIS pigments to the latent heat of evaporation and thus to the primitive Archean water cycle (Michaelian, 2012b)[213], just as the air-water interface today couples the heat of dissipation of the solar infrared light to the water cycle. This is an example of Onsager's principle of coupling of irreversible processes which occurs if this coupling increases the rate of global entropy production (see chapter 6.3.1).

There are also a number of other reasons why the ocean surface would have been an ideal nursery for incipient life. Studies using robots which skim the ocean surface indicate that the upper microlayer (150 μm) is home today to a large assortment of organic molecules such as lipids, amino acids (Kuznetsova et al. 2004)[173], proteins, free floating RNA and DNA, virus, bacteria, and neustron, in concentrations of between 10^2 and 10^4 times greater than that in water slightly below (Hardy, 1982)[124] (see Fig. 10.3). The ocean surface nanolayer contains an even greater concentration of organic material richer in lipids (Laß and G. Friedrichs, 2011)[182]. The high organic density at the surface has been attributed to surface tension and the effect of raindrops producing ris-

Fig. 10.1. The Earth's surface at the begining of life, and for approximately one billion years thereafter during the Archean, was exposed to approximately 5.5 Wm^{-2} of UV-C light at noon at the equator. Sea surface temperatures were hot, about 80 °C and pH values were slightly acidic ~ 6. The atmosphere contained N_2, CO_2, H_2O, some H_2, and H_2S from frequent volcanic erruptions. There would also have been frequent meteorite impacts, possibly contributing to the supply of primordial molecules. Image credit: Adapted from NASA's Goddard Space Flight Center Conceptual Image Lab. Creative Commons.

ing air bubbles which scavenge organic material and bring it to the surface. Furthermore, molecules with a hydrophobic tail will tend to remain at the surface since this is where their interaction energy with their environment is minimized.

The large number of free floating RNA and DNA found at the air-water interface today is rather surprising since RNA and DNA are hydrophilic and have higher density than that of water and would therefore tend to sediment to the bottom of the ocean where they would have been of little use during the Archean for UV-C photon dissipation. However, RNA and DNA have chemical affinity to molecules that can intercalate between consecutive base pairs, or with a stereochemical fit to either the major or minor groove, and having a electropositive tail (as is the case of most pigment molecules - see section 16.5). In such a complex the hydrophobicity of RNA or DNA is increased, causing them to adhere to the air-water interface. The same effect is also found, but to a lesser degree, when their exists a high density of positively charged ions in the water, for example Mg^{+2}.The most probable explanation of these effects is that the electropositive tails, or the positively charged ions, are attracted to and fit within the major groove of the DNA reducing its electronegativity and thus expelling the water from the groove leading to greater hydrophobicity overall for the DNA (Eickbush and Moudrianakis, 1977)[92]. Dai et al. (2013)[69] have also found that double strand DNA can form complex networks amongst themselves over time which makes the network amphipathic, giving them a hydrophobicity which allows them to adhere to the air-water interface.

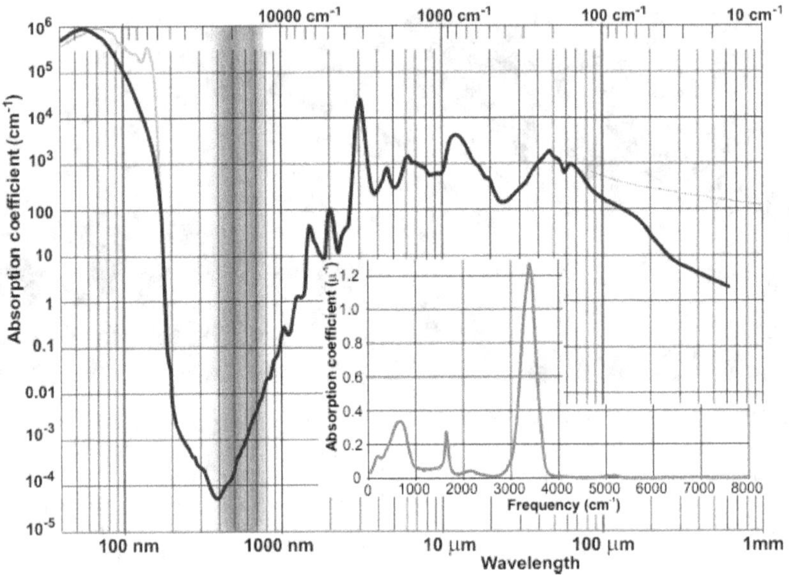

Fig. 10.2. Light absorption in water as a function of wavelength. Light at 260 nm (in the UV-C) is absorbed weakly in pure water, about as weakly as green visible light. Maximum photon dissipation and the coupling of the resulting heat to the Archean water cycle would suggest an origin of life at the ocean surface. Today, the valleys in water absorption spectrum are completely filled in by organic pigments and this speaks in favor of a thermodynamic dissipation theory of the origin and evolution of life. Image credit: Martin Chaplin (2016)[42]. Creative Commons 2.0.

A layer of loosely connected lipid molecules also forms at the air-water interface, due to the hydrophobicity of one of their ends. This layer is fairly resistant to disturbance from ocean waves and rain (Hardy, 1982)[124]. The ocean surface is also much richer than deeper water in metal ions (such as $Mg^{2+}$and Zn^{2+}) which, as noted above, adds to the hydrophobicity and have known catalytic activity in many biological reactions. Divalent ions are also known to promote the formation of stable DNA-phospholipid complexes (see figure 15.3), another manner in which RNA or DNA would be prevented from sedimentation, and the beginnings of encapsulation in a protocell (see chapter 15). Of particular importance here is the presence of Mg^{2+} ions, which today are employed by the enzyme polymerase for RNA and DNA extension, and which have even been shown to promote extension on their own (Szostak, 2012)[371].

The ocean surface microlayer is also subject to relatively large diurnal changes in temperature (see Fig. 10.4), salinity[1], and pH (Wootton, 2008)[411]. Furthermore, it is known that the ocean microlayer today is a fertile region for

[1] During the Archean, the oceans probably had a salinity of between 1.5 to 2.0 times greater than that of today (Knauth, 2005)[164] and a lower pH (more acidic) due to an atmosphere of high CO_2 concentration (see footnote on p. 153 and the following text).

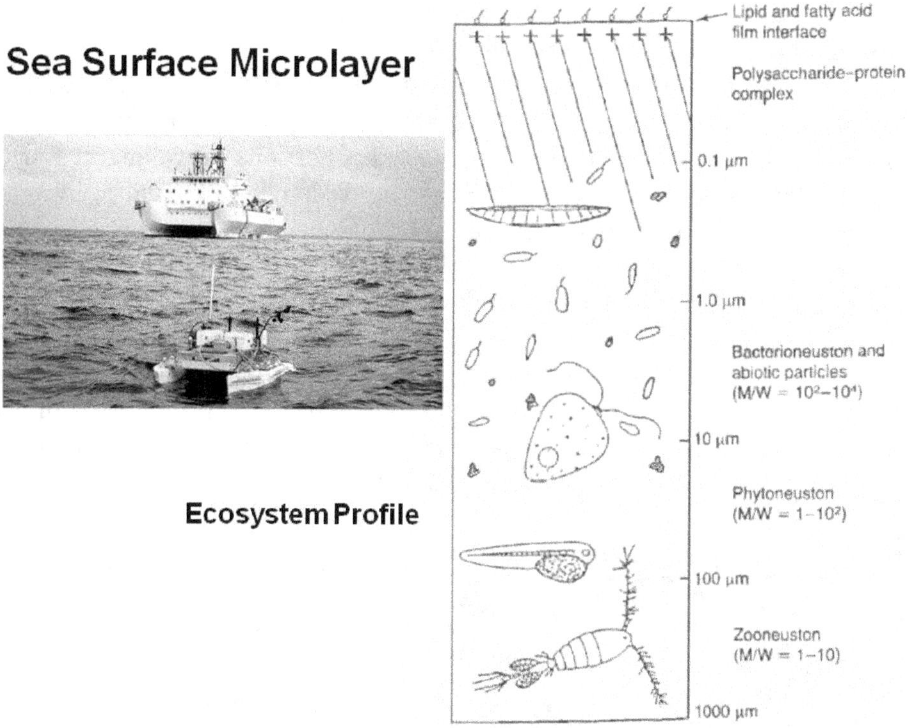

Fig. 10.3. Robots designed to skim the ocean surface reveal an extraordinary ecosystem. Analysis shows that the top microlayer (<150 μm) is very rich in organic material compared to regions just slightly below. It is also where many photochemical reactions take place today and where there is an important component of circularly polarized light. From the perspective of the thermodynamic dissipation theory of the origin of life, these, and many other characteristics (see text) make the ocean surface an ideal nursary for incipient life. Image credit: (left) Soloviev and Lukas, 2006[352], US Navy. Public Domain, (right) Hardy (1982)[124].

photochemical production of organic molecules (Zhou and Mopper, 1997)[428] and there is thus every reason to suspect that this was also the case at the beginnings of life. Very short photon wavelengths (< 220 nm) with the ability to photolyse or cause significant photo-ionization of the molecules would have been mostly eliminated by the Archean atmosphere due to CO_2 and N_2 absorption (see section 4.1 and figure 4.7) and, anyway, are mostly absorbed by a very thin, < 1 μm, layer of surface water (see figure 10.2) through water disassociation and thus could be rendered harmless to the fundamental molecules which would most probably have been found underlying a more stable lipid layer (Laß and G. Friedrichs, 2011)[182]. Finally, the ocean microlayer is also where there exists an important component of circularly polarized submarine sunlight, and this will be shown (see chapter 14.6) to be crucial to a theory I have proposed (Michaelian, 2010b)[209] concerning the accumulation of the homochirality of life through a physical, enzymeless, replication mechanism that fits neatly within this thermodynamic dissipative framework.

Before consideration of the here proposed enzymeless replication mechanism for RNA and DNA operating at the Archean ocean surface microlayer, it is first necessary to emphasize once again the physical conditions prevailing at Earth's surface during the Archean, since, without the existence of enzymes, life's primordial replication would have been much more dependent on the physical conditions of the environment. Based on oxygen isotope ratios found in sediments of the era (3.8 thousand million years ago), Earth's surface temperature was likely hot, close to 80°C (Knauth, 1992; Knauth and Lowe, 2003)[163, 165], probably due to atmospheric methane and carbon dioxide greenhouse gasses. As there exists today, during the Archean there would most likely also have been a diurnal cycling of ocean surface water temperature of amplitude of between 3 to 5 °C (see Fig. 10.4).

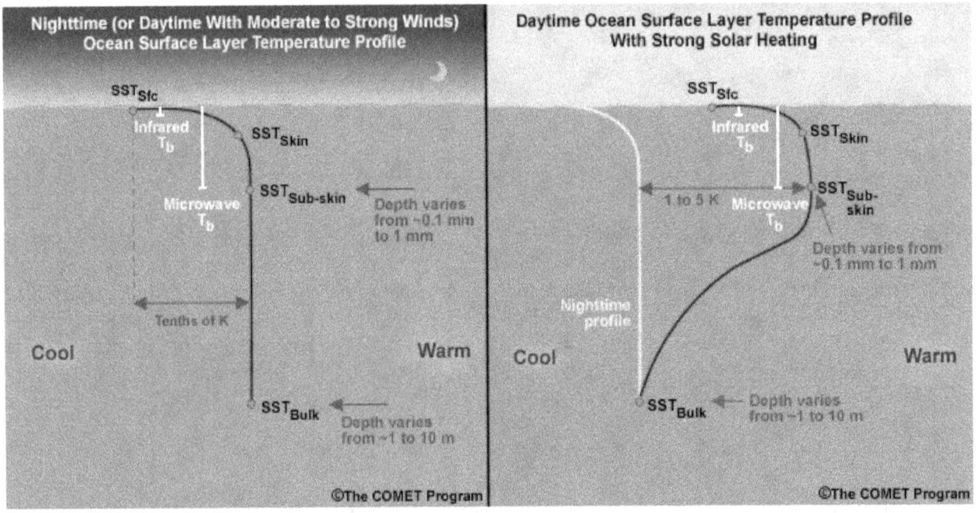

Fig. 10.4. Ocean surface day and night temperature profiles. Within the surface skin layer of ~1 mm from the surface, the day-night temperature variation on average is around 3.0 °C but variations as large as 5 °C have been measured for some oceans. A similar diurnal variation would have ocurred at the Archean ocean surface. Image credit: The COMET Program[55]

The ocean surface during the Archean would probably have been more acidic, close to pH 6, due to the high concentration of CO_2 in the atmosphere but countered somewhat by the lower solubility of this gas in water of higher temperature. The amount of CO_2 in the water affects the pH because it combines with the water to produce carbonic acid which releases H^+ ions. Diurnal variations of ocean surface pH today are related to temperature variations but mostly to photosynthesis and respiration occurring in these layers and can be very large, particularly at near shore regions of high biotic content where diurnal variations are as much as one unit of pH, with values increasing during the day to a maximum in the afternoon, and decreasing at night to lowest values just before dawn (Cornwall et al., 2013)[60]. Far from shore, however, where biotic content is low, diurnal pH variations are much less significant (~ 0.025, see Hofmann et al. (2011)[132]). An important diurnal variation in pH, given a much higher CO_2 content of the Archean atmosphere, > 2500 ppm (Pavlov et al., 2000)[270] compared to 400 ppm today, would probably have occurred on the Archean sea surface. Maximum values of pH would have occurred in the late afternoon when the surface temperature is highest, as would be expected from the CO_2 solubility argument.

11. A Thermophilic or Hyperthermophilic Origin of Life?

Models of the Archean atmosphere suggest hot surface temperatures of between 85 and 110 °C due to the large amount of greenhouse gases in the Archean (Kasting and Ackerman, 1986; Kasting, 1993)[150, 151] which is consistent with geochemical evidence in the form of $^{18}O/^{16}O$ ratios found in cherts of the Barberton greenstone belt of South Africa indicating that the Earth's surface temperature was around 80 °C at 3.8 Ga (Knauth, 1992)[163] (Knauth and Lowe, 2003)[165], falling to 70±15 °C during the 3.5–3.2 Ga era (Lowe and Tice, 2004)[193]. If life arose on the ocean surface, it must therefore have started out hyperthermophilic (optimal temperatures > 80 °C) and stayed at least thermophilic (optimal temperatures > 50 °C) for at least 500 million years, sufficient time for primitive enzymes to have evolved under these high temperature conditions.

Recently, experiments have been performed to test the thermophilic or hyperthermophilic origin of life hypothesis. There are some highly conserved segments in ribosomal RNA (rRNA – small segments of RNA found in cellular organelles called *ribosomes* responsible for converting information, delivered in messenger RNA (mRNA), into proteins and enzymes) found in all domains of life. It is with conserved segments of RNA or DNA that reliable phylogenetic trees can be constructed (see figure 19.3). In fact, it was with these rRNA segments that Woese, Kandler and Wheelis (Woese et al., 1990)[412] first constructed a detailed genetic tree for all living organisms that suggested that life should most naturally be divided into three domains; bacteria, archea, and eukaryote (instead of the older division into prokaryote and eukaryote). This analysis also suggested a thermophilic last universal common ancestor (LUCA). A similar tree analysis conducted with proteins, led to a hyperthermophilic LUA (Di Giulio, 2001; 2003)[79, 80]. More recent analysis of the ribosomal RNA of many newly discovered hyperthermophiles (both Archea and Bacteria) suggest that the base of the phylogenetic tree may indeed be hyperthermophilic (see figure 19.3)(Stetter, 2006)[358].

As a test of the hyperthermophilic hypothesis of the last universal common ancestor, Shimizu et al. (2007)[337] looked at how ancestral mutations of the translation enzyme *glycyltRNA synthetase* affected the thermostability and activity of this enzyme. They noted that phylogenetic trees of the translation system show the same topology as the phylogenetic tree of ribosomal RNA and that since all organisms share a translation mechanism with the same function, there is therefore little probability of modification of function or lateral gene transfer between organisms having occurred. A number (one or two) of amino acid residues of conserved sections of the *glycyltRNA* enzyme were substituted with their more ancestral mutations and their temperature denaturing curves measured by observing the circular dichroism spectra[1]. Shimizu et al. (2007)[337] found that six out of eight of the ancestral mutated enzymes had greater thermal stability (enzyme denaturing temperatures raised by up to ~ 5 °C) and seven out of the eight had improved activity at higher temperatures. This was taken as evidence in favor of a hyperthermophilic LUCA. Earlier, similar studies, but using metabolic enzymes (Miyazaki et al., 2001; Iwabata et al., 2005)[228, 137] also gave results of increased thermostability under ancestral mutation but leading the authors to conclude a thermophilic, rather than a hyperthermophilic, LUCA.

Although the conclusions on the ancestral mutations only apply to the last common universal ancestor, and not to the origin of life itself, and many unlikely but possible scenarios can be envisioned in which the temperature associated with the origin of life was colder than that associated with LUCA, taken together with the geophysical/chemical data of the era, these results paint a consistent picture of a hot origin of life.

Finally, a hot ocean surface would certainly favor enzymeless metabolic reactions, in particular our Ultraviolet and Temperature Assisted Repñication (UVTAR - see chapter 12) proposal for enzymeless reproduction of RNA and DNA which associates reproduction with dissipation. A hot ocean surface would also provide the necessary permeability of the phospholipid enclosures of the first protocells for the diffusion transport of the required nucleotides and amino acids across the cell wall. It is certainly very difficult to imagine effective diffusion transport through phospholipid walls operative in environments at near freezing temperatures, as proponents of the "eutectic concentration"[2] theories on the origin of life advocate. Enzymeless binary fission of the phospholipid enclosure also requires high temperatures, and this temperature, in fact, must be

[1] Circular dichroism (CD) is a measure of the asymmetry as a function of wavelength in the extinction of right- versus left-handed circularly polarized light incident on a molecule. Denatured proteins give a different CD spectrum than natured proteins and thus the technique can be used to construct a melting curve for the molecule. More on this technique of using CD as a tool for studying physical changes in a molecule is given in chapter 14 where we look in detail at the homochirality of the molecules of life.

[2] Eutectic concentration refers to the phenomena of the exclusion of impurities as water freezes into its ice crystal structure, augmenting the concentration of the impurities in regions still in the liquid state. Some have suggested this as a relevant mechanism for increasing the concentrations of the fundamental molecules, assumed to be necessary to bootstrap the autocatalytic set of chemcal reactions that supposedly gave rise to life. However, the evidence is overwhelmingly in favor of a hot origin of life.

higher than that of DNA or RNA denaturing if enzymeless protocellular division is to occur (see chapter 15).

12. Ultraviolet and Temperature Assisted Replication (UVTAR)

12.1 Introduction

In this chapter I shall describe a simple mechanism based on UV-C photon dissipation for enzymeless reproduction of duplex RNA or DNA, or hybrid RNA+DNA (see chapter 9), which I have called Ultraviolet and Temperature Assisted Replication (UVTAR), first published in 2009 (Michaelian, 2009)[207]. The mechanism associates replication with dissipation which is a prerequisite for thermodynamic selection and evolution. Experimental evidence in favor of the postulate will be given in section 12.2.

The temperature of approximately 80-85°C of the ocean surface during the early Archean when it is generally assumed that life arose is strikingly close to the short strand DNA melting temperatures,[1] the temperature at which double strand DNA denatures into single strands, in water at neutral pH[2] and present ocean salinity (see Fig. 7.6). This similarity in temperatures provides an important clue as to the nature of enzymeless reproduction at the origin of life, an idea that will be explored in this chapter.

The temperature at Earth's poles would have been colder, perhaps closer to the denaturing temperature of RNA[3] under similar salt and pH ocean conditions (40 to 50 °C), while the temperature at the equator would have been warmer. There would also have been a temperature profile with depth in the ocean similar to that of today (see Fig. 10.4) but of higher absolute temperature.

The Earth was cooling gradually as the greenhouse gas CO_2 was being consumed in silicate carbonates formed through erosion of the newly forming continents (Lowe and Tice, 2004)[193], and once the local surface temperature fell

[1] DNA or RNA denaturing temperature increases with concentration of salt, closeness to neutral pH, with segment length, and with G-C content.

[2] It is probable that Earth's oceans during the Archean would have been more acidic (pH $\sim 6-7$) than today due to the great amount of carbon dioxide that was probably in the atmosphere and necessary in order to adequately address the faint young sun paradox given the evidence for the existence of liquid water during the Archean. Any movement away from neutral pH (a reduction or increase) tends to decrease the denaturing temperature of DNA (Williams et al., 2001)[407].

[3] RNA is today usually found in single strand form with significant self-adhesion between complimentary segments, but it can exist in double helix form, just as with DNA. In fact, double-strand RNA viruses have been identified.

below their respective denaturing temperatures, DNA and RNA single strand segments (having been formed through the autocatalytic photochemical and polymerization routes described in Chapters 6 and 8) would eventually find and hydrogen bond with short complementary segments and would then normally be unable to separate again, thus excluding the possibility of their reproduction. However, through absorption by RNA and DNA of UV-C light during the day and by dissipating this light into heat deposited locally at the site of the molecule, plus the absorption of some solar infrared light on the ocean surface, the local temperature at the site of the molecule may have been raised sufficiently, and for a time sufficiently long, for denaturing to occur, a process that we call photon-induced denaturing (see figure 12.2).

This enzymeless denaturing using UV-C light is not hypothetical, we have measured it quantitatively for DNA for the first time in our laboratory (see figure 12.1 and Michaelian and Santillán (2014)[216]) and it is even operative for water temperatures well below nominal DNA denaturing temperatures, albeit with decreasing efficiency as the solvent temperature is lowered. In fact, such a photon-induced local heating mechanism has been detected before on other systems. In experiments on heme proteins, Genberg (1988)[102] has shown that photoexcitation produced successive "heating" of the heme, the protein globule and the medium, on the time scale from tens of ps to several nanoseconds, suggesting a similar effect would occur for photoexcitation of RNA or DNA. Our experiments have also shown that UV-C light-induced denaturing of DNA is reversible, meaning that no, or very minimal, damage is inflicted on the molecule by the UV-C light and that the molecule can renature again without difficulty (see figure 12.1).

The characteristic sharpness in the temperature of the denaturing curve of DNA, particularly for larger segments and for acidic pH values ~ 5 (Dubey and Tripathi, 2005)[86], facilitates photon-induced denaturing since, once the ambient temperature fell slightly below the denaturing temperature, only the small amount of energy available in a single UV-C photon (about 4.7 eV) would have been sufficient to rupture all of the hydrogen bonds between the two strands and separate the strands completely. Double strand RNA has a lower denaturing temperature and a less sharp denaturing curve and, from this perspective of the thermodynamic dissipation theory of the origin of life, this may be an argument against the popular belief that an RNA World existed before DNA (see chapter 19.4). During overnight periods of approximately 7 hours (the Earth rotated faster at the origin of life), the sharp denaturing curve of DNA would mean that the small decrease in ocean surface temperature would have been sufficient to allow for Mg^{2+} mediated extension of the separated single strands, completing the reproduction cycle (see Fig. 12.2).

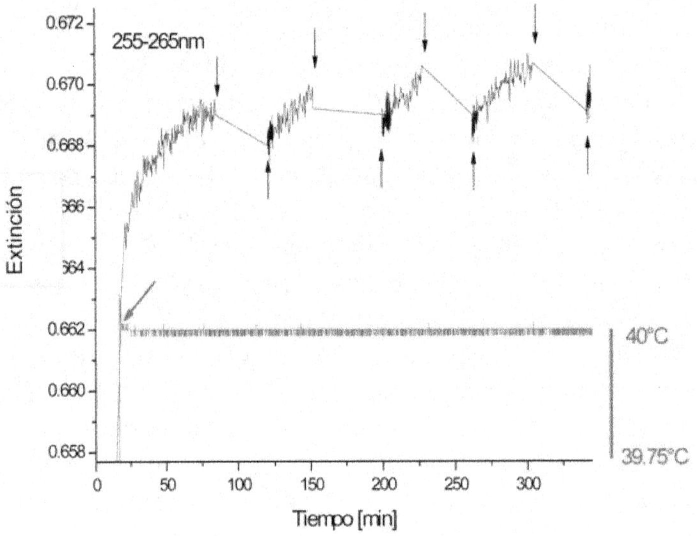

Fig. 12.1. Experimental demonstration of UV-C light denaturing of Salmon sperm DNA of average length 100 Kbp in pure water (no free ions). The graph plots the extinction of the UV-C light (due mainly to absorption with a small amount of scattering) by DNA in the wavelength range 255 to 265 nm against time as the UV-C light was turned on and off. The temperature of the bath was raised over a time period of 20 minutes to 40 °C (lower arrow) and later maintained at this value (±0.01 °C) for the duration of the experiment. The arrows pointing downward mark the time at which the UV-C light was blocked from reaching the sample by a shutter and the arrows pointing upwards mark the times at which the light was allowed on sample by removing the shutter. It can be seen that while UV-C light is on sample, the extinction increased gradually (to about 0.3% of the total absorption, or about 9% of the increase in the absorption due to temperature denaturing at 40 °C) due to light-induced denaturing. While the light was off, the segments renatured. The amount of denaturing depends on the intensity of the UV-C light, the temperature of the bath, and the amount of DNA segments in the sample with a temperature of denaturing somewhat above 40 °C. Image credit: Michaelian and Santillán (2014)[216].

It turns out that single strand RNA and DNA is on average 30-40% more efficient at absorbing and dissipating UV-C photons than is double strand RNA or DNA or the hybrid RNA+DNA duplex (see Fig. 7.3). This effect is know as hyperchromism and is the result of the shadowing or screening of the bases when they are tightly stacked one above the other in a double strand native arrangement (Vekshin, 2005)[386]. In single strand RNA and DNA, the bases are free to take on arbitrary orientation with respect to the long central axis of the RNA or DNA molecule and so there is less shadowing and therefore greater photon absorption. Denaturing by UV-C light would thus increase the rate of dissipation of the solar photon potential in the UV-C by about 30-40%

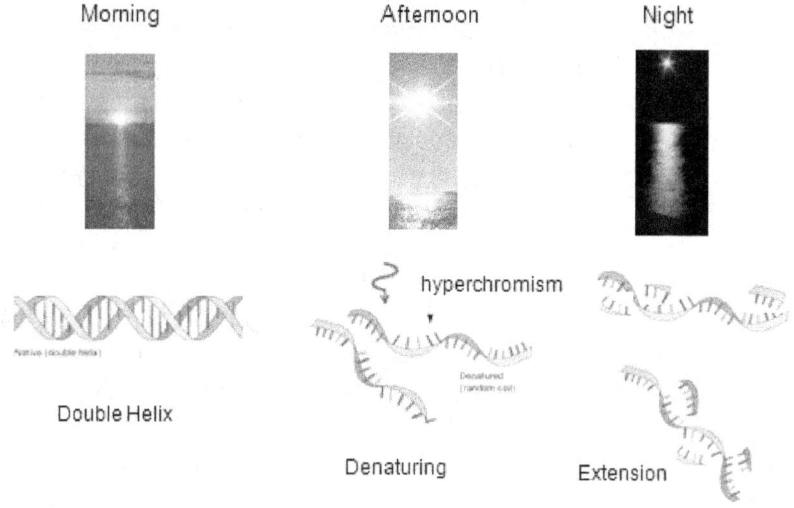

Fig. 12.2. Ultraviolet and Temperature Assisted Reproduction of RNA and DNA (UVTAR). Dissipation mechanism proposed for the enzyme-less reproduction of RNA and DNA assisted by the absorption of the prevailing UV-C light flux and the high temperatures of the ocean surface during the Archean, together with a day/night diurnal cycle of warming and cooling of the ocean surface due to the absorption of solar infrared light. Most denaturing would occur in the afternoon while ocean surface temperatures were highest. "Hyperchromism" refers to the increase (of up to 40%) in the absorption of photons at UV-C wavelenghts (~260 nm) when RNA and DNA denature into single strands (see figure 7.3). The important aspect of this mechanism is that replication is tied to dissipation. Proliferation then has a thermodynamic reason to be.

compared to native double strand and, therefore, RNA and DNA that could have remained in the denatured state during the day through this mechanism of UV-C absorption and dissipation and would have been available for extension during the night, and thus, in this manner, be "thermodynamically" selected. If indeed this UVTAR mechanism were operative in the Archean, it would have endowed incipient life with a *dissipation-replication relation*, which, as we will argue in chapter 17.3, remains to this day in living systems and is essential to placing evolutionary theory on a non-tautological physical-chemical foundation.

Through this UVTAR mechanism, even as the surface temperature of Earth fell below the melting temperature of DNA or RNA, given an incident UV-C component in the solar flux available during daylight hours, RNA and DNA segments could have denatured into single strands without the aid of enzymes. The problem of enzymeless RNA denaturing at temperatures below its melting temperature is considered one of the most difficult problems to resolve in an "RNA World" (Szostak2012)[371].

Although enzymeless photon-induced denaturing may be part of the solution, it is only one half of the complete enzymeless replication problem, the other half being extension (see figure 12.2). However, there exists experimental evidence that extension of a complimentary strand (the formation of a new strand using the single strand as a template) can occur at high temperatures (around 80 °C, similar to the surface temperature of the Archean) using chemically activated nucleotides (see chapter 8.1) in an aqueous solution containing Mg^{2+} or Zn^{2+} ions (Szostak, 2012)[371]. There also exists experimental data indicating that enzymeless DNA extension can be speeded up orders of magnitude by employing planer intercalating molecules which act as a kind of glue (Horowitz et al., 2010)[133]. In our proposed UVTAR scenario, nucleotide activation would have been induced by the UV-C light during daylight hours (see chapter 8.1) and extension would occur overnight when the ocean surface water had cooled by about $3-5$ °C (see figure 10.4) and there would be no UV-C light inhibiting the extension. As noted in chapter 10, the ocean surface microlayer would have been rich in Mg^{2+} and Zn^{2+} ions and planer intercalating molecules such as the amino acid tryptophan. Although the experiments are still being carefully analyzed (Michaelian and Santillán, 2016)[219], all indications, from our light-induced denaturing data and this previously published data concerning enzymeless extension, are that a rudimentary enzymeless reproduction would have been possible utilizing day-night UV-C light cycling, under the physical conditions of Earth's ocean surface during the Archean (high temperatures with $\sim 3-5$°C diurnal cycling, and pH values of around $6-7$).

This UVTAR mechanism connects thermodynamic selection based on photon dissipation rates with RNA and DNA replication. It therefore addresses the current debate of whether replication or metabolism came first by suggesting that they came together under this replication-dissipation relation. The primordial metabolism being the dissipation of the photons into heat, and this is, by far, the most important metabolism of life occurring still today. It is precisely this relation between thermodynamic dissipation and replication which is necessary to remove the tautology – What is natural selection selecting? – from Darwin's theory of natural selection. From the perspective of the thermodynamic dissipation theory of the origin and evolution of life, natural selection is selecting irreversible processes (including the consequent microscopic dissipative structuring of material, e.g. pigments which are the fundamental molecules of life and their associated complexes) and hierarchies of these, coupled with abiotic irreversible processes which increase the dissipation of the generalized chemical potentials in the environment, in particular, the dissipation of the largest of all generalized chemical potentials; the solar photon potential. This has important implications to selection at all biological levels, including at the level of the biosphere (Michaelian, 2012; 2015)[212, 218] and this will be discussed at greater length in Chapter 17.

12.2 Experiments on UV-C Induced Denaturing of DNA

One of the most difficult problems related to the origin of life is explaining enzymeless reproduction of RNA or DNA (Szostak, 2012)[371]. A common theme throughout this book is that pigment proliferation is tied to photon dissipation. Replication is certainly connected to proliferation and here we look in more detail at the proposed dissipation-replication mechanism for RNA and DNA described in the previous section. In particular, in this section we provide experimental evidence for the proposal that the heat resulting from the dissipation of absorbed UV-C photons could have denatured RNA and DNA. The experimental work reported in this section was done in collaboration with my student Norberto Santillán at the Institute of Physics of the National Autonomous University of Mexico (UNAM) and has been previously published (Michaelian and Santillán, 2014)[216].

The experiment was designed to test the proposal that UV-C photon absorption and dissipation in RNA and DNA would lead to enzymeless denaturing once the ocean surface temperature descended to below their relevant melting temperature (Michaelian, 2009; 2011)[207, 210]. Subsequent extension is envisioned to occur during overnight dark periods facilitated by Mg^{2+} ions (Szostak, 2012)[371] and UV-activated phosphorylated nucleotides (Ponnamperuma et al., 1963; Simakov et al., 2005)[286, 343]. This proposed ultraviolet and temperature assisted replication (UVTAR) is similar to polymerase chain reaction (PCR) (Mullis, 1990)[239] but in which the heating and cooling cycle is partially substituted by a day/night UV-C light cycling, and in which Mg^{2+} ions, or their complexes with some hypothetical common Archean molecule, played the role of the extension enzyme polymerase.

The UVTAR mechanism described in this chapter also provides a plausible explanation of the homochirality of life (Michaelian, 2010)[208] as a result of the positive circular dichroism of RNA and DNA at its maximum absorption at 260 nm which happens to be at the center of the Archean atmospheric window (Sagan, 1973; Cnossen et al. 2007)[309, 53], and a small prevalence of right over left handed circularly polarized submarine light in the late afternoon (Angel et al., 1972; Wolstencroft, 2004)[3, 409] when surface water temperatures were highest and thus most conducing to denaturing. This explanation for the origin of homochirality will be detailed in chapter 14.

The UVTAR mechanism also provides a possible explanation for the beginnings of information accumulation in RNA and DNA [207, 210] since the aromatic amino acids that absorb in the UV-C have known chemical affinity to their codons or anti-codons (Yarus et al., 2009; Majerfeld and Yarus, 2005)[418, 196]. RNA or DNA coding for these "antenna" amino acids would dissipate more UV-C light and thus experience greater local heating, endowing them with greater reproductive success through the UVTAR mechanism in increasingly colder seas (see chapter 13). Antenna aromatic amino acids intercalating between consecutive base pairs would not only help in this manner with denaturing but also aid with template extension during dark overnight periods since intercalating molecules act as a kind of "glue" which significantly increases extension efficiency (Horowitz et al. 2010)[133].

In the following sections we present unequivocal experimental evidence for reversible UV-C light denaturing of DNA for four different DNA samples; short 25 bp and 48 bp synthetic DNA, and relatively large (~100 Kbp) salmon sperm and yeast DNA. A quantitative analysis of the hyperchromism and scattering in our extinction data, as well as a careful analysis of the *difference absorption spectra* of the DNA sample obtained at fixed temperature at the beginning and end of a lengthy UV-C light on sample period, all suggest that UV-C light induced denaturing is significant and sufficiently large to support the proposal of an UV-C dissipative origin of life and dissipative thermodynamic explanations of the accumulation of information and homochirality in DNA (see chapters 13 and 14 respectively).

12.2.1 Method

Salmon sperm and yeast DNA samples of varied lengths (average 100 Kbp) were obtained from the Institute of Cellular Physiology at the National Autonomous University of Mexico (IFC-UNAM). Complementary synthetic DNA oligonucleotides of 25 and 48 bases were synthesized at the IFC-UNAM. The oligos were designed to be free of adjacent thymine to avoid complications of dimer formation induced by light of around 280 nm wavelength (and dimer reversion by light of around 239 nm) since dimer formation also causes local denaturing (Setlow and Carrier, 1963)[334]. The oligos were also designed to have convenient denaturing and priming temperatures in order to facilitate a possible UVTAR mechanism operating under only a very small simulated diurnal temperature cycling. The sequences chosen for the synthetic 25bp and 48bp oligos were

5'-CTATGGAGCGGATATACCATGGACG-3'

3'-GATACCTCGCCTATATGGTACCTGC-5';

and:

5'-GGTCGCAGAGGCACATGGGTATATATATACGTATGCGGCGTGGCATCC-3'

3'-CCAGCGTCTCCGTGTACCCATATATATATGCATACGCCGCACCGTAGG-5'

having 52% and 54% G-C content, respectively.

To form the double helices of 25 and 48 bp, complementary oligos at equal concentration were mixed in a Dulbecco PBS buffer (pH 7.3) solution containing 2.7 mM potassium chloride (KCl), 136.9 mM sodium chloride (NaCl), 1.5 mM potassium phosphate monobasic (KH_2PO_4) and 8.9 mM sodium phosphate dibasic (Na_2HPO_4), and heated rapidly to 85 °C and kept there during 10 min before being brought to ambient temperature at a rate of 1 °C/min.

The yeast DNA was dissolved in a Dulbecco PBS buffer and the salmon sperm DNA dissolved in purified water (Mili-q). The resulting concentrations of double helix DNA were determined from their absorption at 260 nm to be 2.2, 0.7, 0.0015 and 0.00023 μM for the 25 bp, 48 bp, yeast and salmon sperm DNA respectively (assuming average lengths of 100 Kbp for the latter two).

Two ml of the corresponding solution was placed into a tightly stoppered standard 3.5 ml quartz cuvette of 1 cm light path length. The cuvette was placed inside a precise (± 0.01°C) Ocean Optics@ temperature control unit operating via the Peltier effect with water flow stabilization of the temperature; important for precise and rapid temperature control when removal of heat from the cuvette was required. The temperature was monitored via a probe located in one of the four supporting towers of the cuvette and in direct contact with it. A magnetic stirrer provided assurance of temperature equilibration throughout the sample, insurance against DNA sedimentation, and homogeneous passage of the DNA through the approximately 3.7 mm diameter light beam on the sample.

UV light from a 3.8 W (9.4 μW on sample) deuterium source, covering continuously the range 200 to 800 nm, or light from a 1.2 W (7.0 μW on sample) tungsten-halogen source (used as a control) covering 550 to 1000 nm, was defocused onto the DNA sample via UV light resistant 30 cm long optical fiber of 600 μm diameter and a quartz lens. After passing through the sample, the surviving light was collected by a second lens and focused onto a similar optical fiber which fed into an Ocean Optics@ HR4000CG charge-coupled device spectrometer. The spectrometer covered the range 200 to 1200 nm with a resolution of 0.3 nm. Spectrometer integration times were of the order of 1.5 s and the data were averaged over five wavelength channels (1.5 nm).

After allowing approximately 1/2 hour for lamp stabilization, sample extinction (absorption plus scattering) spectra, E(λ), were obtained by comparing the measured intensity spectrum, I(λ), of the light through the DNA sample with a reference spectrum, I_R (λ), obtained at room temperature with a similar quartz cuvette containing either purified water or PBS buffer as required, but no DNA. A dark spectrum, I_D (λ), obtained with a shutter blocking the deuterium light to the sample at room temperature, was subtracted from the spectra to correct for stray light and electronic noise. Extinction was thus calculated as;

$$E(\lambda) = -log_{10}\frac{(I(\lambda) - I_D(\lambda))}{(I_R(\lambda) - I_D(\lambda))} \tag{12.1}$$

The synthetic DNA samples were heated rapidly (5 °C/min) to 90 °C and held there for approximately 10 minutes to ensure complete denaturing. The temperature of the sample was then lowered (1 °C/min) to that desired for the particular run. The longer DNA samples were brought directly to the temperature of the run from ambient temperature (~24 °C). Runs consisted of allowing the sample to equilibrate (between denatured single and natured double strands) for approximately 40 minutes at the given temperature of the run and then inserting and retracting a shutter to cycle the deuterium light on and off through the sample for different time periods while continuously monitoring the wavelength dependent extinction, Eq. (12.1), during light on periods.

12.2.2 Results

Figure 12.3 shows a typical extinction spectrum using the deuterium light source for (a) 48 bp synthetic DNA and (b) yeast DNA, at three different temperatures. The extinction peak at 260 nm is due predominantly to absorption on the nucleic acid bases but also includes a small amount of Rayleigh and Mie scattering. An increase in absorption with temperature, known as hyperchromism, is observable at the absorption peak at 260 nm and is the result of the nucleobases becoming more exposed (less shielded) to the UV-C light due temperature denaturing of the tightly stacked, and therefore, shielded nucleobases in double helix form (Vekshin, 2005)[386]. The inset shows an amplified view of the longer wavelength region for the three different temperatures. The extinction in this region is due to Mie scattering dominating in the short wavelength region (~200 to 350 nm) and Rayleigh scattering dominating in the long wavelength (>350 nm) region for the 25 bp and 48 bp synthetic DNA (physical size ~2 x 17 nm). For the much longer yeast and salmon sperm DNA, Mie scattering is predominant over the whole wavelength region (Fig. 12.3(b) inset).

Scattering is a complicated function of the number of scattering centers, the ratio of the indices of refraction DNA/buffer, and shape of the scattering centers (Johnsen and Widder, 1999)[141]. However, for the small 48 bp (inset, figure 12.3(a)) and 25 bp (not shown) DNA the scattering increases with denaturing (correlated with temperature), while the scattering is constant or only slightly increasing with denaturing for the large yeast (inset, figure 12.3(b)) and salmon sperm DNA (not shown). This result is interpreted as being due to complete denaturing for at least some of the small synthetic DNA (even at low temperature) where the separated strands act as separate scattering centers, but only partial denaturing for the large DNA, where the still united strands act as single scattering centers. Light scattering is thus an additional (to hyperchromicity)

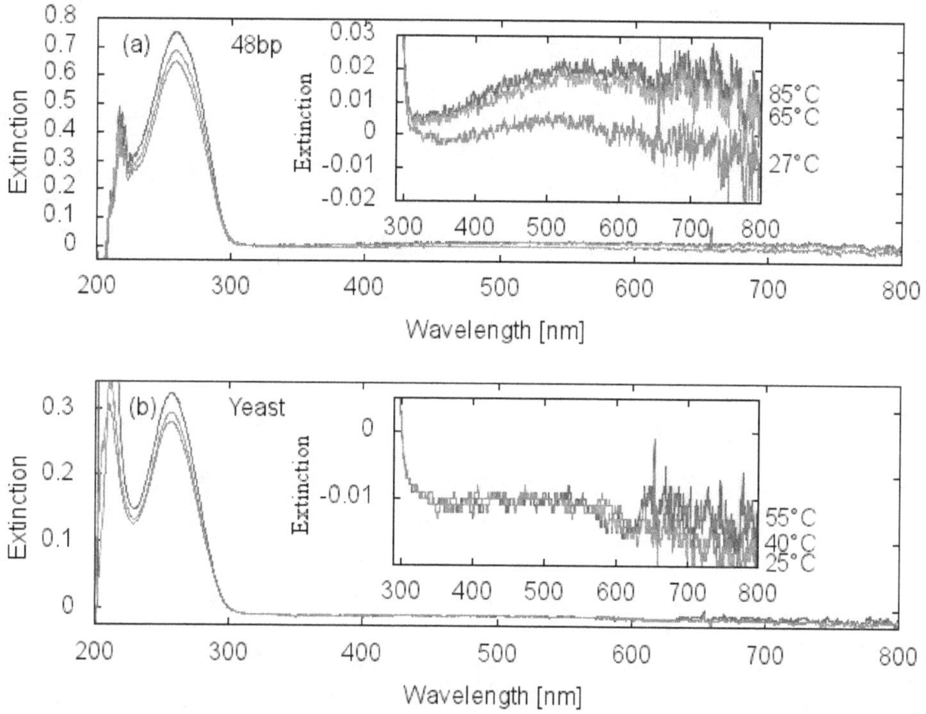

Fig. 12.3. Extinction spectra. a) 48 bp synthetic DNA. b) 100 Kbp (average size) yeast DNA. Hyperchromism is observable at 260 nm in both samples. The insets show amplified views of the scattering regions at three different temperatures.

sensitive indicator of denaturing for the small synthetic DNA, while it has little utility for the long DNA, except at high temperature where complete denaturing also occurs. For the short synthetic DNA, therefore, a measure of photon induced denaturing was obtained by integrating the extinction over wavelength from before the peak in absorption at 260 nm up to 495 nm.

Figures 12.4, 12.5, 12.6 and 12.7 plot the extinction, Eq. (12.1), over the duration of the run, integrated over 245-295 nm and 255-275 nm, for the yeast and salmon sperm DNA respectively, and integrated over wavelengths 225-485 nm for the short synthetic DNA samples. The periods of deuterium light cycling on and off (through manual control of a shutter) at the given constant temperature are noted on the graphs by up and down arrows respectively. It can be observed that for all DNA samples, the extinction rises during the light on periods and decreases or stays constant during the light off periods. Given that extinction increases on denaturing due to hyperchromism, this data is putative evidence for UV-C light-induced denaturing of DNA.

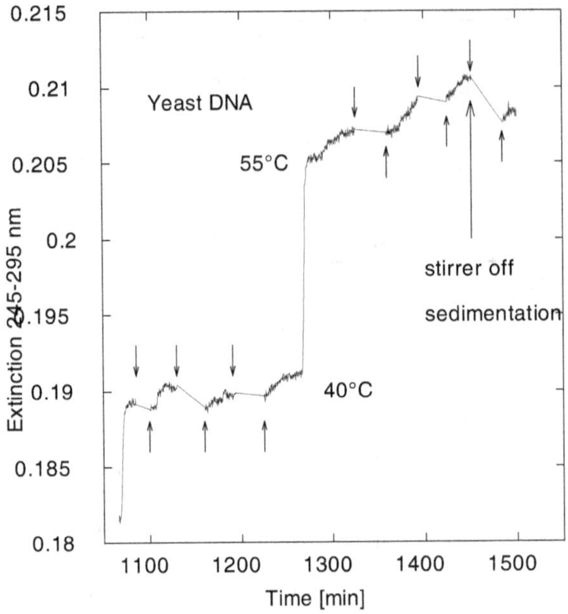

Fig. 12.4. Extinction for Yeast DNA: a) Extinction (mainly absorption) as a function of time in minutes. Deuterium light on an off changes are noted on the graphs by up and down arrows respectively. Turning the stirrer off, as noted on the graph for yeast DNA, causes a decrease in extinction over time due to sedimentation of the DNA.

However, before considering these results as conclusive evidence for UV-C induced DNA denaturing, it is important to consider other physical or instrumental effects which may have influence on the result and need to be looked at in detail. An observed increases in extinction may, in fact, be due to any combination of all of the following possibilities;

1. absorption hyperchromicity due to light-induced denaturing (the sought after effect),
2. increased Rayleigh or Mie scattering, particularly for the short synthetic DNA, due to light-induced denaturing changing the number and shape of the scattering centers (also useful for verifying the existence of the sought after effect),
3. evaporation of the solvent, thereby increasing the concentration of the DNA solute,
4. sedimentation of the DNA, thereby decreasing the concentration of the DNA solute,
5. variation over time of the light output from the light source, or variation of the detector response.

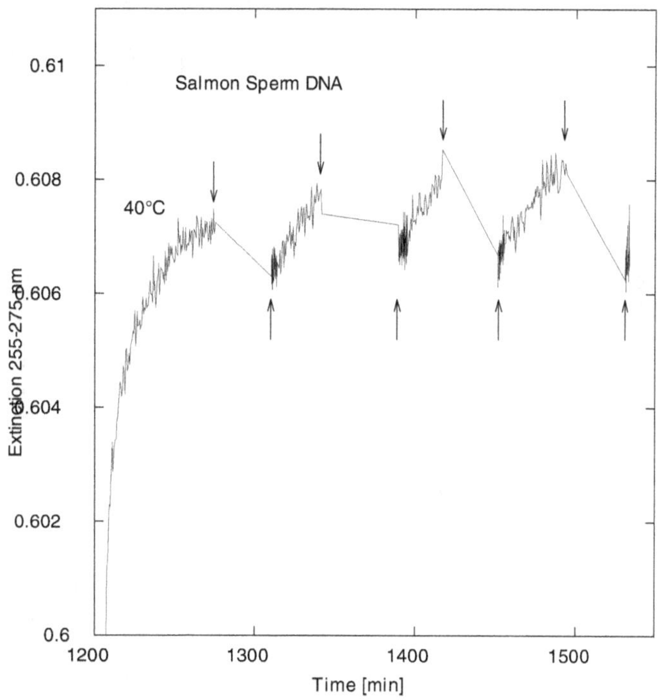

Fig. 12.5. Extinction for Salmon Sperm DNA. See Fig. 12.4 for an explanation.

With respect to point 3), even though the cuvette was tightly stoppered, evaporation of the solvent occurred into the approximately 1.5 ml air space above the 2 ml solvent volume of the 3.5 ml cuvette volume. Evaporation of the solute into the airspace could lead to a rise in extinction as DNA concentrations would grow as the liquid volume decreased. Evaporation would be greater at higher temperatures and should be independent of the light on or off condition. In some of the runs at high temperature, evaporation indeed occurred and the water was observed to condense around the colder Teflon cuvette stopper, accumulating into a large drop ($\sim 50\ \mu$l) that would eventually fall into the solvent, rapidly reducing the DNA concentration to nearly its initial value at low temperatures and leading to a sudden (within 30 seconds) and noticeable drop in our measured extinction data (see "condensation drop" on figure 12.6). This continuous evaporation of the solvent into the air space above liquid solvent is also responsible for the general increase in the observed extinction in time, even during light off periods, particularly for the high temperature runs at 85 °C (see Fig. 12.6 and 12.7).

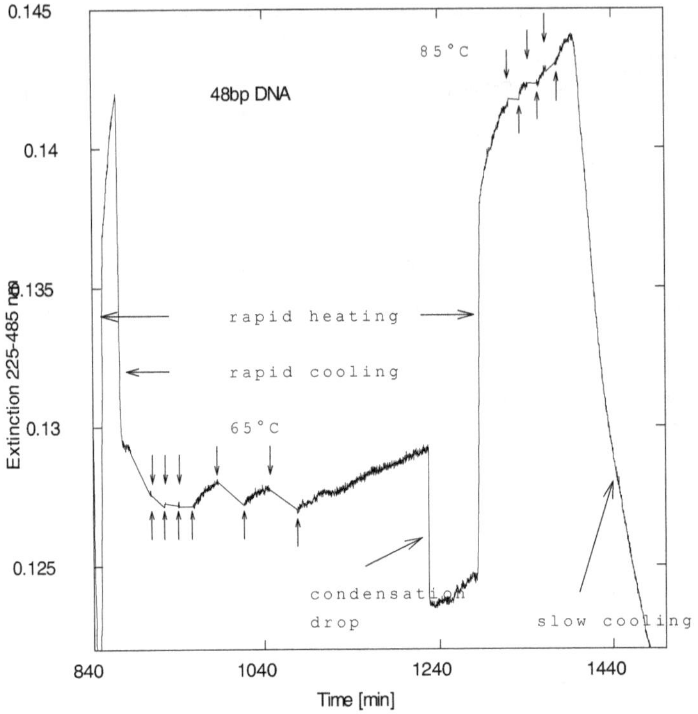

Fig. 12.6. Extinction for 48 bp Synthetic DNA. See Fig. 12.4 for an explanation.

Concerning point 4), sedimentation of the DNA was observed (see Fig. 12.4) after a magnetic stirrer within the quartz cuvette was turned off, however sedimentation was completely avoided as long as the stirrer was kept on.

Concerning point 5), the relative drift of the deuterium light source at different wavelengths convoluted with the detector response function has been measured by the supplier of both these instruments using a similar spectrometer (Ocean Optics, 2015)[255]. Drift of both the light source and detector is influenced mostly by variations of the ambient temperature and line voltage. Drift is seen to be small (<0.05%) and uniform over a 4 hour run (Ocean Optics, 2015)[255] and should also be independent of the light on or off condition since an electronic shutter enacted this condition. Since neither native double strand nor single strand DNA absorb beyond 310 nm, it was possible to correct the absorption for drift by normalizing to the extinction measured at 315 nm for each spectrum. However, at wavelengths > 310 nm there is an important component due to light scattering differences between single and double strand DNA, especially for the short strand DNA (see inset figure 12.3(a)). Therefore, in order not to obscure this useful scattering signal the normalization was ap-

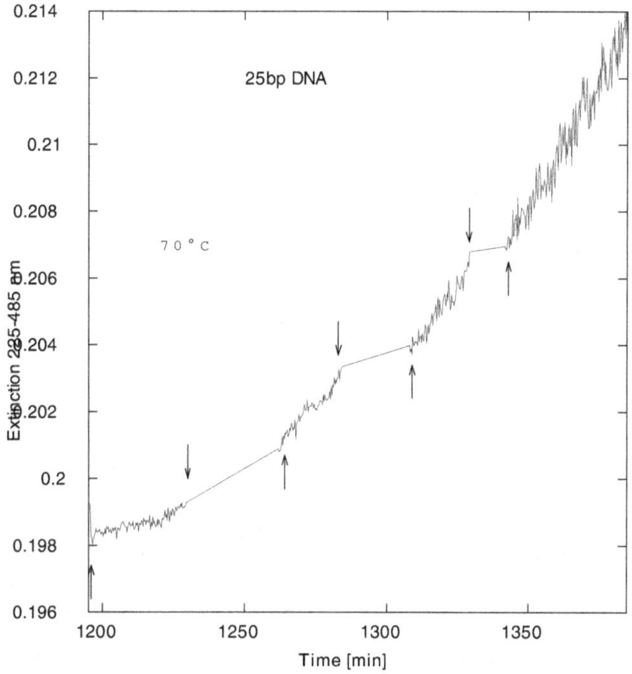

Fig. 12.7. Extinction for 25 bp Synthetic DNA. See Fig. 12.4 for an explanation.

plied only for a definitive check of UV-C induced denaturing using the difference spectra around 260 nm which we now proceed to describe.

Fortunately, a rather simple and definitive method exists to distinguish the sought after effect of light-induced denaturing from an instrumental (e.g. evaporation, sedimentation, or electronic drift), or any other non-denaturing effect, which could give rise to a change in extinction. This method is based on the very particular difference in the extinction spectrums between natured and denatured DNA, especially for the short 25 bp synthetic DNA (see below and figure 12.9). By comparing the difference spectrum obtained for UV-C light induced denaturing with that obtained for temperature-induced denaturing at around the fixed temperature of the UV-C light-on run, it is possible to definitively verify light-induced denaturing and to quantify the amount of denaturing due to it.

In figure 12.8 we show the extinction data of a particular run for the 25 bp synthetic DNA as a function of time overlaid with a graph showing how the temperature of the sample was varied as a function of time.

The difference in the absorption spectra between denatured single strand DNA and natured double strand DNA is unique for different DNA (Mergny et al., 2005)[204]. Obtaining an expected difference spectrum is therefore a clear signature of denaturing. Such a signature can thus be used to verify that we in-

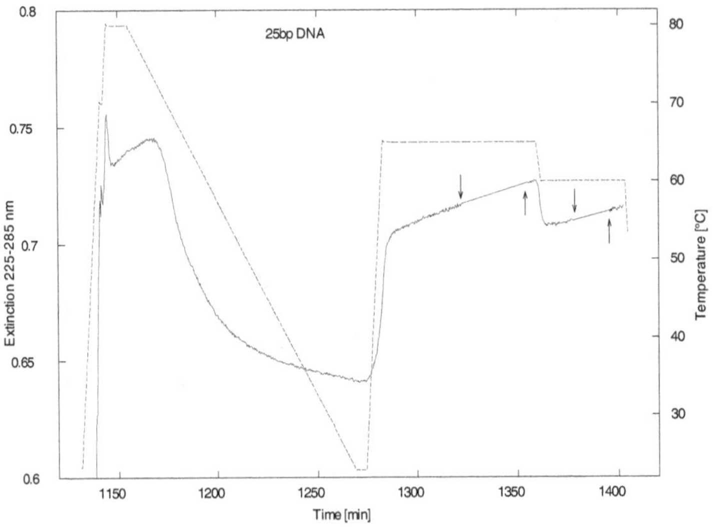

Fig. 12.8. Extinction for 25 bp DNA (solid line) and the variation of the temperature of the sample (dashed line) as a function of time. Light on and off conditions are marked on the graph by up and down arrows respectively. Note that at the high constant temperature periods at 80 °C, 65 °C and 60 °C the extinction increased due to evaporation of the solvent (increasing the concentration of DNA), and this occurred irrespective of the light being on or off. However, the slope of the increase in the light on period is greater than that due to evaporation and this difference is due to UV-C light induced denaturing.

deed have UV-C light induced denaturing during the light on runs. Figure 12.9 plots the extinction curves of 25 bp DNA at two temperatures, 80 °C and 25 °C, corresponding to temperature denatured single strand and natured double strand DNA respectively and the difference spectrum, which is unique to this particular DNA (of particular G-C content and even sequence). Figure 12.10 plots the contributions to the difference spectra for different temperature bins of 3 °C. This figure shows that at around 60 °C we begin to get important contributions from G-C denaturing as can be seen by comparing to the absorption spectra of the individual nucleotides as given in figure 12.11.

We now take a look at the difference spectrum for the extended light on period at the constant temperature of 65 °C from time 1289 min to 1322 min (see figure 12.8). The extinction spectra at these two times are averaged over 11 distinct graphs each and are normalized at 315 nm to remove instrumental and other non-denaturing effects (such as evaporation and electronic drift). The result is given by the pink curve in figure 12.12.

In figure 12.6 we have plotted the extinction for the 48 bp DNA while varying the temperature during the run. In figure 12.13 we plot the contribution to the denaturing difference spectrum for temperature differences of 3.2 °C at the

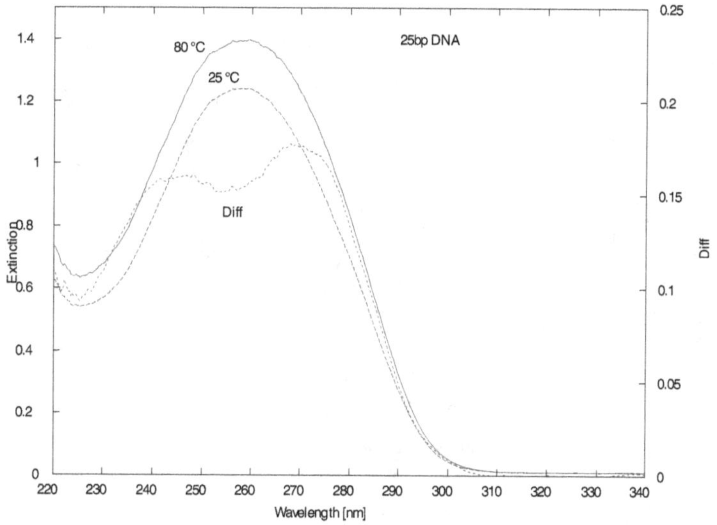

Fig. 12.9. Extinction spectra for the 25bp DNA at 80 °C and at 25 °C (corresponding to times 1167 min and 1270 min respectively on the graph of Fig. 6) normalized at 315 nm to remove any instrument or other non-denaturing effect (including electronic drift and evaporation). The difference spectrum (dotted curve) of the two extinction curves is also plotted and is a unique signal for 25bp DNA denaturing and has two peaks at 244.4 nm and 269.2 nm. The difference spectrum is, in fact, just the wavelength dependent hyperchromicity.

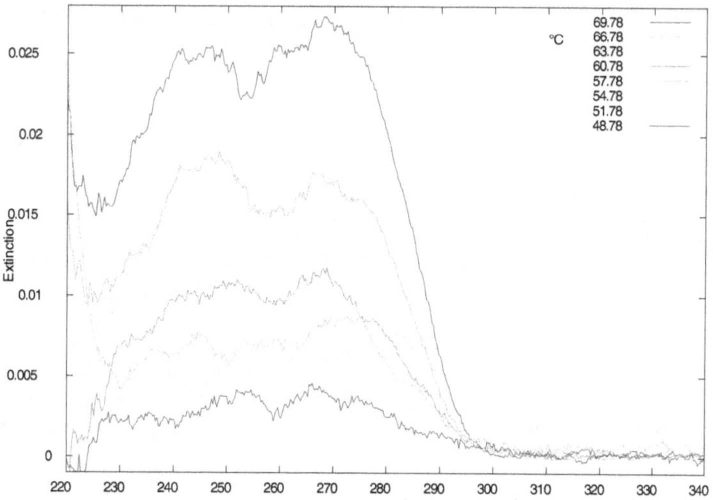

Fig. 12.10. Contributions to the difference spectrum for different temperature bins of 3 °C. Each difference spectrum is an average over approximately 10 distinct extinction spectra at both the beginning and end of the temperature bin.

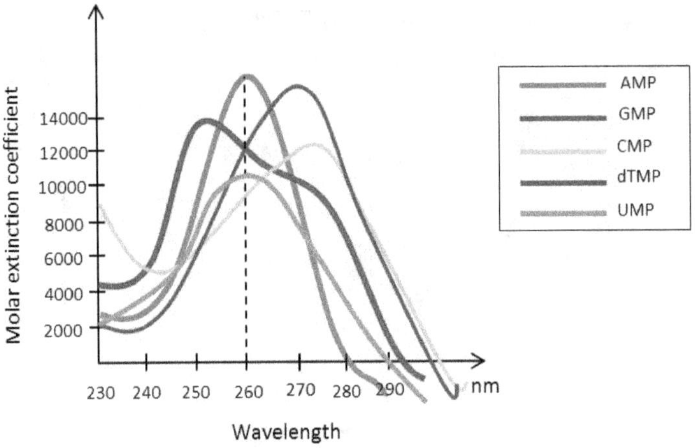

Fig. 12.11. Absorption spectra of the different nucleotides.

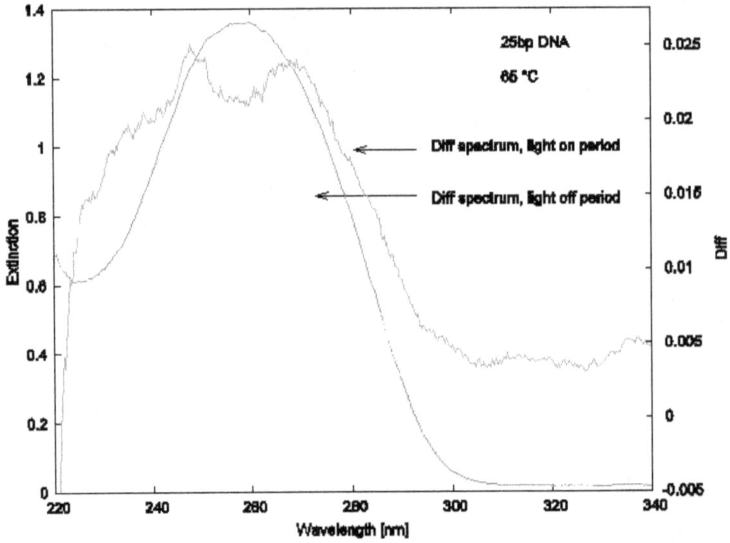

Fig. 12.12. Extinction spectra for 25 bp DNA obtained at the beginning and end of the light-on period (green dotted and red dashed lines respectively) at the constant temperature of 65 °C (see Fig. 6). Difference spectrum of these two extinction spectrum for UV-C light-induced denaturing at constant temperature of 65 °C is also plotted (pink curve, scale on right). The two peaks are at 248.5 nm and 268.6 nm and result from hyperchromicity of UV-C light-induced denaturing of mainly the G-C bonds (most A-T bonds are already broken at this temperature). Renaturing of the G-C bonds can be seen by the inversion of both the two peaks and the valley at around 260 nm in the difference spectrum for the light-off period (yellow curve).

temperature given, obtained during the slow cooling part of the run from time 1400 min to 1500 min for the 48 bp DNA (Fig. 12.6).

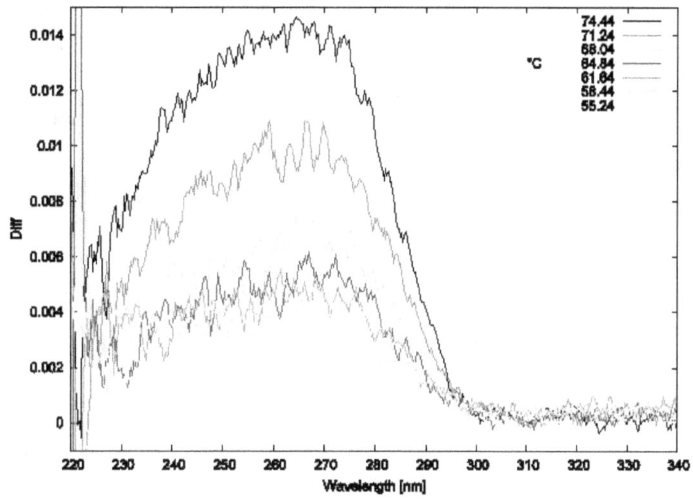

Fig. 12.13. Contributions to the difference spectrum for a temperature difference of 3.2 °C centered on the given temperature for the 48 bp DNA. Each difference spectrum is an average over approximately 17 distinct extinction spectra at the beginning and end of each temperature 3.2 °C bin.

Figure 12.14 plots the difference spectra obtained at constant temperature (65 °C) for the UV-C light on period of the 48 bp DNA run from 1076 to 1226 min (Fig. 12.6) after normalizing for evaporation. The shape of the difference spectrum is similar to that of temperature denaturing of 3.2 °C around 65 °C (see Fig. 12.13). This provides further evidence, in addition to the increase and decrease in extinction with light on and off respectively as seen in figure 12.6, for UV-C light-induced denaturing of the synthetic 48 bp DNA.

Renaturing of completely separated DNA proceeds via chance encounter of a few complementary base pairs and then rapid re-zipping of the remaining pairs. It is thus second order in the concentration but there is also a volume or steric effect since for longer DNA the folding of single strands implies that some base pairs are inaccessible to first encounter. Renaturing is thus a diffusion plus steric limited process but has some peculiarities that depend on the nucleotide complexity of the strand, the ionic nature of the solvent, and also the viscosity of the solvent (Wetmur et al., 1968)[400]. Renaturing is readily observed as a decrease in extinction after the end of the light off period in the photon induced partial denaturing of yeast and salmon sperm DNA (Figs. 12.4 and 12.5) and to some extent in the 48 bp DNA (Fig. 12.6) but not readily observed in the extinction data for the short 25 bp synthetic DNA (Fig. 12.7) although observed

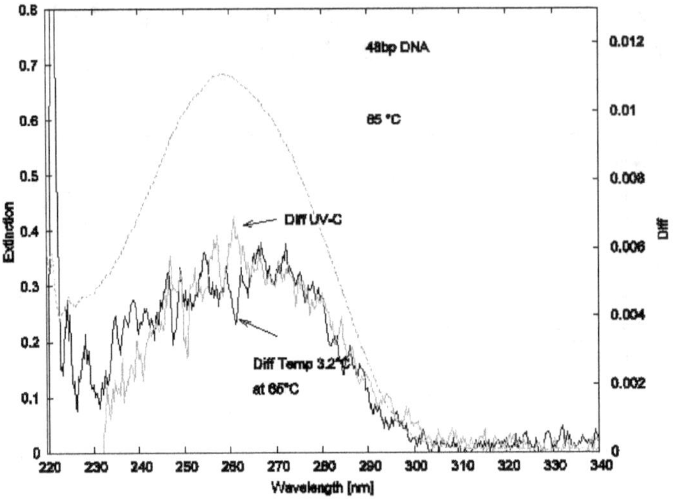

Fig. 12.14. Extinction spectra obtained at the beginning and end of the UV-C light on period (green dotted and red dashed lines respectively) and the difference spectrum (blue line, right scale) for the 48 bp DNA. The similarity in the shape of the UV-C light-induced difference spectrum (blue line, right scale) with the difference spectrum obtained for temperature denaturing of 3.2 °C centered on 65 °C (black line, see Fig. 12.13) is further evidence of UV-C light-induced denaturing.

in the difference spectrum for 25 bp DNA (yellow curve, figure 12.12). This can be attributed to a number of effects, i) the difficulty of chance encounter for renaturing completely separated single strands, ii) the effects of evaporation increasing the concentration of DNA exposed to the light and therefore increasing the extinction even during the light-off periods, iii) increased probability of heat denaturing once UV-C light-induced partial denaturing of G-C bonds has reduced the effective melting temperature of the surviving double strand segments.

As an experimental control, to exclude the possibility of other mechanical, optical, electronic, temperature, or other artifacts giving rise to the changes in extinction, we carried out an identical experiment with 48 bp DNA (Fig. 12.6) by replacing the deuterium UV-C lamp with the control halogen-tungsten visible lamp. In this case, no increase in the extinction (which could only be observable in scattering) was found during the light on periods (Fig. 12.15) and no decrease was found during the light off periods, confirming that the increase in extinction observed during the light on periods (Fig. 12.6) is indeed UV-C light-induced.

As a final check of the theoretical feasibility of UV-C light-induced denaturing, we calculated the increase in local temperature at the site of the DNA due to the conversion of a single UV-C photon of 260 nm (4.8 eV) into heat

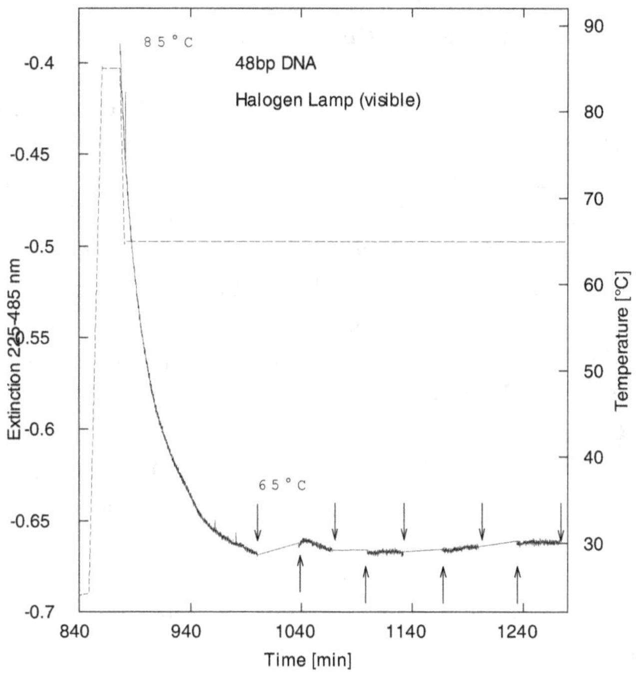

Fig. 12.15. Extinction for 48 bp synthetic DNA under halogen (only visible) light. The up and down arrows mark the beginning of the light on and off periods respectively. No increase in extinction is observed during the light-on periods, nor a decrease in extinction during the light-off periods, indicating that the variations in extinction observed in Fig. 12.6 are indeed UV-C light-induced, and not an instrumental artifact. The temperature of the sample is given by the dashed line corresponding to the scale on the right.

with the DNA in a water solvent environment. We assume that the photon is absorbed on one or a few contiguous bases (Crespo-Hernández et al., 2005)[62] and the electronic excitation energy is rapidly dissipated (within picoseconds) into heat (Pecourt et al., 2000)[271] which diffuses from the interaction site into a spherical volume centered on the site. Given the heat capacity of water ($C_V=$ 4200 J kg^{-1}K^{-1}) the calculated diameter of the volume at which the average temperature of the solvent increases to 3 °C over ambient temperature at 65 °C (a rise in temperature significant to cause a considerable amount of denaturing at these temperatures, see Fig. 12.10), is 4.9 nm, which corresponds to a volume containing 16 base pairs assuming the double helix as a rigid cylinder. This must be considered as a conservative estimate since DNA longer than 15 bp are not rigid cylinders but are folded, allowing more base pairs to be within this volume, and, furthermore, it is most probable that the heat of photon dissipation will not diffuse into a spherical volume centered on the base, but rather diffuse preferentially along the axis of the DNA following the high vibrational modes of the strong covalent bonds of the molecule. We are presently performing molecular dynamic simulations to understand this process of heat diffusion along DNA in greater detail.

A quantitative measure of the UV-C light-induced denaturing can be obtained by comparing the hyperchromicity obtained during the light on periods with the maximum temperature induced hyperchromicity of completely denatured DNA, assuming a linear relation between hyperchromicity and denaturing. For the salmon sperm DNA in purified water at 40 °C (at the start of the denaturing curve), a half hour light on period is sufficient to increase the extinction in the 255-275 nm region by 0.002 units (see Fig. 12.5), which corresponds to 1.7% of the maximum temperature induced hyperchromicity at 90 °C, implying a UV-C light-induced denaturing rate of approximately 3.4% per hour. A similar value was found for yeast DNA at 40 °C. An analogous calculation for the 48 bp DNA in PBS buffer (see Fig. 12.6) shows that at 65 °C the rate of UV-C light-induced denaturing is approximately 4.6% per hour. For the 25 bp DNA in PBS buffer at 65 °C (somewhat into the denaturing curve, see Fig. 12.8) the UV-C light-induced denaturing was determined to be approximately 24% per hour after correcting for evaporation as evaluated from the increase in extinction during the light off period (see Fig. 12.8).

Finally, it is pertinent to compare our experimental conditions of light, temperature, and DNA concentrations to those likely existent during the Archean at the beginning of life on Earth (~3.8 Ga). Sagan (1973)[309] has calculated an integrated UV-C flux during the Archean over the 240-270 nm region where DNA absorbs strongly (see Fig. 12.3) of 3.3 W/m^2 while we have estimated a somewhat larger flux of 4.3 W/m^2 (Michaelian and Simeonov, 2015)[217]. Our light on sample was estimated to have a flux over this same wavelength range of 2.9 W/m^2. However, the beam volume on sample was only 0.107 cm^3 while the total volume of the sample was 2.0 cm^3 (2 ml) and uniform mixing of the DNA sample due to the magnetic stirrer ensured homogeneous passage of the DNA through the whole volume, giving an effective UV-C light flux of only 0.155 W/m^2 during light-on periods, less than 1/20 of what it may have been at the ocean surface during the Archean.

Isotopic geologic data suggest that at 3.8 Ga the Earth was kept warm by a CO_2 and CH_4 atmosphere, maintaining average surface temperatures around 80 °C (Knauth, 1992; Knauth and Lowe, 2003)[163, 165] and falling to 70±15 °C at 3.5–3.2 Ga (Lowe and Tice, 2004)[193]. Of course, polar regions would have been colder and equatorial regions warmer. The light-induced denaturing rates presented here were obtained at lower temperatures and therefore shows that at these high temperatures at the beginning of life, UV-C light-induced DNA denaturing could have been even greater.

Miller (1998)[224] has determined adenine concentrations of 15 μM in the prebiotic soup using calculations of photochemical production rates of prebiotic organic molecules determined by Stribling and Miller (1987)[363]. Although these estimates have been considered as overly optimistic by some, we emphasize that Miller was not aware of non-equilibrium thermodynamic routes to nucleotide proliferation based on photon dissipation (see chapter 6.1), for example, the Ferris-Orgel photochemical route to the nucleobases, or the purine production in UV-C irradiated formamide solutions (Barks et al., 2010)[12], nor of the existence of an organically enriched sea surface skin layer (Hardy, 1982; Grammatika ad Zimmerman, 2001)[124, 114], so Miller's determinations of nucleotide concentrations at the beginnings of life may, in fact, be overly conservative. Adenine (and the other nucleobase) concentrations in our sample DNA were of the order of 50 μM.

The above comparisons, particularly our much lower on-sample UV-C light flux relative to estimates of the flux at Earth's surface during the Archean, suggest that our determined UV-C induced denaturing rates should probably be taken as a conservative lower limit to rates that could have occurred on the Archean sea surface at the beginning of life.

12.2.3 Discussion

Through an analysis of extinction (absorption at UV-C wavelengths plus scattering at longer wavelengths) and considering the hyperchromicity plus Rayleigh and Mie scattering characteristics of DNA, we have demonstrated that absorption and dissipation of UV-C light in the range of 230 to 290 nm denatures DNA in a reversible and apparently benign manner. Analysis of the scattering component suggests complete denaturing for short 25 bp synthetic DNA while partial denaturing for 48 bp DNA, and long salmon sperm and yeast DNA. We are presently investigating whether higher light intensities, closer to those incident on the ocean surface during the Archean, may be able to completely denature DNA significantly longer than 25 bp. We have conclusively verified that what we see in the extinction data is indeed UV-C light-induced denaturing by comparing our light on difference spectra with difference spectra obtained for temperature induced denaturing at the temperatures of the light-on run. Our light off difference spectra for 25 bp DNA clearly shows renaturing. Further verification of this comes from the fact that we see no increase in scattering at long wavelengths if we replace our UV-C deuterium light with a tungsten- halogen visible light, indicating that the increase in scattering seen at longer wavelengths as obtained with the deuterium light must come from UV-C induced denaturing with the $\sim$ 260 nm component of the deuterium light. Finally, we have confirmed the theoretical possibility of UV-C induced denaturing using a simple heat flow calculation given the heat capacity of water.

Although our postulate of denaturing through energy transport (heat flow) along the chain of the DNA has been challenged on the basis of the internal conversion process of the electronic excitation energy being on the order of picoseconds, much too short to support a collective phenomena of denaturing (Werner, 2011)[398], there are now known mechanisms for energy transfer along double helix RNA and DNA and these mechanisms will be discussed in relation to UV-C induced denaturing in the following section.

Indirect evidence of UV light-induced denaturing of DNA was suspected, although not confirmed, previously in a very different context (Hagen et al., 1965)[122]. Our results are the first conclusive evidence that UV-C light denatures DNA and we presented it within the context of the thermodynamic dissipation theory of the origin of life (Michaelian, 200; 2011)[207, 210]. We have quantified this effect under conditions that may be considered conservative with respect to those expected at the beginnings of life on the surface of the Archean seas.

Since RNA has similar optical and chemical properties as DNA, it is plausible that such ultraviolet and temperature assisted replication (UVTAR), homochirality, and information acquisition mechanisms driven by photon dissipation would have also acted similarly over duplex RNA and hybrid RNA+DNA duplexes, and even over triplexes (see chapter 9). We are presently conducting experiments to verify this. If the latter is borne out, then DNA and RNA may have been contemporaries in the primordial soup, evolving independently with other fundamental UV-C absorbing molecules (Michaelian and Simeonov, 2015)[217] towards increased photon dissipation before finally forming a symbiosis which augmented further still this dissipation. Evidence for a stereochemical era in which amino acids and other UV-C and visible absorbing fundamental molecules of life had chemical affinity to both DNA and RNA (Michaelian and Simeonov, 2015; Govil et al., 1985)[217, 112] supports this assertion. This would have allowed continued replication and proliferation, and thus increases in photon dissipation, despite the cooling of the ocean surface and the eventual attenuation of the UV-C light brought on by the accumulation of oxygen in the environment after the invention of oxygenic photosynthesis at around 2.9 Ga and the delegation of UV-C dissipation to life derived ozone in the upper atmosphere.

Regarding the second part of the UVTAR mechanism, that of enzymeless extension during dark and cooler, overnight periods, or during cooler early morning periods, there exists experimental data obtained using chemical activation indicating that long random sequence linear DNA oligonucleotides can be template synthesized from a random pool of short oligonucleotides with high fidelity (with fidelity increasing with temperature) (James and Ellington, 1997)[138]. In our proposed UVTAR scenario, day-time UV-C light activation of the nucleotides (Ponnamperuma et al., 1963)[286] would replace chemical activation.

Our result, analyzed within the context of the non-equilibrium thermodynamic dissipation theory of the origin of life, sheds new light on some of the formidable difficulties related to the origin of life. The association of growth, replication, and evolution with dissipation is a necessary thermodynamic prerequisite of any irreversible biotic or abiotic process and in this context our result could reconcile "replication first" with "metabolism first" scenarios for the origin of life. Photon dissipation would certainly have been the most primitive and still is the most important and extensive of all "metabolisms".

The significant magnitude of the UV-C light-induced denaturing results presented here also lends plausibility to the proposal of the acquisition of homochirality and information content in DNA through dissipation (Michaelian, 2010)[208] and this will be discussed in the following chapters. The thermodynamic dissipation theory of the origin of life (Michaelian, 2009; 2011)[207, 210] suggests that all biotic, and coupled biotic-abiotic, evolution has been driven by increases in the entropy production of the biosphere, mostly through increasing the global solar photon dissipation rate, and that life started out dissipating in the UV-C region before gradually evolving dissipation towards the visible where photon fluxes were greater but biosynthetic pathways necessarily had to evolve to greater complexity because of the lower photon energy available for making and breaking covalent bonds.

12.3 Energy Dissipation and Transfer in RNA and DNA

This section provides a look at the microscopic experimental and theoretical foundation for the UVTAR mechanism of enzymeless replication of RNA and DNA which, we believe, may have occurred under the prevailing environmental conditions existing during the Archean. In particular, here we look at UV-C induced energy dissipation and energy transfer or dispersion mechanisms in single nucleobases, in single strand, and in the duplex A- or B- form of DNA and RNA. The dissipation time scales and the various mechanisms for base to solvent, or base to base, energy transfer so far recognized differ in importance depending on the particular macromolecular structure and base sequence, and it is quite probable that other pathways have yet to be delineanated.

Kang and co-workers (2002)[148] as well as Kohler and collaborators (Kohler, 2010)[167] have studied single nucelobase non-radiative de-excitation after excitation with a pumped femtosecond laser operating in the UV-C region. They find extremely rapid de-excitation times of the electronic excited states on the sub picosecond time scale (200 fs) for all the single natural bases (see figure 7.4) which they attribute to one or more conical intersections (see figure 7.5).

Calculations have shown that the conical intersections are accessed through an out-of-plane nuclear movement while in the excited state structure. In all these cases, the intermolecular vibrational energy transfer to the solvent (vibrational cooling) is found to be the rate limiting step of the dissipation process which leads to overall ground state recovery times for the bases of about 2 pico seconds (2×10^{-12} s), an order of magnitude larger than the observed time for depopulation of the excited state.

Multiple conical intersections may be present in a single base because the nucleobases have several close-lying $^1\pi\pi^*$ and $^1n\pi^*$ states, leading to the existence of several decay pathways for each of the bases. Although the $^1n\pi^*$ states are not significantly populated by light absorption, they can be reached via internal conversion from the intense $^1\pi\pi^*$ states (Kohler, 2010)[167]. The decay traces, particularly for pyrimidines, thus often require the fitting of two or more exponentials. For example, it is found that thymine has a 10% chance of branching to a $^1n\pi^*$ state which has a larger decay time constant of about 30 ps (see figure 7.4). The $^1n\pi^*$ states have not yet been detected in the purines and these, in fact, demonstrate decay only through a single rapid component. This is probably the reason that purines are not as prone to UV-C light induced lesions as are the pyrimidines. It has been determined that a small fraction of the $^1n\pi^*$ population can undergo intersystem crossing to the lowest triplet state in competition with vibrational cooling, and this could explain the higher triplet yields observed for pyrimidine with respect to purine bases (Hare et al., 2007)[125]. Non-natural tautomers of the bases were found to have depopulation times of the excited state orders of magnitude greater than that of the natural bases (Serrano-Andres and Merchán, 2009)[333], ostensibly because they lack a conical intersection.

The stacking of the bases, even in single strand RNA and DNA, a result of the non-covalent but strong π-stacking interactions between consecutive bases allows for the possibility of energy migration along the axis of the DNA. Such a phenomena would certainly be of relevance to a UVTAR mechanism for enzymeless denaturing of DNA or RNA. It was known, in fact, since the 1980's that there must exist some form of energy migration along DNA since lesions in DNA were induced at a relatively large distances, up to microns, from the site of energy deposition created by neutron-induced nuclear stopping[4]. The evidence for energy transfer at large distances along DNA in a water solvent environment comes in the form of the production of nucleobase radicals far from the localized cite of energy deposition for DNA irradiated perpendicular to the DNA axis

[4] Nuclear stopping referes to the process in which a neutral particle (such as a neutron or an initially charged ion at the end of its stopping curve, when it has picked up electrons from the medium) looses energy to the medium by a direct collision with a nucleus of an atom from the stopping material. This deposits a large amount of energy over a very small region at the end of the neutrons (or ions) track.

(Arroyo et al., 1986)[7]. It is important to note in this regard that one of the products, protonation of thymine, can be thermally stimulated. This observation, of damage occurring far from the site of energy deposition, was interpreted as possibly resulting from the formation of soliton-like stable energy packets that can travel large distances along the quasi-one dimensional molecule of the DNA (Baverstock and Cundall, 1988)[14]. It was suggested that energy dissipation into the solvent in this case would be offset by the non-linearity through endothermic coupling of the vibrational modes and lattice phonons, allowing the disturbance to propagate large distances and last for nanoseconds (Baverstock and Cundall, 1989)[15] .

More recent work by Buchvarov et al. (2007)[29] using femtosecond time-resolved broadband spectroscopy, has traced the electronic excitation in both time and space along the stack of bases in single-stranded and double-stranded DNA oligonucleotides. A femtosecond broadband pump-probe spectroscopy allows them to monitor the temporal evolution of the excited-state absorption, including the absorption of "dark states" (the non-radiative states of interest with negligible fluorescence quantum yields), which are, in fact, dominant in excited DNA. The combination of broadband spectral probing (from 300 to 700 nm) with a time resolution of 100 fs allowed them to monitor a variety of spectroscopic transitions simultaneously. They looked at single strand homoadenine DNA $(dA)_n$ (with $n = 2, 3, 4, 5, 6, 12, 15, 18$ consecutive 2 -deoxyadenosine (dA) residues), as well as double-stranded oligonucleotides $(dA)_n\cdot$ $(dT)_n$ (with $n = 12, 18$) and a hetero duplex $(dAT)_9\cdot(dAT)_9$. By measuring the ratio of exciton absorption intensity at 435 nm to monomer absorption at 330 nm (see figure 12.16), they found that in both the single strand DNA and in the double strand oligos, delocalized Frenkel exciton states were formed in which the UV-C induced excitation energy became rapidly dispersed over a number of consecutive bases. For the single strand homoadenine DNA they found that delocalization of energy typically shows an exponential decrease with distance with a 1/e scale of 3 to 4 bases, implying that up to 8 bases could be sharing a significant amount of energy of the incident photon. The A·T duplexes were found to have an exciton distribution corresponding to somewhat greater delocalization of the energy and this was attributed to the larger electronic coupling between bases because of shorter base–base distances and/or a more rigid stack structure in duplexes as compared to single-stranded sequences. Although the present author has not been able find experimental confirmation in the literature, the same argument would lead to the supposition of even greater delocalization of the exciton in A-DNA which has a shorter distance between the bases, greater overlap of the bases, and a more rigid structure than B-DNA (Ussery, 2002)[385].

Buchvarov et al. (2007)[29] found a clear indication of a temporal decay of the electronic exciton structure, with a time constant of about 8 -10 ps (see figure 12.16), indicating that the exciton is dynamic and that changes in the delocalization length occur during the lifetime of the excited base stack. Given the length of the observed time scale, it was suggested that these changes must be "induced by nuclear motions within the base stack and/or in the immediate surroundings of the DNA". The alternating $(dAT)_9{\cdot}(dAT)_9$ duplex had lower values of exciton delocalization than their homoadenine counterpart and did not demonstrate temporal decay of exciton structure on the time frame of the experiments. The authors associate this with the greater possibility for charge transfer reactions in these heterogeneous macromolecular systems that would not participate in the exciton state wavefunctions.

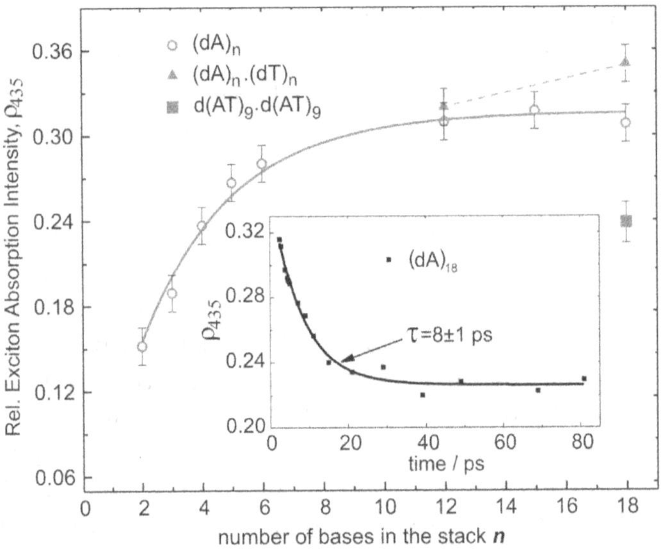

Fig. 12.16. Spectral intensity ratio between the exciton absorption (in a 30-nm interval, centered at 435 nm) and the monomer absorption at 330 nm ($\rho435$), 3 ps after excitation, as a function of the stack length n. For the (dA)n series, a single exponential fit (solid curve) was used to extract the "1/e delocalization length" d. The $\rho435$ values for the A·T duplexes are connected with a dotted line. (Inset) Time dependence of $\rho435$ for (dA)18. The decay of $\rho435$ is characterized by time constants of 8–10 ps in all DNA systems studied, except for (dAT)9·(dAT)9 where no decay of $\rho435$ was observed. Image credit: Buchvarov et al. (2007)[29]. Reproduced with permission by the National Academy of Sciences.

Charge transfer is thus another possible mechanism of energy dispersion and dissipation and this has been studied in some detail by numerous groups. The study of charge migration (either electrons or protons) in DNA is receiving a lot of recent attention in the fields of both biology and bioengineering since it has been suspected, for example, that electron or hole transfer in DNA may promote oxidative damage at particular sites known as a mutation hot-spots remote (up to 200 bp) from the location of initial energy deposition (Núñez et al., 1999)[252]. Another important biological reason for studying electron transfer is its role in the repair of ultraviolet-induced cyclobutane pyrimidine dimers in DNA in which an electron is transferred from FADH- to the dimer lesion, thereby playing an important role in photoreactivation of DNA (Tashiro et al., 2006)[373].

When the bases are stacked in multimers, either in single or double strand, for example in short oligos, in addition to the very fast picosecond decay components reminiscent of the monomers, there appears a much slower component which lasts on the order of 100's of picoseconds to nanoseconds (see figure 12.17). These long lived states in duplex DNA are now receiving a lot of attention but the microscopic details have yet to be sorted out. Some workers attribute them to the reduced ability to access the conical intersections due to a restriction on the out-of-plane movement when the bases are stacked. However, there is an increasing consensus that this slower (by at least 2 order of magnitude) component in DNA de-excitation is due to charge transfer, either electron or hole propagation through a hopping like mechanism (Kohler, 2010)[167]. For this slower component attributed to charge transfer, the photon-induced excitation energy is distributed and dissipated over a large region of DNA, of over 4.0 nm (Murphy et al., 1993)[240], corresponding to about 15 to 20 base pairs in B- or A-DNA respectively.

There is also the possibility of decay through multiple steps, for example, excitons that trap to long-lived charge transfer or exciplex states, or interstrand proton transfer which can occur since it is coupled to intrastrand electron transfer because the latter reduces the energetic barriers to proton transfer within base pairs (Kohler, 2010)[167]. There is still significant uncertainty regarding the dynamics of the long-time component of DNA de-excitation and other experiments are needed before the complete picture becomes clear. For example, little is now known about the dependence of a particular decay channel on the energy or intensity of the incident photon flux. There is also a need for more information on the sequence dependence of energy transfer, which is known to be important (Xu and Nordlund, 2000)[417], and on the differences between A and B form DNA and also RNA/DNA hybrid duplexes which normally take on the A form.

There is no doubt, however, of the existence of a long lived decay component in single strand and duplex DNA. Coorborating the evidence mentioned above are experimental observations of fluorescent emission from the DNA long lived states extending up to the nanosecond time scale. Even if the energy dispersion were limited to no more than 8 bases or so, at the high ocean surface temperatures existing during the Archean, this could be sufficient to fray the small oligos at one end or produce a sufficiently large bubble in mid regions and thereby promote the unraveling of the macromolecule to the denatured state.

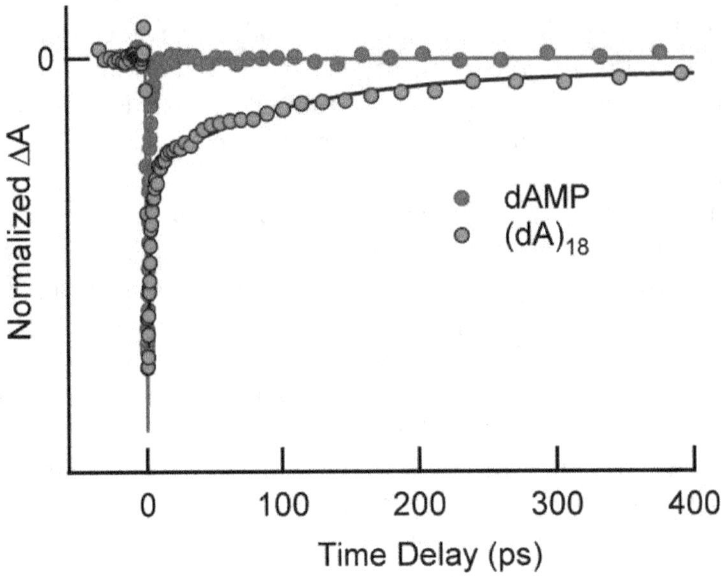

Fig. 12.17. By pumping the system with ultraviolet light at 267 nm from a femptosecond laser, and then probing the system with light at 250 and 570 nm, Kohler and coworkers have been able to determine the dynamics of the decay of the exicted state for different nucleobase structures. For single bases, the decay is ultrafast on the order of 2 pico seconds (blus dots), while for stacked bases such as occurs in double strand DNA, there appear an additional second component to the decay dynamics (red dots are for double strand 18-mer poly A DNA, $(dA)_{18}$). It is postulated that this second component is due to charge migration (electron or hole hopping) along the length of the DNA molecules. Image credit: Kohler (2010)[167]. Reproduced with permission by the American Chemical Society.

A microscopic description of UV-C induced denaturing is therefore at hand but understanding the details will require further work to delineate all the DNA excited state decay pathways and their dynamics. Molecular dynamic simulations coupled with time-dependent density functional theory is being employed to illuminate the decay pathways, but the size of the macromolecules present a real practical problem given the computational resources presently available. Another difficult problem is modeling the photon-induced initial excited state of the macromolecule (Marquetand et al., 2016)[200].

There is, in fact, prior experimental evidence of UV-C induced denaturing of DNA. Marmur and Doty (1959)[198] first noted a lowering of thermal denaturing temperature of DNA after exposure to even low intensities of UV-C light. As mentioned in section 12.2.3, (Hagen et al., 1965)[122] have observed how the absorption amplitude of initially double strand DNA on exposure to intense UV-C light begins to take on an identical absorption cross section as heat denatured single strand DNA and they attributed this to UV-C induced denaturing of DNA. Marmur and Grossman (1961)[199] described how double strand DNA exposed to UV-C light reacts preferentially with homologous antibodies as well as does heat denatured DNA, but not double strand DNA.They suggest that the implications of this data is that one of the effects of UV light is to cause breakage of hydrogen bonds exposing antigenically reactive groups. Some of the UV-C light induced denaturing has been attributed by these workers to the formation of cyclobutane pyrimidine dimer formation which may cause local denaturing at specific sites resulting in an observed difficulty of renaturing (Marmur and Grossman, 1961)[199]. All of these experiments were performed on long natural DNA sequences and at a specific wavelength of incident light (254 nm) corresponding to a mercury line.

In our experiments on short synthetic DNA, we designed the oligos to be free from having adjacent thymine bases, thereby reducing the possibility of denaturing due to cyclobutane lesion formation. Our data clearly show significant UV-C induced denaturing at intensities as low as 1/20 that expected on the Archean ocean surface (see section 12.2.2).

13. Information Accumulation in RNA and DNA

The non-equilibrium thermodynamic principle of material organization into structures and processes leading to an increase in dissipation provides novel insight into the meaning of genetic information and how information could have began to accumulate in RNA and DNA. Today, RNA and DNA are not considered as photon dissipating molecules, and they certainly cannot act as such since there is no longer any UV-C light at Earth's surface to dissipate[1], but rather are known as information containing molecules. However, it is apparent that the photon dissipation attributed to life today, across the near UV and visible, is still the predominant thermodynamic utility of the information stored within these molecules.

The information stored within the collective gene pool of all the organisms on Earth today defines the biosphere (including apparently abiotic and unrelated factors such as the transparency of the atmosphere and the size of the water cycle) and the principal thermodynamic function performed by the biosphere is that of the dissipation of the solar photon potential. Genetic information, instead of being associated with the entropy of the dissipative structure itself should, therefore, more correctly, be associated with the rate of entropy production derived from the dissipative structure which was produced with the aid of this information and which arises in response to the applied external photon potential. But, how could the relation between information and dissipation have arisen at the dawn of life?

The complex enzymes of today are the result of complex biosynthetic pathways, and therefore could not have existed at the beginnings of life. There was thus a need for a purely physical mechanism for enacting RNA and DNA reproduction, and this reproduction necessarily must have been related to dissipa-

[1] However, DNA of about 150 kbp is known to exist in the chloroplasts of plants (called plastid DNA), thought to be the remnants of the DNA of once free living cyanobacteria before the endosymbiotic event that led to the eukaryote cell. Most of the engulfed cyanobacteria DNA appears to have been transfred to the nuclear DNA of the eucaryote plant cell. However, the question arises as to why not all of the chloroplast DNA has been transferred to the nucleus. It may, in fact, be that plastid DNA was, or still is, acting as an aceptor molecule for dissipating rapidly into heat the electronic excitation energy of some particular extinct or even existent donor pigment. If this could be verified, it would indicate a clear vestige of DNA's original thermodynamic function and would add to the evidence in favor of the thermodynamic dissipation theory of the origin and evolution of life.

tion (see chapter 12). The stereo-chemical affinity of a particular base sequence of RNA or DNA other molecules to that could have aided in the dissipation-replication process must have been the first information selected for. Here we give two particular examples, the first being antenna UV-C absorbing molecules that could have increased the photon capture rate and therefore the probability of denaturing in ever colder sees, and the second being molecules which increase the hydrophobicity of RNA and DNA, causing them to adhere to the air-water interface thereby reducing substantially sedimentation and exposing them to the largest UV-C flux which is available at the surface.

Polymers of nucleotides which absorbed in the UV-C could have continued to replicate through the simple enzymeless mechanism described in the previous chapter which we have called Ultraviolet and Temperature Assisted Replication (UVTAR). Since the temperature of the Earth's surface continued to cool during the Archean, there would have come a time when the dissipation (metabolism) of one, or a few simultaneous, UV-C photons absorbed on RNA or DNA would not have been sufficient to cause denaturing. However, if by chance some segments of native DNA or RNA had chemical affinity to other UV-C absorbing chromophore molecules, such as, for example, the aromatic amino acids, or one of the other fundamental molecules which also absorbs in the UV-C (Michaelian and Simeonov, 2015)[217] (see figures 4.7 and 8.4, and table 4.1), then the DNA or RNA with affinity to these complexes could continue denaturing in the colder sea because of greater local heat deposited due to the greater cross section for UV-C absorption afforded by the complex. These particular RNA or DNA segments could thus be template copied during overnight dark, or colder, periods, thereby increasing their representation in the population of RNA or DNA strands. They would, in essence, be *thermodynamically selected* based on their dissipation capacity over segments without chemical affinity to the fundamental UV-C absorbing molecules.

It is a corroborating fact that out of the 22 amino acids used by life today, only those with significant UV-C absorption properties (the four aromatic amino acids, Tryptophan, Tyrosine, Phenylalanine, Histidine plus Alanine, Cystine, Methionine) have strong affinity to their DNA codons or anti-codons (Yarus et al., 2009)[418]. In this manner, information, which allowed for an increase in dissipation and the continued replication during periods of environmental change (in particular the cooling of the oceans), would become selected for and would become programmed through this thermodynamic selection mechanism into RNA and DNA (see Fig. 13.1).

The formation and proliferation of a codon for one of the UV-C absorbing amino acids in this manner is another example of microscopic dissipative structuring under the prevalent external UV-C flux. The persistence of the codon until today, even though there is no longer a UV-C photon flux at Earth's sur-

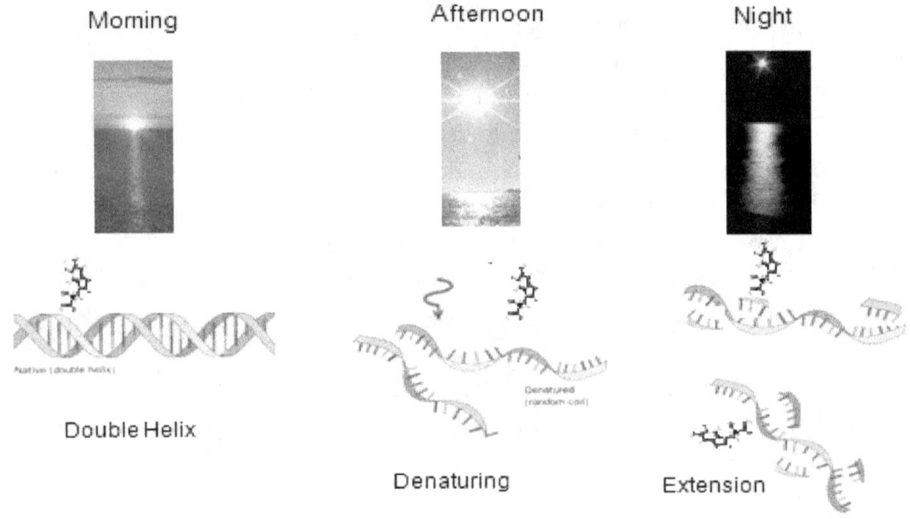

Fig. 13.1. RNA and DNA which stereochemically coded for particular aromatic amino acids, or other UV-C absorbing antenna molecules, could continue reproducing under conditions of a cooling ocean surface while non-coding segments would not be able to denature during daylight hours and so could not be template copied during cooler, overnight periods. This provides a relation between replication and dissipation which provides a mechanism for thermodynamic selection.

face, is an example of the persistence of nanoscale structuring even after the removal of the imposed generalized chemical potential due to atomic mobility issues (see chapter 3.4).

Not only do DNA and RNA have affinity to aromatic amino acids and other fundamental molecules that dissipate in the UV-C, but certain sequences of RNA and DNA are known to have affinity to some fundamental enzymes that appear to be intimately connected with the primitive energy metabolism and nucleotide production routes (Krypides and Ouzounis, 1995)[172]. Many of the cofactors of these enzymes are, in fact, themselves, UV-C dissipating molecules (Krypides and Ouzounis, 1995; Michaelian and Simeonov, 2015)[172, 217]. We thus have the beginnings of hierarchal structuring based on dissipation.

It was described in chapter 10 how certain planer aromatic molecules with an electropositive tail could have intercalated between base pairs with their tails occupying the major groove of RNA and DNA, thereby making these molecules more hydrophobic and thus causing them to adhere to the air-water interface. There is an interesting correlation between the ranking of amino acid hydrophobicity and their ranking of chemical affinity to their anticodon (Weber and Lacey, 1978; Lacey and Mullins, 1983)[394, 181] which supports our suggested mechanism for the accumulation of information – those segments of DNA being selected which could promote the UVTAR mechanism (in this case the hydrophobicity exposing DNA to greater intensity UV-C light at the ocean surface). Another possibility may have been the amino acids with hydrophobic tails which have been found to fit stereochemically into the pocket left by removal of the intermediate base of the anticodons (Hendry et al., 1981 and Erives, 2011)[128, 95]. This would have given RNA or DNA+molecule complex; i) a larger photon absorption cross section, ii) made them less susceptible to sedimentation , iii) exposed them to the maximum surface photon flux, iv) given them a lower denaturing temperature (by canceling the negative charges on the phosphate groups of the backbone), v) increasing the efficiency of overnight extension through the intercalations acting as a kind of glue (Horowitz et al., 2010)[133], and vi) reduced pyrimidine dimer formation known as cyclization (Horowitz et al., 2010)[133]. RNA or DNA sequences coding for these intercalating, or nucleotide substituting water expulsing molecules would have therefore been thermodynamically selected for.

Replication associated with dissipation and thus entropy production, for which there exists both a well defined thermodynamic sense and thermodynamic imperative, instead of replication associated with a mysterious "will to survive" and an ill defined "fitness" which is, in fact, inherently tautological, suggests how the Darwinian paradigm needs to be revised according to this physical-chemical basis of dissipation. By adopting the thermodynamic dissipation basis of selection in evolution, selection is seen to operate at the molecular level, and indeed at all levels up to the biosphere. Just as Boltzmann argued for contemporary life over one hundred and thirty years ago, it is not competition for energy or material, and it was not competition for a particular molecular reactant species as once suggested by Oparin and still commonly argued in the contemporary origin of life literature, but competition for entropy production through photon dissipation which drove, and still drives, evolution through natural selection, or better said, through *thermodynamic* selection.

14. Homochirality

One of the salient characteristics of the fundamental molecules of life which has resisted a plausible explanation thus far is their homochirality; that they are found in only one particular handedness (chirality) (see Fig. 14.1), right-handed RNA, DNA, and sugars, and left-handed amino acids, even though both chiralities have identical Gibb's free energies and thus identical probabilities of formation and degradation under near equilibrium conditions. "Right-handed" DNA are so named because they absorb preferentially more right-handed circularly polarized light over left-handed circularly polarized light (see Fig. 14.2) around their absorption maximum wavelength of 260 nm (see Fig. 12.3). The circular dichroism (CD) spectrum of a material is a plot of the difference in the molar extinction coefficients between left- and right-handed circularly polarized light as a function of wavelength. It is defined as,

$$\Delta\varepsilon(\lambda) = \varepsilon_L(\lambda) - \varepsilon_R(\lambda) \tag{14.1}$$

where $\varepsilon_L(\lambda)$ [M^{-1} cm^{-1}] is the molar extinction coefficient for left-handed circularly polarized light and $\varepsilon_R(\lambda)$ [M^{-1} cm^{-1}] is the same for right-handed circularly polarized light (see Fig. 14.3).

The homochirality is a fundamental characteristic of life today. Since it occurs at the level of the nucleotides, and since the template copying of a mixed chiral chain of nucleotides is fatally frustrated (Joyce et al., 1984)[142], any theory purporting to explain the origin of life must simultaneously explain the origin of homochirality.

The homochirality of life is usually attributed to some initial asymmetry in the amounts of primordial molecules of different chirality resulting from coincidental external factors, such as a region of gas in space exposed to the circularly polarized light emitted by only one hemisphere of a nearby pulsar, resulting in the selective photolysing of one chirality over the other. Some evidence for this comes in the small ($\sim 8\%$) differences from a racemic distribution of the α-methyl amino acids, the most stable amino acids, in both the Murray and Murchison meteorites (Pizzarello and Cronin, 2000)[281]. This small enantiomer excess (in favor of the L-chirality amino acids) led some to suggest that these space derived amino acids could have biased the homochirality towards the L-amino acids on Earth because only a small asymmetry in the initial chiralities

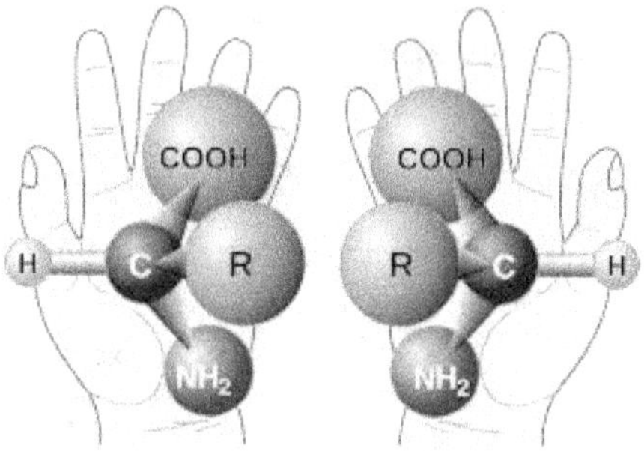

Fig. 14.1. The two distinct chiral versions (left- and right-handed) of a given amino acid with functional group R. The isomers have identicle Gibb's free energies so both have equal formation and degradation probability under near equilibrium conditions. However, left-handed amino acids, and right handed nucleic acids and ribose sugars, predominate in biology and this fact must be explained by any theory purporting to explain the origin of life. Image credit: NASA, Public Domain.

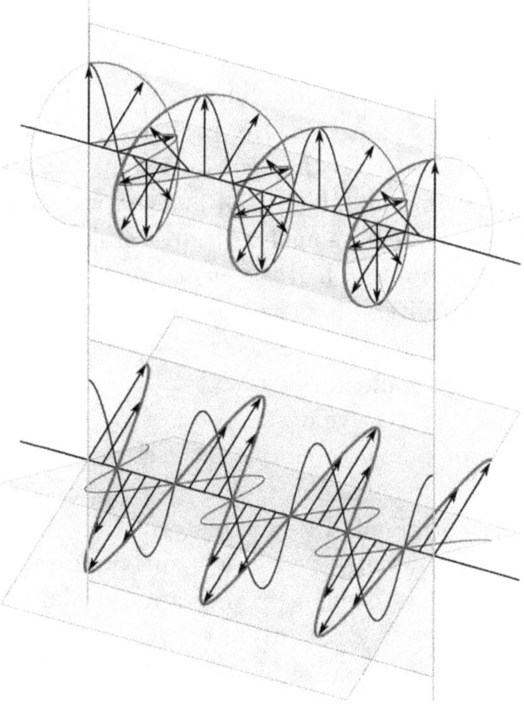

Fig. 14.2. Circularly polarized light. The wave vectors, corresponding to the electric and magnetic fields, rotate around the direction of propagation. If the observer is viewing the incident light and the wave vector is rotating clockwise (like in this figure), the light is called "right-handed circularly polarized". Image credit: Dave3457, Public Domain.

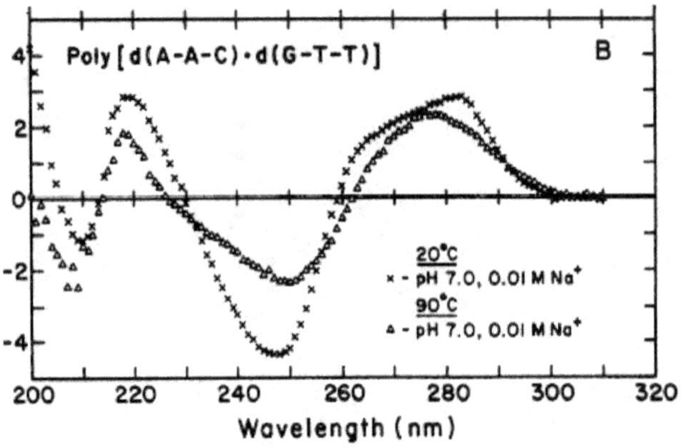

Fig. 14.3. Circular dichroism (CD) spectra of synthetic poly [d(A-A-C) . d(G-T-T)] DNA at neutral pH; native polymer (crosses) and heat-denatured polymer (triangles). DNA absorbs strongly from about 230 to 280 nm (see figure 12.9). The integrated CD spectrum over this range is negative, indicating that DNA absorbs more right-handed circularly polarized light than lef-handed. Image credit: Gray et al. (1978) [115].

would be needed if there existed an amplification mechanism such as a chirality based autocatalytic chemical reaction. However, no detailed description of such an autocatalytic chemical reaction has yet been put forward. Another suggestion, first proposed by Ulbricht (1959)[384] has the very small parity violation seen in the weak force as being somehow responsible for the initial asymmetry in the chirality of the material. However, a comparison of the weak energy difference of the two chiralities to thermal energy at the Earth's surface gives $\Delta E/k_B T \simeq 10^{-17}$ (Cline, 2005)[52], much too small to be a plausible solution in itself to the problem of homochirality.

Explanations based on selective destruction of a given chirality, such as that given above, are also not plausible since they require exotic conditions such as the shading of a particular hemisphere of a nearby pulsar and require almost 100% destruction of the originally racemic material in order to obtain even nominal amounts of chiral asymmetry in the few surviving molecules (Michaelian, 2010b)[209]. There has thus been no satisfactory explanation of homochirality, certainly none in which the resolution of the problem fits neatly within a proposed mechanism for the origin of life itself.

The thermodynamic dissipation theory of the origin and evolution of life argues that homochirality developed from an asymmetry, not in the initial chiral concentrations of the fundamental molecules of life, but rather in the dynamics of replication due to an asymmetry in the external conditions imposed on incipient life, particularly in the submarine light environment at the sea surface which, according to this theory, life arose to dissipate. This theory suggests that homochirality could have gradually been incorporated into the fundamental molecules of life through the repeated application of an asymmetric UVTAR reproduction mechanism (see chapter 12) driven by thermodynamic dissipation.

The mechanism I am suggesting can be described as follows; due to the large heat capacity of water compared to that of air, the warmest water temperatures of the ocean surface occur in the late afternoon around 15:00 (see figure 14.4) and therefore this time period would have been most suitable for denaturing DNA or RNA through the UVTAR mechanism described in chapter 12.

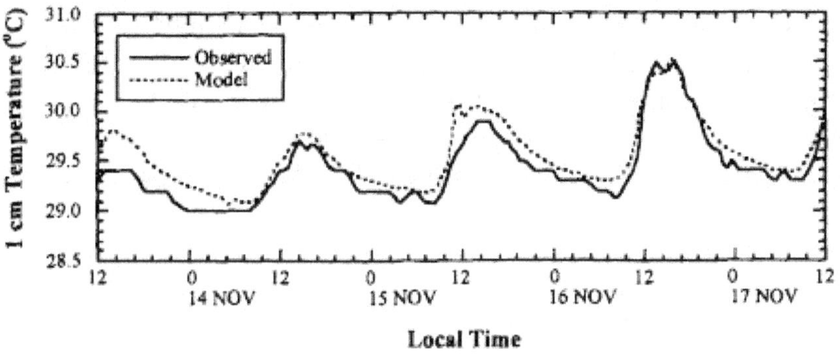

Fig. 14.4. Diurnal temperature cycling of the western tropical ocean surface at a depth of 1 cm as a function of local time. Note that maximum temperatures occur in the afternoon at about 15:00. Note that the temperature variations at the surface microlayer would be much greater, perhaps as large as 5 °C (see figure 10.4). Most of the surface heating is due to the infrared light from the sun. This diurnal asymmetry would be expected to occur similarly on the Archean ocean surface and would introduce an asymmetry into the UVTAR mechanism for replication (see chapter 12). Image credit: Webster et al. (1996)[395]. Copyright 1996, American Meteorlogical Society.

It turns out that light scattering on water molecules (which linearly polarizes the light) being followed by total internal reflection of the light at the water-air interface, for a particular viewing direction of scattering with respect to the incident photon beam, and very close to the ocean surface, produces a small but important excess of right-handed circularly polarized light at the sea surface in the late afternoon, whereas there is a similarly slight excess of left-handed circularly polarized light in the early morning (Wolstencroft, 2004)[409] (see Fig. 14.5). This effect becomes more prominent the shorter the wavelength of light and the more confined the observer is to the ocean surface (Wolstencroft, 2004)[409].

Segments of RNA and DNA can come in two chiral versions, those that absorb preferentially right-handed circularly polarized light and those that absorb preferentially left-handed circularly polarized light at 260 nm where absorption is strongest. It is therefore apparent that the UVTAR mechanism would give those right-handed RNA or DNA oligos, that absorb preferentially right-handed circularly polarized light, a slight overall advantage of denaturing during the day because of the higher water temperature in the afternoon when there was an important excess of right-handed circularly polarized light, and thus these would have had slightly higher probability of becoming templates for extension during over night cooler periods.

Given the specific RNA and DNA circularly polarized photon absorption cross sections for the two molecular chiralities, and a quantitative measure of the asymmetry in the circular polarization of sunlight at the sea surface in the afternoon today (about 5%), I have shown through simulations that such a mechanism would be sufficient to produce practically 100% homochirality of right-handed RNA and DNA within only 400 Archean years (see figure 14.6). The details of the simulation have been given in Michaelian (2010b)[209].

A possible reason as to why the amino-acids are left-handed has also been given in Michaelian (2010b)[209]. It turns out that left-handed amino acids have greater chemical affinity to right-handed RNA and DNA (a "stereochemical" effect) and together, as a complex, they present greater absorption cross section over the region 230 to 290 nm and also together present an even greater molar extinction coefficient for right-handed circularly polarized light over left-handed circularly polarized light (a greater negative circular dichroism with respect to

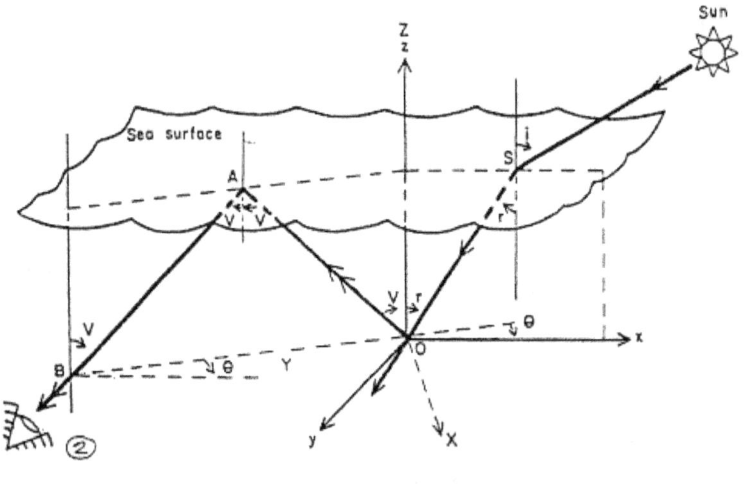

Fig. 14.5. Circularly polarized light is produced through light scattering on water molecules (at O), which linearly polarizes the light, and then being totaly internally reflected at the water-air surface (at A). The amount of circular polarization increases towards the surface (greater angle V) and for shorter wavelength light. For a given viewing direction (angle θ)it is left-handed in the morning and righ-handed in the afternoon, or vice versa. Image credit: Wolstencroft (2004)[409].

DNA alone) around this region of the spectrum where together they absorb strongly (Arcaya et al., 1971)[5] and where dissipation is thus greatest for the complex. The circular dichroism spectra of L-tryptophan with the nucleosides is given in figure 14.7. Arcaya et al. (1971)[5] report that "DNA-tryptophan displays the same average pattern found for the mixture of tryptophan with nucleosides, as if a similar interaction occur when DNA or nucleosides interacts with tryptophan", suggesting that this supports the idea "that the interaction of the amino acid with the DNA could occur through an intercalation of the amino acid between the bases of the macromolecule without changing greatly its tertiary structure.".

Thus, for the same reason of a greater denaturing probability due to larger right-handed circularly polarized photon absorption cross section and ensuing dissipation, and therefore higher reproduction probability of the RNA- or DNA+L-amino acid complexes under the UVTAR mechanism, right-handed RNA or DNA with stereo-chemical affinity to left-handed amino acids would have been thermodynamically selected over left-handed RNA or DNA with stereo-chemical affinity to right-handed amino acids.

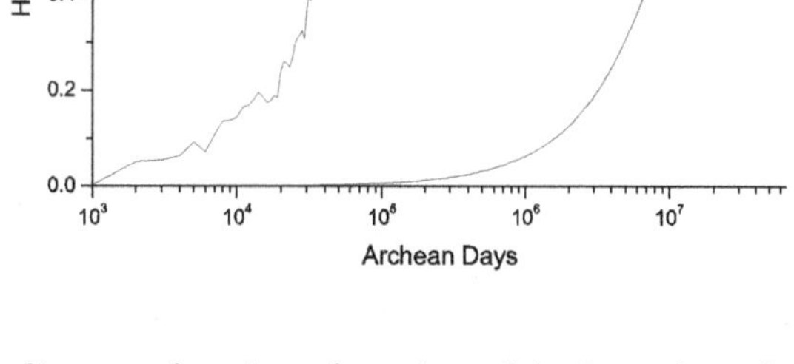

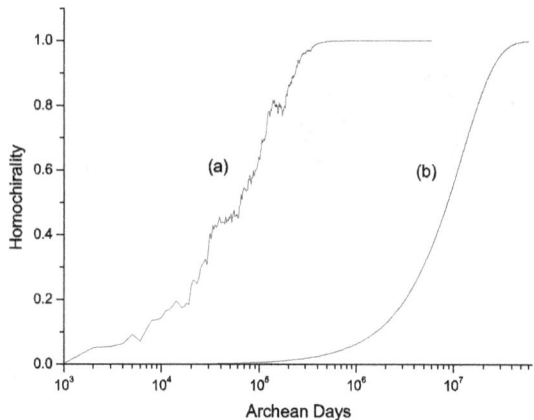

Fig. 14.6. Homochirality as a function of number of Archean days simulated using an asymmetric UVTAR mechanism due to right-handed circularly polarized light being more prevalent at the ocean surface in the afternoon when water temperatures are higher. (a) Including an energy threshold for denaturation related to the complementary strand binding energy, (b) assuming denaturation probability is simply proportional to the number of photons absorbed (see Michaelian (2010b)[209] for details).

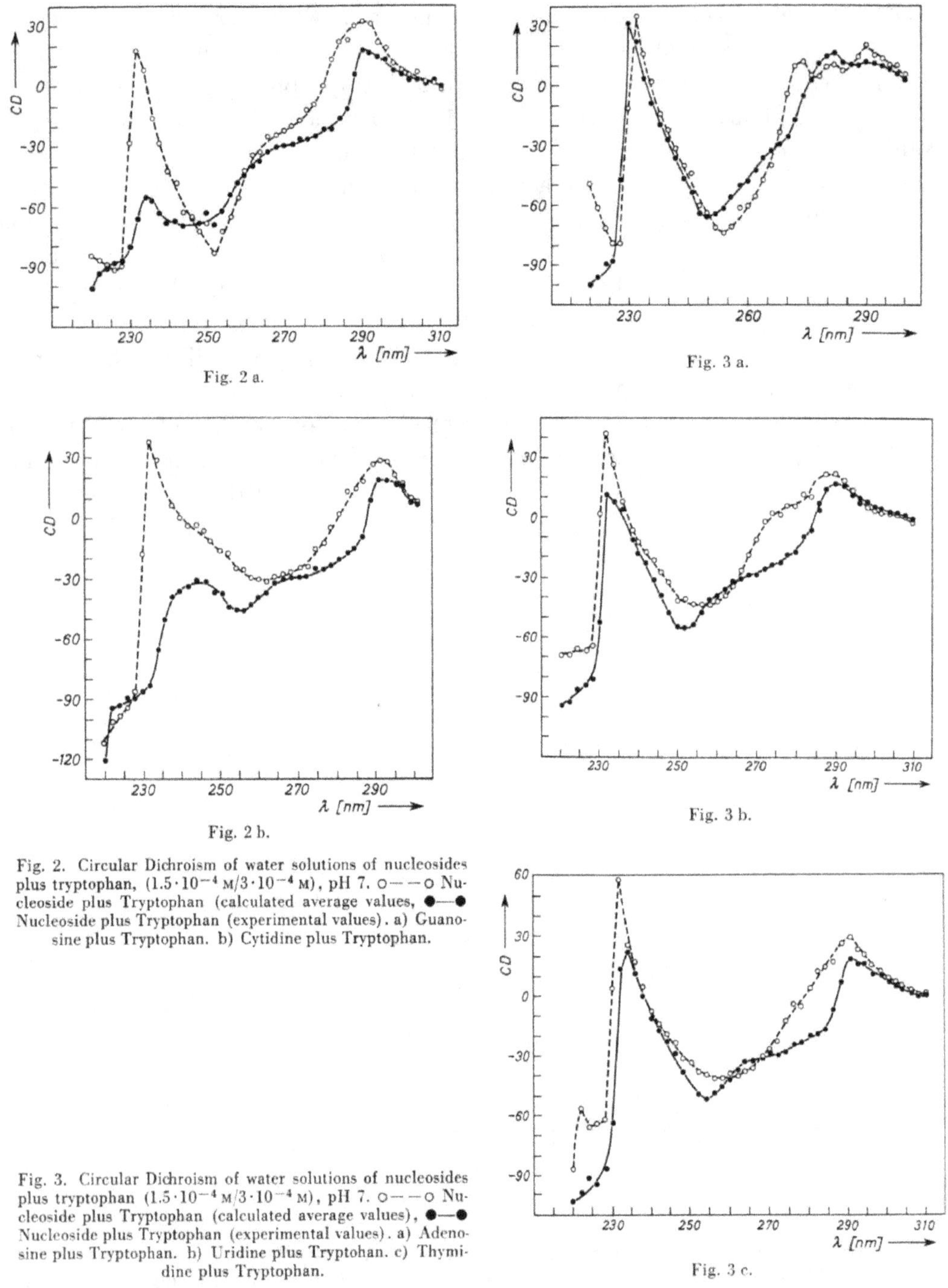
CD
30
0
-30
-60
-90
230
250
270
290
310
λ [nm]
Fig. 2 a.
Fig. 3 a.
-120
Fig. 2 b.
Fig. 3 b.
60
Fig. 2. Circular Dichroism of water solutions of nucleosides plus tryptophan, (1.5·10⁻⁴ M/3·10⁻⁴ M), pH 7. o—o Nucleoside plus Tryptophan (calculated average values, ●—● Nucleoside plus Tryptophan (experimental values). a) Guanosine plus Tryptophan. b) Cytidine plus Tryptophan.
Fig. 3. Circular Dichroism of water solutions of nucleosides plus tryptophan (1.5·10⁻⁴ M/3·10⁻⁴ M), pH 7. o—o Nucleoside plus Tryptophan (calculated average values), ●—● Nucleoside plus Tryptophan (experimental values). a) Adenosine plus Tryptophan. b) Uridine plus Tryptohan. c) Thymidine plus Tryptophan.
Fig. 3 c.

Fig. 14.7. Circular dichroism spectra for L-tryptophan with the different nucleosides. In all cases there is negative circular dichroism (the complex absorbs more strongly right-handed circularly polarized light compared to left-handed) where there is strong absorption (~230 to 290 nm). Arcaya et al. (1971)[5] find the same average pattern (i.e. strong negative CD spectra) for DNA with L-tryptophan. With a UVTAR mechanism operating (see chapter 12), this would provide a thermodynamic selection for right-handed DNA with left-handed Tryptophan. Image credit: Arcaya et al. (1971)[5]. (CC BY-NC-ND 3.0)

Part IV

Evolution is Driven by Dissipation

15. The First Protocell

The ocean surface temperature was descending steadily throughout the Archean as the greenhouse gas CO_2 was being consumed in the formation of silicon carbonates as a result of weathering of the newly forming continents. Schwartz and Chang (2002)[328] estimate that the surface temperature of the Earth descended below 100 °C about 4.4 billion years ago. Giant impacts, extending into the "late lunar bombardment era" of ca. 3.9 Ga, may have periodically reset ocean temperatures to above the boiling point (Zahnle, 2007)[426]. There is geochemical evidence in the form of $^{18}O/^{16}O$ ratios found in cherts of the Barberton greenstone belt of South Africa indicating that the Earth's surface temperature was around 80 °C at 3.8 Ga (Knauth, 1992)[163] (Knauth and Lowe, 2003)[165], falling to 70±15 °C during the 3.5–3.2 Ga era (Lowe and Tice, 2004)[193].

As the temperature of the ocean surface descended further, another adaptation that would have allowed continued RNA and DNA denaturing during the day could have been a UV-C transparent, but infrared opaque, enclosure surrounding RNA or DNA that would have allowed high energy photons to enter, be absorbed on RNA or DNA, and then dissipated into heat which could not easily escape the enclosure. It is known that phospholipids today form a loosely connected but surprisingly stable layer on the sea surface (see chapter 10) and that they can self-assemble into a number of distinct forms, the most common of which are given in figure 15.1. Lipds are also extremely diverse, with distinct molecular forms of lipids suited to almost all pH and temperature environments existent on Earth (Nes and Nes, 1980)[246].

DNA, RNA and phospholipids are *fundamental* molecules found in all three domains of life and phospholipids are readily produced in Miller-type experiments from precursor molecules such as HCN and CO_2 interacting with UV-C light or spark discharge (Miller, 1955)[223]. Experiments have shown that DNA, in the presence of divalent cations, has chemical affinity to neutral phospholipids and that the combined DNA-lipid system has an increased UV-C absorption of about 15% at 260 nm (Ranjnohova et al., 2010)[296]. The lipids have known low heat conduction properties, being the molecules of the insulating tissue of most organisms (see figure ??). Trapping heat from photon dissipation within the lipid bilayer enclosure for some time would have allowed DNA and RNA to

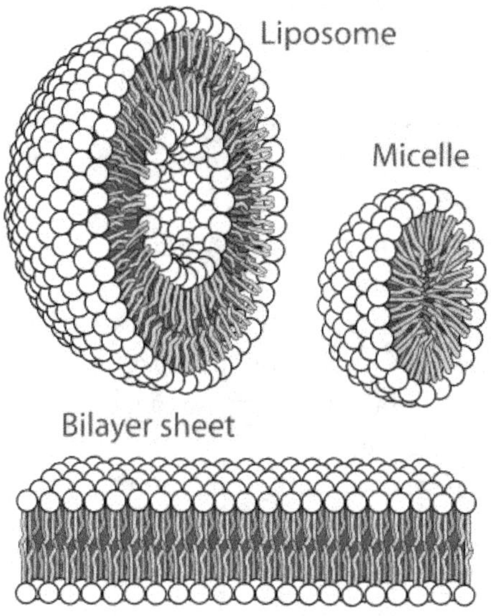

Fig. 15.1. Self-organization of common phospholipid geometries: a spherical liposome, a micelle, and a lipid bilayer sheet. The white spheres represent the hydrophilic ends of the molecule which like to stay in contact with water. The tails are hydrophobic and avoid contact with water. These hydrophilic and hyrdophobic ends of the molecule give it its unique properties of self-assembly into the given geometries. Image credit: Mariana Ruiz Villarreal, LadyofHats, Public Domain.

denature and remain in their higher photon absorbing and dissipating, single strand, mode (see figure 7.3) during the day, even though sea surface temperatures descended below their denaturing temperatures. This would have been particularly important for DNA since it has a significantly higher denaturing temperature than RNA under similar conditions.

Today, the focusing of incident light through the phospholipid layer onto the inner surface of the cell membrane of bacteria has been shown to be operative in the phototaxis of bacteria (Schuergers et al., 2016)[326]. It is therefore reasonable to assume that such lensing was operative in protobacteria during the Archean with UV light being focused onto RNA or DNA (or other pigments) adsorbed onto the inner surface of the phospholipid bilayer membrane (see figure 15.3). A spherical enclosure of this kind would serve to increase the capture rate of photons and to direct the absorbed energy onto RNA or DNA, or onto other chromophores employing RNA and DNA as acceptor quencher molecules for the dissipation of the photon energy into heat through their conical intersections.

In order to facilitate reproduction within such an enclosure, some requirements on the nature of the enclosing lipid wall would have to be met. These are;

1. transparency to, or absorption of (Zabelinskii et al., 2005)[425], short wavelength UV-C light (230-300 nm) but some reflectivity to long wavelength infrared light, thereby raising the internal temperature of the vesicle,
2. permeable to short segments of RNA or DNA required for extension, but impermeable to the long segments being reproduced within the enclosure,
3. permeable to the aromatic amino acids and other UV-C absorbing fundamental molecules of life,
4. flexible enough to allow a binary fission of the enclosure, that would have to accompany any replication of the internal RNA or DNA,
5. stable at the high surface temperatures existing during the Archean.

An enclosure having all such characteristics are the vesicles formed from lipid bilayers, such as the phospholipid monolayers or bilayers (see figure 15.2), which, in fact, today form the cell walls of all organisms. Although at low temperatures the hydrophilic heads of the lipid bilayer prevent polar (hydrophilic) molecules, such as some amino acids (serine, threonine, asparagine, glutamine, histidine and tyrosine), nucleic acids, and ions, from diffusing across the membrane, this is not the case at high temperatures or for the hydrophobic amino acids (alanine, valine, leucine, isoleucine, proline, phenylalanine, tryptophan, cysteine and methionine).

The enclosure would also have allowed specific conditions of temperature, pH, and ionic concentration to be maintained within the phospholipid liposome, independently of the values in the external environment (Nes and Nes, 1980)[246] and thus would have permitted the RNA or DNA dissipating molecules within their liposome enclosure to conquer new inhospitable environments and so increase their dispersion and proliferation over the entire surface of the Earth exposed to sunlight.

Experiments and theoretical calculations (Antipina and Gurtovenko, 2015)[4] have shown the propensity of DNA to adsorb on phospholipid bilayers when mediated by Ca^{2+} ions (see Fig. 15.3). Other experiments have shown how DNA-amino acid complexes (for example, polylysine) can be efficiently incorporated into micelles, the negatively charged DNA complex and positively charged micelle form a stable system (Zintchenko, 2005)[430].

Lipid vesicles could also have provided a more confined environment for promoting resonant energy transfer between UV-C absorbing donor chromophores and RNA and DNA acceptor quencher molecules. For resonant energy transfer, donor and acceptor molecules do not have to be in physical contact (i.e. demonstrate chemical affinity), excitation energy can be passed efficiently to the acceptor dissipater (in this case DNA or RNA) through fluorescent resonant energy transfer if the molecules are separated by less than 10 nm (see figures 8.2 and 8.3).

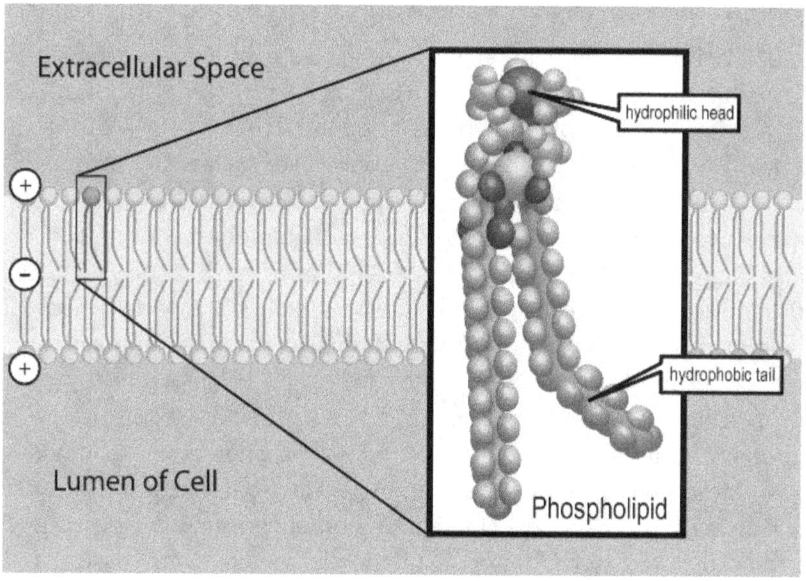

Fig. 15.2. Phospholipid bilayers form the cell walls of all organisms. The inner layer is hydrophobic while the outer layers are hydrophilic. Phosphoipids have ideal characteristics for providing a conducive environment for the UVTAR mechanism which ties replication to dissipation. Image credit: TvanBrussel, Public domain.

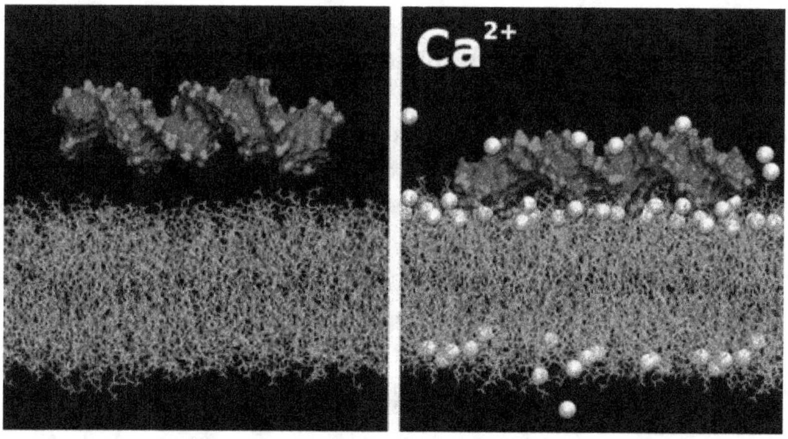

Fig. 15.3. Interaction of DNA with phospholipids. Divalent cations can promote formation of stable DNA–phospholipid complexes. This figure is the result of computer simulations to gain atomistically resolved insight into the kinetics of calcium-induced adsorption of DNA on zwitterionic phosphatidylcholine membrane. This is an important long-standing problem in the field of liposome-based gene delivery. Findings show that calcium ions play a dual role in DNA–phospholipid systems. First, binding of divalent cations to the lipid–water interface turns the surface of the zwitterionic membrane positively charged, promoting thereby the initial electrostatic attraction of a polyanionic DNA molecule. Second, it was shown that calcium ions are crucial for stabilizing the DNA–lipid membrane complex as they bridge together phosphate groups of DNA and lipid molecules. Image credit: Antipina and Gurtovenko (2015)[4].

It is known that vesicle fission can be induced by heating (Hanczyc and Szostak, 2004)[123] and in our suggested scenario of non-enzymatic protocell division, vesicle division may have been induced by internal heating as a result of UV-C photon absorption and dissipation on RNA or DNA and the other fundamental UV-C absorbing molecules contained within the protocell. Such a division would have to occur at internal temperatures higher than the denaturing temperature of RNA or DNA so that each daughter vesicle could contain a separate single strand DNA or RNA. This high temperature requirement for vesicle fission does not present a practical problem since today many forms of lipid-bilayers are known with sufficiently high transition temperatures[1] (Silvius, 1982)[342], for example, those forming the cell walls of the Archea, which are the most common extremophiles and thought to be the closest relatives to the last universal common ancestor[2] (LUCA). In fact, it is known that today's microorganisms subjected to thermal stress can alter the lipid composition of their cell membrane to reduce this stress (known as homeoviscous adaptation). The permeability and propensity to fission of the membrane can thus be adjusted within today's organisms in response to the external environment (Gennis, 1989)[103]. At the beginnings of life in the Archean, these properties would have been thermodynamically selected to give the greatest photon dissipation rates.

It is commonly believed that among the oldest group of cellular organisms known today are the archea (see Fig. 15.4). These have cell membranes of phospholipid bilayers quite unlike either that of eukaryote or bacteria. The archea have membranes composed of glycerol-ether lipids while the eukaryote and bacteria have membranes composed of glycerol-ester lipids (de Rosa, 1986)[76] (see Fig. 15.5). The ether bonds are more chemically stable than the ester bonds which helps the archea to survive as extremophiles at high temperature or in very acid or alkaline environments. Since there is much evidence that the Earth surface was hot at the origin of life, and that the oceans were somewhat acidic ($5 < pH < 7$) due to a high concentration of CO_2 in the atmosphere, and since the UVTAR mechanism requires a high temperature for replication of the RNA and DNA chromophores, it is thus probable that the first cellular membranes were of the glycerol-ether lipid composition and this is consistent with such isoprene lipids being detected in rock in the Isua district of Greenland which have been dated at 3.8 Ga (Jürgen and Huag, 1986)[145], very close to the beginnings of life.

[1] The lipid transition temperature is that temperature at which the integrity of the lipid walls are compromised.

[2] The Last Universal Common Ancestor (LUCA) is that organism that gave rise to the three domains of life, Archea, Bacteria, Eukaryote, existing about 3.5 - 3.8 billion years ago before branching off into the three domains. The Eukaryote are thought to have formed through an endosymbiosis of different bacteria (see figure 19.3).

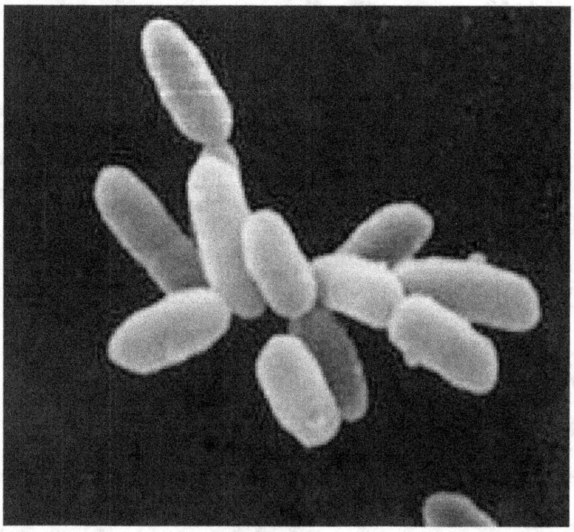

Fig. 15.4. Halobacteria, from the domain Archea. Their cell membranes are made of glycerol-ether lipid bi-layers (see Fig. 15.5). These cells are 5 μm long. Phospholipid bilayers may have formed around RNA and DNA to increase the UV-C photon capture and dissipation rate. The bilayer would also have increased the probability for denaturing during the day in cooler ocean surface environments and provided an optimal physical environment for enzymeless replication which allowed the proliferation of the photon dissipating molecules into new, inhospitable, environments. Image credit: NASA, Public Domain.

It has been shown experimentally that chromophore groups consisting of polyunsaturated acids of phospholipids participate in the absorption of photons in the range of 260–280 nm, which leads to excitation of valence electrons of multiple (double) bonds, and that the energy of such electrons can be used in interactions with other molecules, in particular, for energy transfer inside the membrane (Zabelinski et al., 2005)[425]. It is also relevant that lipid bilayers are today structurally and functionally associated with the energy conversion units of the cell. For example, in the photosynthetic system of plants and bacteria, the lipid bilayers provide the required gradient for the proton potential which drives the production of ATP (Kritsky et al., 2013)[171].

Thermodynamic selection in securing continued replication and proliferation of photon dissipating complexes in ever colder seas or in new environments could thus have provided the stimulus for the incorporation into early life of the lipid based cell wall, providing a selective, semi-permeable, boundary between replicating pigments (the fundamental molecules of life, see table 4.7) and their external environment when changes in the external environment proved too antagonistic for replication of free floating RNA and DNA through the UVTAR mechanism.

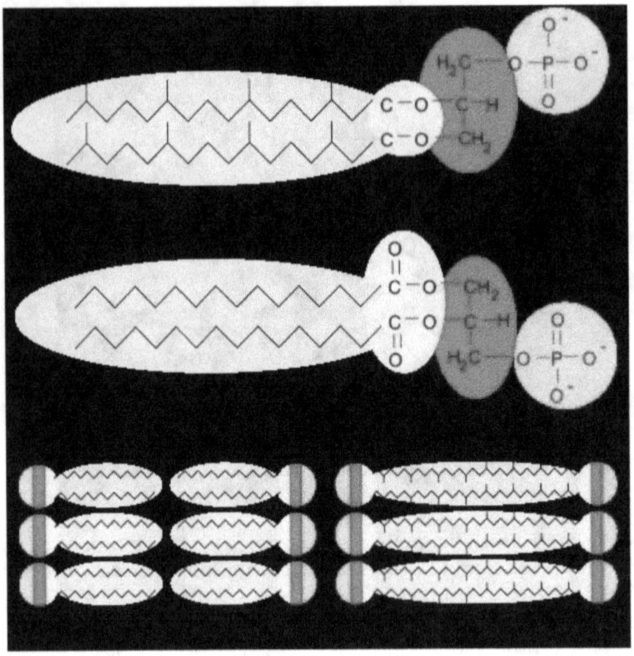

Fig. 15.5. Phospholipid membrane structures. Top, an archaeal phospholipid: in white, isoprene chains; yellow, ether linkages; red, L-glycerol moiety; green, phosphate group. Middle, a bacterial or eukaryotic phospholipid: white, fatty acid chains; yellow, ester linkages; red, D-glycerol moiety; green, phosphate group. Bottom left: lipid bilayer of bacteria and eukaryotes; bottom right: lipid monolayer of some archaea. Image credit: Franciscosp2, Public Domain.

16. Evolution of Photon Dissipation Efficacy

It is often incorrectly assumed that the overwhelming interaction of photons with organic material has to do with photosynthesis and that it is photosynthesis which Nature optimizes under varying environmental conditions and over evolutionary time. However, today, after more than 3.8 Ga years of biological evolution, less than 0.1% of the free energy in the solar photons captured by the plants or cyanobacteria is channeled into the fixation of carbon in the process known as photosynthesis (Gates, 1980)[100]. It therefore makes little sense to base biological arguments on the supposition that Nature optimizes the photosynthetic process, given the very poor energy conversion efficiency of this process today. For example, humans, in only 25 years, have obtained close to 40% efficiencies in converting the free energy in sunlight into usable electrical energy.

Through Earth's biosphere, most of the free energy available in the solar photon flux arriving at Earth's surface is simply dissipated directly into heat by an enormous quantity and variety of organic pigments, of which chlorophyll is only one. For example, the carotenoids are known acceptor quencher molecules for the rapid dissipation (within 200 ps) of excited chlorophyll donor molecules, the two pigments being held close together in high-light inducible proteins (Hilps) which are the cyanobacterial predecessors of the light harvesting complexes (LHC) found in plants (Staleva, 2015)[356]. Even with such a formidable array of evolved pigment complexes, the wavelength integrated flux of photons from the Sun arriving at Earth's surface is so copious ($\sim 10^{22}$ $m^{-2}s^{-1}$) that only approximately 26% of the solar photons arriving at Earth's surface are absorbed and dissipated by today's evolved plants, diatoms and cyanobacteria (Michaelian, 2015)[218]. Well over one half of the solar photons incident on the leaves, principally those with wavelength greater than the red-edge ($\sim$700 nm), are simply reflected with little dissipation[1] by contemporary organic material.

The thermodynamically most important biotic-abiotic coevolutionary process has been that of increasing the photon dissipation efficiency of Earth in its solar

[1] Reflection of a directed incident beam of photons also amounts to dissipation (entropy production) due to the fact that the energy after diffuse reflection is distributed over a larger solid angle of space. See Michaelian (2012b)[213] for the calculation of this component of the dissipation for the Earth and neighboring planets.

environment and this evolution has continued ever since the formation of Earth in the Hadean until the present era. In the opinion of the author, there is no better or succinct description of biological evolution than this observation. Over the short wavelength UV and visible region, evolution of ever greater photon dissipating efficacy of the organic pigments has incurred all of the following strategies: Evolving,

1. a greater photon absorption cross section per pigment size, for example by increasing molecular electric dipole moments,
2. faster dissipation of the photon-induced electronic excitation energy through the evolution of conical intersections so that the molecular complex promptly returns to the ground state, ready to process another photon,
3. quenching of fluorescent and phosphorescent radiative decay channels, through, for example, Förster resonant energy transfer,
4. quenching of electron transfer reactions, except where required for specific metabolic processes, e.g. photosynthesis,
5. stability against photochemical reactions and electron transfer reactions which could destroy the molecule,
6. new pigments absorbing at different wavelengths, covering ever more completely the solar spectrum,
7. mechanisms for promoting the dispersal of pigments over ever more of Earth's sunlit surface and into new inhospitable environments, through, for example, the evolution of cell membranes, mechanisms such as viruses for horizontal gene transfer, and insects and other animals for dispersal of pigments and nutrients,
8. greater coupling to other biotic and abiotic irreversible processes, for example the water cycle, winds and ocean currents,
9. transparency of the gases of Earth's atmosphere so that the most intense high energy part of the Sun's spectrum can arrive at the surface where photons can be best intercepted by organic pigments in liquid water and thereby be dissipated most efficiently.

In the following sections, each of the above evolutionary strategies towards fomenting photon dissipation will be discussed in detail.

16.1 Molecular Photon Absorption Cross Section

Organic molecules can be ionized, and therefore photochemically destroyed at photon wavelengths of less than approximately 200 nm (> 6.2 eV). There is also a problem with photochemical stability for chromophores that absorb at the other end of the visible spectrum at > 700 nm where the low-lying singlet or triplet electronic excited states increase the reactivity of the molecule with solvent impurities, such as dissolved oxygen (Sauer et al., 2011)[320]. Therefore, the absorption wavelength region where organic molecules could retain greatest chemical stability is between 200 and 700 nm, and this is basically, in fact, just where most organic pigments are found to absorb today. It also happens to be the region where water does not absorb strongly.

Chromophores with conjugated bonds (with alternating single and double covalent bonds) form planer structures with the electrons delocalized over the framework of the "conjugated" system. The atoms are linked by either σ-, or π-bonds. Those π-electrons have a node in the plane of the molecule and form a charge cloud above and below this plane along the conjugated chain.

The probability of the absorption of a photon in the UV or visible on a particular molecule depends not only on the energy differences of the electronic molecular orbitals of the molecule being similar to that of the photon energy ($\Delta E = h\nu$) and the size of the transition dipole moment $\boldsymbol{\mu} = \sum_n e_n \mathbf{r}_n$ (where e_n and $\mathbf{r}_n$ are the charge and position vector of the nth atom of the molecule), but also on selection rules that can be either spin or symmetry dependent. These selection rules derive from the time-dependent quantum mechanical formalism.

For example, the spin selection rule is that the spin multiplicity[2] should not change in a transition. The spin multiplicity for a quantum system is given by $(2S + 1)$ where S is the spin quantum number of the system. For example, for a molecule with all electrons (individual electrons have spin 1/2) paired the net spin S is zero and the spin multiplicity is therefore 1 and this state is called a singlet state. If there are two unpaired electrons, then the spin quantum number is one, $S = 1$, and the spin multiplicity is 3, and the state is referred to as a triplet state. Transitions from a singlet to a triplet state, or vice-versa, are strongly forbidden. Allowed transitions are therefore those in which $\Delta S = 0$, i.e. there is no change in the spin quantum number.

The required symmetry of the wavefunctions also forbids certain transitions. For example, a quantitative measure of the intensity of a transition is given by the transition dipole moment,

$$\mu_{ij} = \int \psi_i^* \mu \psi_j d\tau, \tag{16.1}$$

where μ is the dipole moment vector connecting the two molecular states ψ_i and ψ_j, and the integration is over all coordinates, with $d\tau$ being the volume element. Since μ is an odd function of the coordinates (see above paragraph), ψ_i and ψ_j must be either odd and even, or even and odd, respectively, to obtain an even function for the intensity of the transition. Therefore, a $1s$ to a $2s$ transition

[2] The spin multiplicity represents the number of ways the spin vector (internal angular momentum) for the system can be aligned in an external magnetic field. Quantum mechanics gives it as $(2S + 1)$ where S is the total spin of the system.

is not allowed but a $1s$ to a $2p$ transition is allowed. Here, s and p refer to the orbital angular momentum quantum states. The symmetry selection rule is thus $\Delta l = \pm 1$, where l is the orbital angular momentum quantum number.

As equation (16.1) makes clear, the greater the dipole moment vector connecting the ground state and the excited state (called the transition dipole moment), the greater the transition intensity or the photon absorption cross-section (for photons of the required energy of the difference between the ground and excited states) respecting, of course, the spin and symmetry selection rules. It is an interesting fact that most of the fundamental molecules of life (those found in all three domains of life), including the nucleic acid bases and the aromatic amino acids, have large transition dipole moments and the energy difference between the ground state and first electronic excited state corresponds to energies in the UV-C to UV-B region. These characteristics make them excellent absorbers of photons in this wavelength region, just that region for which there existed a window in the early Archean atmosphere (Sagan, 1973)[309]. The dissipative microscopic structuring of the fundamental molecules leading them to have large dipole moments and electronic excitation energy levels corresponding to the UV-C is not fortuitous, but instead explained by thermodynamic principles which govern the structuring such that the same solar photon potential that produced them can be efficiently dissipated (see chapter 6). The first step in the dissipation process is the absorption of a photon, so large photon absorption cross-sections are important to efficient dissipation.

Some evidence indicating that electric dipole moments of the fundamental molecules (pigments) have been thermodynamically selected to be large[3] is that electric dipole moments of pigment molecules are much larger than those of similar sized non-biological molecules.

16.2 Excited State Lifetimes and Photochemical Stability

A photon dissipation event begins with the absorption of a photon on a DNA or RNA nucleobase (A, T, G, C, U) through the allowed $^1(\pi \rightarrow \pi^*)$ singlet state transition, calculated using time dependent density functional theory to be at

[3] Note that not all molecular structures with large electric dipole moments are optimal for dissipation since the molecule must also de-excite rapidly through non-radiative channels, dissipating the electronic excitation energy into heat, for example, through a conical intersection. This requires the absence of energy barriers along the reaction path from the excited state. It has been shown through quantum mechanical calcuations that, except for the natural and methylated bases, all other tautomers or substitutions of the bases have energy barriers along the pathway and are thus orders of magnitude slower in non-radiative de-excitation, or else they de-excite radiatively with high quantum efficiency (Serrano-Andrés and Merchán, 2009)[333]. Only the natural nucleobases are thus efficient dissipaters and this is the reason they have been thermodynamically selected by Nature over other, in fact more tightly bound, tautomers (see following section).

5.02 (U), 4.89 (T), 4.41 (C), 5.35 (A), and 4.93 (G) eV, with related oscillator strengths ranging from 0.053 for C to 0.436 for U (Serrano-Andrés and Merchán, 2009)[333], in near agreement with gas-phase experimental data, 5.1 (U), 4.8 (T), 4.6 eV (C), 5.2 (A), and 4.6 (G) eV (Crespo-Hernández et al., 2004)[61]. The purine (adenine and guanine) and pyrimidine (thymine, cytosine and uracil) nucleotides are extraordinarily rapid in dissipating this photon-induced electronic excitation energy into heat on sub-picosecond time scales (Pecourt, 2001)[272]. There exist other nucleotide substitutes for the bases, isomers, that are equally, or even more strongly, hydrogen bonded, for example, the isomer 2-aminopurine of adenine (6-aminopurine) which has a similar association constant for pairing with uracil, but the lifetime of the excited state of 2-aminopurine is 10's of nanoseconds, or about 10,000 times longer than the lifetime of naturally occurring adenine (Broo, 1998)[27]. There are also bondings other than the Watson-Crick among the nucleotides which, in fact, are stronger, i.e. have lower Gibb's free energies, but under a UV-C photon flux these also have much longer de-excitation times (Serrano-Andrés and Merchán, 2009)[333]. There thus appears to have been a non-equilibrium thermodynamic, rather than equilibrium thermodynamic, selection for the native bases, which can be described in terms of optimizing dissipation efficiency through the dissipation-replication mechanism described in chapter 12.2 and the autocatalytic non-linear, non-equilibrium mechanism described in chapter 6. It is evident that a similar type of dissipation efficiency has evolved in many other pigments over time in accordance with the prevailing photonic environmental condition and this will be considered in section 16.5.

Another important consequence of this evolution of dissipation efficiency is that those nucleotides, and their particular pairings, with very short excited state lifetimes (those that dissipate most rapidly) have a short temporal opportunity to undergo destruction or change through photochemical reactions (which happen most frequently in the excited state) and thus have greater "stability" under the non-equilibrium condition of the imposed UV-C photon flux of the Archean. This type of photo-protection is often argued (Pecourt, 2001; Mulkidjanian et al., 2003; Serrano-Andrés and Merchán, 2009)[272, 238, 333] to be the main reason for the evolution of these dissipative characteristics. However, the case for evolution in favor of photo-protection is not as strong as the case for evolution in favor of photon dissipation since, first, dissipation is the fundamental driver of the irreversible process we call life and must have accompanied life since its very beginnings, and secondly, Nature could have just as easily evolved nucleotide variants and other fundamental molecules of life that did not absorb in the atmospheric UV-C window, which would have made them completely oblivious to the UV-C photons. Rapid non-radiative de-excitation is necessary for dissipation, but it is not for photoprotection. Note that neither the photo-dissipation nor photoprotection explanations for the selection of efficient photon dissipating characteristics of the nucleobases could be employed by those who advocate a hydrothermal bottom of ocean origin of life where UV-C light would have been completely irrelevant. This will be given as strong evidence against the hypothesis of a hydrothermal origin of life in section 19.11.

The efficient photon dissipation characteristics of the RNA and DNA bases have been attributed to internal conversion through a conical intersection. A conical intersection occurs when the highest energy vibrational states of the slightly nuclear deformed electronic ground state are degenerate in energy with the lowest energy vibrational state of the electronic excited state (see figure 7.5). However, besides the existence of a conical intersection, it is important that there are no energy barriers on the reaction pathway to the ground state (Serrano-Andrés and Merchán, 2009)[333]. The molecule can thereby decay extremely rapidly and radiationless through vibrational cooling to the ground state, effectively converting the high energy single photon (electronic excitation energy) efficiently into many infrared photons (heat). As has been argued in chapter 12, such efficient de-excitation into local heat provides the possibility for an enzymeless denaturing mechanism for reproduction of RNA and DNA, leading to a dissipation-replication relation.

Conical intersections are common to conjugated systems (those molecules with alternating single and double covalent bonding) such as in the aromatic molecules (see figure 16.1).

16.3 Quenching of Radiative Decay Channels

Besides ultrafast de-excitation of an electronic excited state through a conical intersection and vibrational cooling to the ground state, covalently bound molecules can also decay through other channels which are slower or less efficient at dissipation. Some of these are; radiative decays such as fluorescence and phosphorescence, charge migration, and photochemical reactions detrimental to the integrity of the pigment. All these alternative de-excitation pathways reduce the efficiency of photon dissipation and would therefore have been thermodynamically selected against in all systems evolving under the non-equilibrium imperative of increasing dissipation.

Würfel and Ruppel (1985)[416] have shown that a steady state of maximal entropy production is produced when a material absorbing an arbitrary photon flux converts this flux into a black-body spectrum at the temperature of the material. The result is independent of whether the material is, or is not, connected to an external heat reservoir. This implies effective coupling of the electronic degrees of freedom of the material to its phonon degrees of freedom. Canonical intersections in organic molecules do exactly this and can thus be considered as

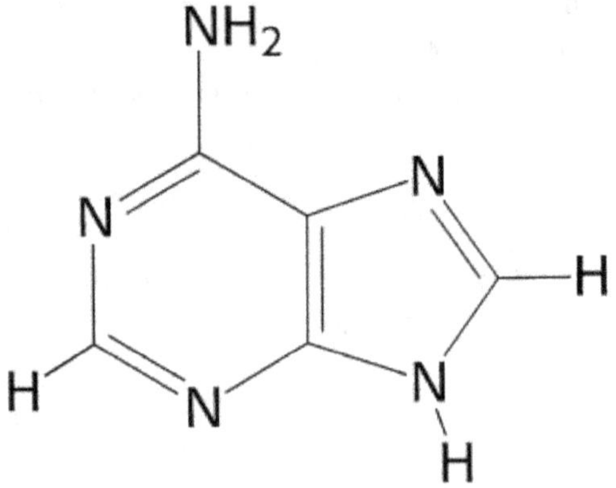

Fig. 16.1. The nucleobase adenine is an example of a "conjugated molecule" having alternating single and double covalent bonds. Conjugated molecules often have conical intersections for the rapid dissipation of electronic excitation energy into heat. Most of the fundamental molecules of life (see table 4.1) are conjugated molecules.

examples of microscopic dissipative structuring of material in response to the impressed photon potential.

Another effective technique that Nature has available to increase dissipation through reducing the quantum efficiency for radiative decay and photochemical reactions is that of resonant energy coupling between an electronically excited molecule and an acceptor molecule that has a conical intersection. As we have seen, DNA and RNA have an extremely rapid de-excitation pathway (sub picosecond) through a conical intersection which they can offer to an excited donor molecule. DNA and RNA also have chemical affinity to most of the fundamental molecules of life and these have large cross sections for UV-C absorption (see table 4.1). For example, the amino acid tryptophan has an intrinsic excited state life-time of 200 ns and a large quantum efficiency for fluorescence of 20%, however, when complexed with DNA or RNA the quantum efficiency for fluorescence of tryptophan goes to zero and the lifetime is on the order of picoseconds. Intercalations of tryptophan with DNA involving the stacking of the tryptophyl ring with the bases is known from monomer studies to lead to the total quenching of both the tryptophan and nucleobase phosphorescence[4] (Montenay-Garestier and Hèlene, 1968; 1971)[232, 233]. In double strand DNA, the quenching persists on intercalation and the double strand opens somewhat at the site of the

[4] Isolated nucleobases also emit phosphorescence between 370 and 450 nm with a quantum efficiency of 0.02, about 10 times greater than that of DNA (see figure 8.5).

tryptophan (Rajeswari et al., 1987)[295]. Toulmé et al. (1974)[376] analyzed the binding of tryptophan-containing small peptides Lys-Trp-Lys to heat-denatured and UV-irradiated DNA, and observed direct stacking of the indole ring of tryptophan to preferentially single-stranded regions of DNA that led to tryptophan fluorescence quenching. Interestingly, the Lys-Trp-Lys peptides showed a preferential binding to thymine dimers which often led to the photosensitized splitting of the dimer, implying that it could have acted as a primordial photolyase enzyme for photoreactivation of DNA. Kim et al. (1992)[159] have also shown that a particular tryptophan containing residue of the *Escherichia coli* photolyase, Trp-277, can alone bind to thymine dimers and split them using light at 280 nm with a quantum efficiency of 0.56. Besides the two known chromophores of *Escherichia coli* photolyase operating in the visible, this chromophore of Trp-277 also operates in the UV-C. This most probably indicates that *Escherichia coli* photolyase is a modern evolved version of a predecessor primordial photolyase operating in the UV-C. Chemical affinity between fundamental molecules of life thus reduced radiative de-excitation thereby facilitating photon dissipation and this could have been the origin of information accumulation through, for example, codon affinity to donor amino acids (this is discussed in detail in chapter 13).

It is also known that flavonoids can protect DNA from photochemical reactions, in particular they are effective against UV-A to UV-C induced covalent dimerizations of adjacent pyrimidines (Stapleton and Walbot, 1994)[357], a photochemical reaction that often leads to mutations and related diseases, such as skin cancer. It has been determined that flavonoid absorption in the visible (see figure 16.2) can provide enough energy to induce cyclobutane pyrimidine dimer reversion, thereby acting as a kind of protective photolyase. With flavonoids present, up to a 5 fold reduction in pyrimidine dimerization has been measured under laboratory conditions (Stapleton and Walbot, 1994)[357]. Flavonoids also absorb radiation at wavelengths of peak DNA dimerization potential (~280 nm, see figure 16.2) giving the complex DNA+flavonoid a larger cross section for absorption and more rapid dissipation of the complex (employing the conical intersection of DNA) in comparison to the molecules existing separately without interaction. This provided a thermodynamic imperative, in terms of dissipating efficacy, for what was perhaps one of the first of a number of "symbiosis" events at the molecular level. The extraordinary dissipation characteristics of RNA and DNA provided the thermodynamic incentive for their symbiotic association with most of the fundamental molecules of life.

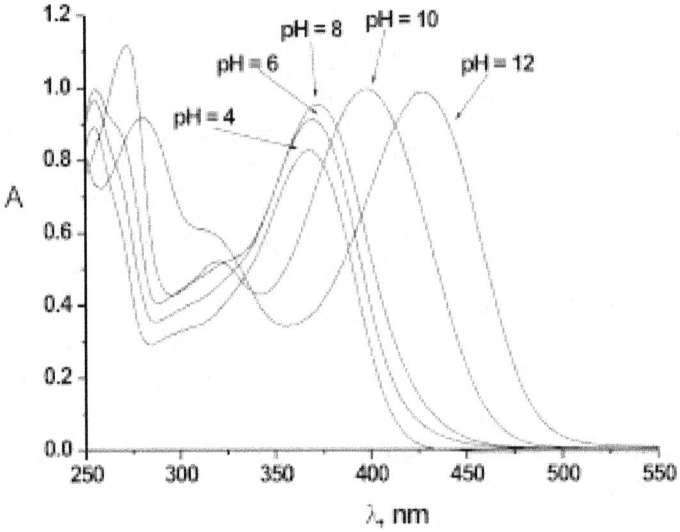

Fig. 16.2. Absorption spectrum of extracted flavonoid Quercetin at different pH values. Flavonoids can act as a photolyase for reversion of DNA thymine dimers through the energy of light absorbed in the visible (Stapleton and Walbot, 1994)[357]. However, flavonoids also absorb strongly at 280 nm at which wavelength the potential for DNA dimer formation is greatest. Flavonoids excited at 280 nm, when in close proximity to DNA, can transfer, through resonant energy transfer, their electronic excitation energy to DNA taking advantage of DNA's or RNA's conical intersection for rapid dissipation. This was the overriding thermodynamic reason for the association of flavonoids, and, in fact, all of fundamental molecules of life, with DNA and RNA at the begining of life. Image credit: Mezzetti et al. (2011)[205]. Reprinted with permission Royal Society of Chemistry.

16.4 Physical Size of Pigments

Reducing the physical size of the pigment while maintaining its photon absorption cross section would allow tighter packing of the pigments and thus greater capture rate of the incident photons. The copious flux of high energy photons from the Archean sun, and its increase in intensity by approximately 30% since then, has meant that there has always been room for improvement in photon capture efficiency. For example, today only 26% of the photons intercepted by organic matter are dissipated, the rest, mainly of wavelength beyond the red-edge, are simply reflected since dissipation processing rates are limited by finite excited state lifetimes, and by finite packing geometries related to the physical size of the pigment and its support structures.

Following Vekshin (2005)[386], the probability for the absorption of a photon of wavelength λ by a molecule of physical cross-sectional area X, is

$$P(\lambda) = \delta(\lambda)/X, \tag{16.2}$$

where $\delta(\lambda)$ is the molecular absorption cross-section for a photon of wavelength λ.The molecular absorption cross section is related to the molar extinction coefficient $E(\lambda)$ by $\delta(\lambda) = 2.3E(\lambda)$, where the factor 2.3 arrises from replacing the natural logarithm by the decimal one. It is noted that a 1 Å^2 molecular absorption cross-section corresponds to a 60,000 $M^{-1}cm^{-1}$ molar extinction coefficient (see table 4.1 for a list of extinction coefficients for the fundamental molecules).There is also an orientation factor q that must be taken into account since absorption is strongest when the oscillations of the electric vector of the light wave (perpendicular to the direction of propagation) coincide with the direction of electronic transition, which lies in the plane of the chromophore (Vekshin, 2005)[386]. This is the reason for the dynamic tracking of the sun by plant leaves during the day. Parallel and random orientations correspond to $q = 3$ and $q = 1$ respectively. For the cross-sectional area X of a molecule, it is appropriate to use the area of the electron density which is responsible for optical transitions, referred to as the cross-section of the valence electrons S. This gives the more useful formula for Eq. (16.2) as

$$P(\lambda) = 2.3E(\lambda)q/S. \tag{16.3}$$

From Eq. (16.3) it can be seen that by increasing the molar extinction, or by decreasing the size of the molecule (the cross-sectional area of the valence electrons), the probability for photon absorption by the chromophore increases. If natural thermodynamic selection is optimizing photon dissipation, it may be assumed that evolution has been working towards increasing the molar extinction coefficient $E(\lambda)$, while decreasing the size of the molecule S. It is also probable that in pigment supporting structural complexes, evolution has been working towards an alignment factor q such that optimal orientation of the planes of the chromophores, perpendicular to the direction of the absorbed photons is attained. This, indeed, seems to be the case, for example, for the chlorophyll chromophores which are stacked in the thylakoids of the chloroplasts in the leaves of plants. Plants have thus developed several mechanisms to ensure the optimal orientation of the leaves with respect to the oscillating field of the incident light flux to give maximum photon absorption. However, in completely parallel arranged stacks there would be screening leading to hypochromism and it seems for that reason the chlorophyll molecules in the thylakoids have a q value closer to 2 than to 3 (Vekshin, 2005)[386]. It is relevant to note that such mechanisms would not be needed if the plant were only interested in photosynthesis.

The size S of the molecule can be accurately determined through quantum mechanical calculations but can also be estimated by summing the number of double bonds, each contributing an effective area of approximately 0.9 Å^2, and the number of electron pairs on nitrogen atoms, each contributing an area of approximately 0.7 Å^2 (Vekshin, 2005, p. 27)[386].

16.5 Pigments Evolved to Cover Ever More of the Solar Spectrum

The photon absorption by today's plants and cyanobacteria covers a very large part of the high energy region of the solar spectrum arriving at Earth's surface. For example, a typical leaf absorbs strongly photons with wavelengths from the UV-C to the near infrared (see Fig. 16.3) However, there is a sharp drop in the absorption spectrum beginning at the near infrared, at about 700 nm, referred to as the red-edge, before absorbing strongly again in the infrared at approximately 1.3 μm, mainly due to the absorption by water within the leaf (see figure 16.4).

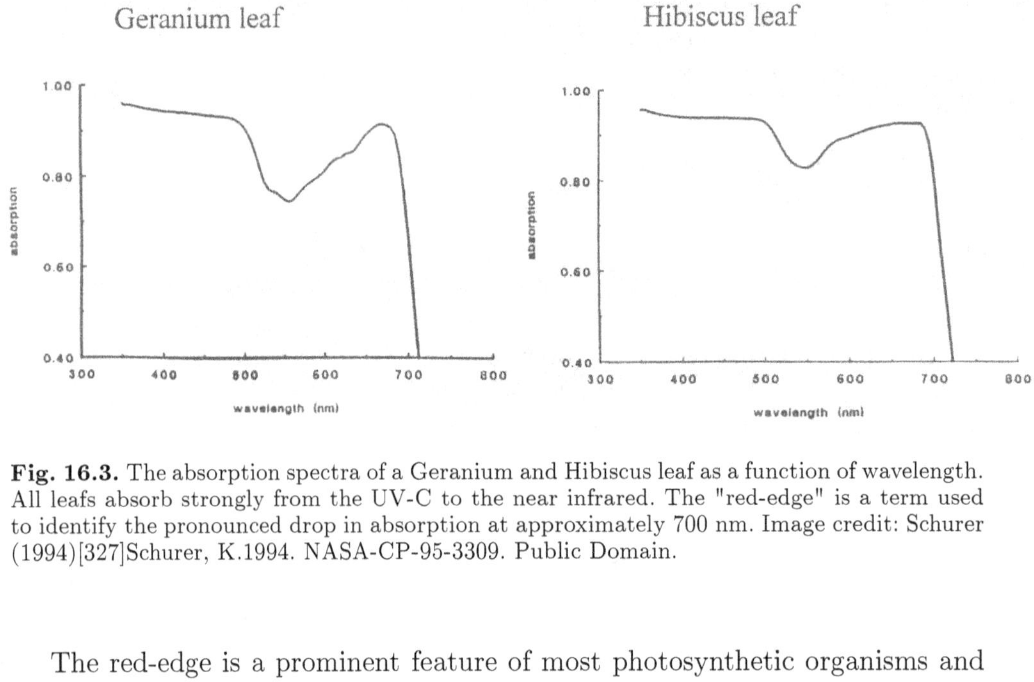

Fig. 16.3. The absorption spectra of a Geranium and Hibiscus leaf as a function of wavelength. All leafs absorb strongly from the UV-C to the near infrared. The "red-edge" is a term used to identify the pronounced drop in absorption at approximately 700 nm. Image credit: Schurer (1994)[327]Schurer, K.1994. NASA-CP-95-3309. Public Domain.

The red-edge is a prominent feature of most photosynthetic organisms and has been suggest to be due to a lack of molecular excited states between the highest energy vibrational state of the electronic ground state and the lowest energy vibrational state of the first electronic excited state of pigment molecules (Gates, 1980)[100]. It has also been assigned to an evolved protection strategy of plants to avoid overheating of the leaves which could compromise photosynthetic efficiency (Gates, 1980)[100]. Both these arguments, however, appear to be incorrect since deep sea bacteria have been found to be performing anoxygenic photosynthesis using wavelengths as long as 1,015–1,020 nm (Trissl 1993; Scheer 2003)[378, 321], well beyond the red-edge at 700 nm, making use of the faint red light given off by hydrothermal vents. To avoid heating of the leaves,

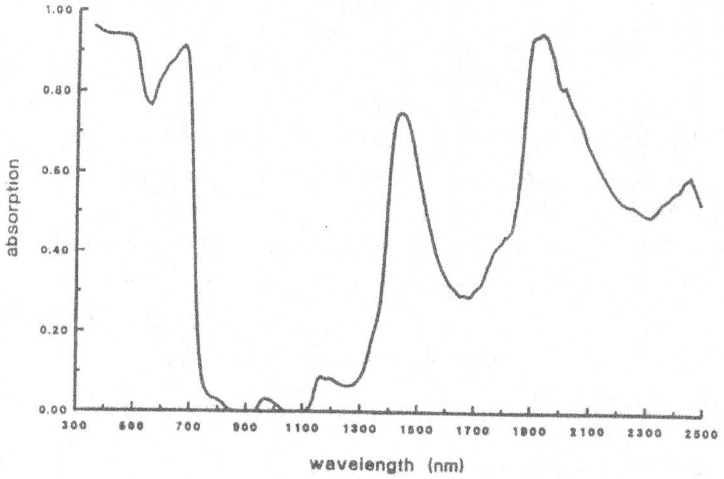

Fig. 16.4. Absorption of a Fuchsia leaf over an extended wavelength region. There is a sharp drop in the absorption spectrum at about 700 nm, referred to as the red-edge, before absorbing strongly again in the infrared at approximately 1.3 μm, mainly due to absorption by water inside the leaf. Image credit: Schurer (1994)[327]Schurer, K.1994. NASA-CP-95-3309. Public Domain.

plants could have avoided absorption in the UV and visible and concentrated their absorption to a limited region around 700 nm where photosynthesis is most efficient. Furthermore, efficient photosynthesis is in fact known for temperatures beyond 75 °C (Nes and Nes, 1980)[246]. These explanations for the existence of a red-edge are therefore not very plausible.

Instead, if plants and other photosynthetic organisms have evolved to optimize dissipation rather than photosynthesis, it is more probable that the reason for the red-edge has to do with dissipation rather than with photosynthesis. The flux of photons from the Sun at Earth's surface is copious ($\sim 10^{22}$ $m^{-2}s^{-1}$) such that given the finite lifetimes of pigment excited states, and finite areas occupied by the pigments, it maybe physically impossible, given the present level of evolution, to dissipate all photons over the whole spectrum. Plants and cyanobacteria would therefore maximize dissipation by devoting their resources to dissipation of the higher wavelength photons from the UV towards the visible, since these wavelengths give larger entropy production per photon dissipated, until a wavelength at which saturation of their dissipating capabilities would define a red-edge. Furthermore, as mentioned in section 16.1, there may also exist a problem with photochemical stability for chromophores that absorb at > 700 nm where the low-lying singlet or triplet electronic excited states increase the reactivity of the molecule with solvent impurities, such as dissolved oxygen

(Sauer et al., 2011)[320]. More details regarding this explanation of the red-edge will be given in chapter 19.15.

On Earth, organic molecules dissipating the solar spectrum are found only in association with water. Water provides a thermal bath at low temperatures into which the molecules can dissipate their excess vibrational excitation energy. As described in chapter 4, water is an efficient dissipative medium since it provides high frequency vibrational modes to facilitate rapid non-radiative de-excitation of the pigment. Without a water (or other solvent) environment, pigments would be less efficient photon dissipaters, remaining for longer times in an electronic excited state and therefore incapable of processing other photons and more vulnerable to destruction by photochemical reactions. This characteristic of the water solvent, of facilitating dissipation, was probably the primordial, and still the fundamental, reason for the association of life with water. Organic pigments complement water in covering the whole solar spectrum. That part of the solar spectrum not absorbed and dissipated by pigments is dissipated directly by water (see figure 4.1 and 10.2).

Organic pigments found in contemporary life on Earth can dissipate efficiently from the mid UV-C ($\sim$ 180 nm) up to the red-edge ($\sim$ 700 nm) and it is likely that evolution is pushing this red-edge limit ever further towards the infrared. There is still room for further evolution in this regard since today only about 26% of the solar photons reaching the Earth's surface and intercepted by living organisms are dissipated by them. There is a region of important intensity in the solar spectrum between the the red-edge at 700 nm and where water starts to absorb strongly at 1300 nm (see figure 16.4) where future evolution of new organic pigments could increase global dissipation, however, there is competition for space, resources, and entropy production with the contemporary pigments which are already dissipating photons of the highest entropy production potential, at the short photon wavelength region.

At wavelengths shorter than 180 nm, or greater than 1300 nm, water itself efficiently absorbs and dissipates photons (see figure 10.2) and there are no known organic molecules which absorb while maintaining their integrity (i.e. avoid being ionized) in these regions. Just as with the organic molecules, the ubiquity of water in the cosmos most likely has to do with its catalytic activity in dissipating the neighboring stellar spectra below 180 nm and above 1300 nm. Buhl (1973) [30] found that the production of water was occurring through photochemical reactions with light at UV-V wavelengths (consistent with 120 nm photon-induced production of H_3^+ which is responsible for the catalysis of many important organic molecules in space, see figure 5.10), around condensing star systems within a few astronomical units from the cores where the temperature reached 600 K and the H_2 density was approximately 10^{10} molecules per cubic centimeter (see section 5.1).

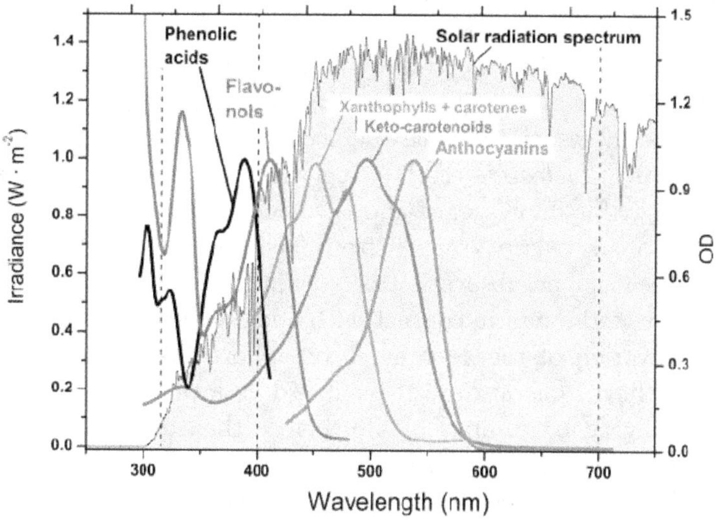

Fig. 16.5. Absorption spectra of representatives of key groups of what are known in the biological community as the "photoprotective pigments". The spectrum of solar radiation at the surface of the Earth is superimposed. The absorption maxima of most phenolics are located in the UV-B and UV-A regions; anthocyanins and flavonoids possess a long-wave maximum absorption in the green region, but also a shortwave maximum in the UV-C (not shown here, but see figure 4.15). The green (550 nm) is the band where the maximum of energy in the solar spectrum has always been located since the beginning of life in the Archean. The band of maximum entropy production per photon dissipated is, however, in the UV-C. The so called "photoprotective" carotenoids absorb in the blue-green region. The spectra are normalized to their absorption maxima. The chlorophyll pigments (not included in the above graph), which also have a much greater quantum efficiency for dissipation than for photosynthesis, fill in the remainder of the solar spectrum out to the red-edge at about 700 nm. For example, Chl *b* has maximum absorption at 650 nm while Chl *a* has maximum absorption at 680 nm (see figure 4.12). Image credit: Solovchenko and Merzlyak, 2008[351]. Reproduced with permission from Springer Science+Business Media.

Common families of pigments in today's plants and cyanobacteria include scytonemins, mycosporines, flavonoids (anthocyanins), carotenoids, porphyrins, phycobilins. The coverage of the solar spectrum of some of these pigments is given in figure 16.5.

Given this complete coverage of the high energy portion of today's solar spectrum by pigments, there can be little doubt that the evolution of chromophores to cover the whole high energy solar spectral region has been a continual process since the origin of life. Vestiges of this evolution of wavelength coverage from the past can be discerned in a number of important molecular complexes. RNA riboswitches[5] bind a range of protein cofactors which absorb in the UV-C, such

[5] Riboswitches are very short segments of RNA, about 100 to 200 nucleobases long, which have been found to act as switches to turn on or off gene expression by altering their

as flavin mononucleotide (FMN), thiamine pyrophosphate, tetrahydrofolate, S-adenosylmethionine and adenosylcobalamin (a form of vitamin B12). The pigment flavin, which also absorbs strongly at 450 nm in the blue (see figure 4.15), has long excited state decay times and a large quantum efficiency for fluorescence, can be found associated with the acceptor quencher nucleobase adenine in *flavine adenine dinucleotide* (FAD). Together, in the FAD complex, both the quantum efficiency for fluorescence and the excited state lifetimes are much reduced. Since resonant energy transfer from flavin excited at 450 nm to adenine is prohibited because the absorption of adenine is in the UV-C, the exciplex, formed by the excited flavin in contact with adenine, makes use of the vibrionic modes of deactivation of the adenine chromophore (Vekshin, 2005)[386]. It is interesting that flavin has also been identified as a possible energy conversion molecule at the very beginnings of life, before the advent of the chlorophyll photosynthetic system (Kritsky et al., 2013)[171].

A very similar situation occurs for reduced *nicotinamide adenine dinucleotide* (NADH) which absorbs strongly in the UV-A at 340 nm (Vekshin, 2005)[386]. The list of similar cofactors, coenzymes, and pigments with chromophores covering the whole solar spectral region is large (see chapter 4.3 and Michaelian and Simeonov (2015)[217] and references therein). Any of these fundamental pigments can couple to the nucleotide vibrionic modes of deactivation, making complexes of the two formidable dissipating systems.

What were the probable mechanisms of evolution of pigments to cover ever more of the enthropically most important region of the solar spectrum? Selection based on the optimization of photon dissipation certainly was responsible for the evolution then as it is today. How such a mechanism for evolution could have worked in practise at the molecular level through a replication-dissipation relation during the Archean was given in chapter 12. However, what explicit change at the molecular level of the pigments occurred to alter or widen their dissipative wavelength region? In chapter 4 it was shown how pigment absorption in the visible can be attributed to a large number of conjugated double bonds (see figure 4.10), the greater the number, the more closely spaced the electronic excitation energy levels of the molecule. Their is a de-localization of charge in these molecules among the p-orbitals, each additional conjugated p-orbital pushes the molecules absorption further towards the red (see figure 4.13). It, therefore, must have been a relatively simple matter for nature to evolve new pigments covering ever more of the UV and visible spectrum starting from molecules that absorbed in the UV-C simply by adding conjugated bonds to the molecule.

secondary structure in response to the binding of a metabolite. These riboswitches can act as terminators or promotors of gene transcription. Thermal riboswitches have also been found; RNA segments which change their secondary structure on temperature change.

There is, however, a way of increasing the width of the size of the spectral region of absorption without changing substantially the molecule. This is due to the natural line width of a spectral line, a consequence of the quantum nature of the cosmos and is related to the Heisenberg uncertainty principle, $\Delta E \Delta t \geq h/4\pi$, where h is Planck's constant. This relation states that the the smaller the uncertainty in the lifetime t of the excited state, the larger the uncertainty in the energy E of that state. Since the uncertainty in the lifetime of an excited state cannot be greater than the lifetime itself, this implies that excited states with short lifetimes will have large widths in the spectral region for absorption. Examples of such mechanisms for increasing the spectra absorption width in molecules are the conical intersections (see chapter 7) in which the energy barriers along the reaction pathways of the radiationless transitions are very small, making the excited state lifetimes exceedingly short. This is the case for the nucleic acid bases in which non-natural tautomers of the bases (variations of the bases with different protonation) have much longer lifetimes. The tautomers of the bases selected by nature hence have the greatest spectral absorption width, even though these are usually not the most tightly bound structures (Serrano-Andrés and Merchán, 2009)[333]. Nature, therefore, takes very seriously her non-equilibrium dissipative program compromising on equilibrium stability in order to gain dissipative efficiency.

Finally, there is yet another way of changing the wavelength region of absorption and even amplifying the absorption of a molecule without changing substantially its structure. This is by adding a functional group of atoms with nonbonded electrons to a chromophore. This alters both the wavelength and intensity of absorption of the original molecule. Such a functional group is known as an *auxochrome*. Typical auxochromes can be either acidic (accepting an electron), for example -COOH, -OH, $-SO_3H$, or basic (donating an electron), for example -NHR, $-NR_2$, $-NH_2$.

16.6 Animals Arised and Evolved to Propagate Photon Dissipating Pigments

The dispersal of photon dissipating pigments today is global, except for regions where there exists little liquid water necessary for efficient dissipation of electronic excitation energy into heat, for example, the Sahara desert or the polar ice caps (see figure 16.6). This dispersal and maintenance of pigments over land and over the ocean surface can be attributed to mobile animals.

For at least 2.5 billion years after its origin, life remained as single celled organisms with little mobility, confined to the ocean surface or confined to land surfaces near the ocean shore. The development of multi-cellular organisms is known from the fossil record (Ediacaran biota – see figure 16.7) to have occurred

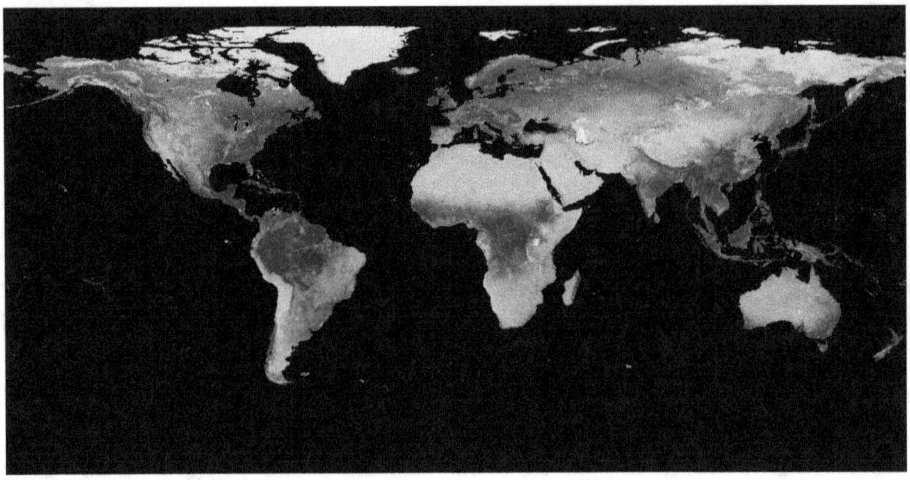

Fig. 16.6. The vegetation index on land on a global scale. The distribution of pigments over the Earth's surface and their maintenence is provided for by the animals. Only regions with little liquid water to disipate the resultant heat of photon disipation are barren of pigments; for example the Sahara desert and the polar ice caps. Image credit: NASA, TERRA/MODIS, Public Domain.

at about 610 Ma, only some 60 million years earlier than the colonization of the land by plants and animals at about 550 Ma. The thermodynamic dissipation theory of the origin of life and evolution suggests that novelty in the evolution of the Earth system is selected based on thermodynamic dissipative efficacy, and this provides a thermodynamic imperative for the appearance of multi-cellular mobile organisms.

At the time of the Cambrian explosion[6] and the colonization of land by plants and animals, about 1/3 of Earth's surface consisted of continental land mass, as today. Before the colonization of the land by plants and animals the water cycle would have been confined to regions close (perhaps up to 1000 Km) to the shores (Michaelian, 2012b)[213]. Inland continental surfaces would have consisted of essentially dry rock and sand dunes, eroded by winds from rock. Barren rock reflects much of the incident sunlight back into space, having a high albedo of, on average, 0.35 in the visible while sand has an albedo even higher of 0.40 (see figure 16.8). This compares with a 0.05 albedo for climax land ecosystems which probably came into being later into the Cambrian. Similarly, pure water has a larger albedo that natural water containing organic pigments, especially for the Sun at large incident angles to the normal (Michaelian, 2012b)[213]. These smaller albedos for life over non-life covered areas decrease even further the shorter the wavelength of the light. Clarke et al. (1970)[49] give measurements

[6] The Cambrian explosion refers to the rapid, on geological timescales, radiation of new animal species which occurred in the Cambrian beginning at 542 Ma.

Fig. 16.7. The first multicellular fossils known are the Ediacarans which appeared at about 610 Ma. This specimen is known as *Dickinsonia*. A scale in cm is given in the upper right. Image credit: Verisimilus, Creative Commons Attribution-Share Alike 3.0.

for the reduction of water albedo at different wavelengths due to the presence of organic material. In general, the smaller the albedo, the greater the photon dissipation since a smaller albedo means greater photon absorption[7].

There would thus have been a strong non-equilibrium thermodynamic incentive to spreading photon dissipating pigments, not only over land areas and ocean surfaces near the shores, but also over far inland and mid ocean surfaces. However, far inland and mid ocean surfaces would have been generally barren of the nutrients necessary for producing organic pigments. A mechanism of nutrient dispersal was need if all of Earth's surface was to be made available for photon dissipation.

In chapter 6 I have presented how proliferation of pigments as microscopic dissipative structures can be explained from non-equilibrium thermodynamic principles if the photochemically formed pigments also dissipate the same photon potential that produced them. However, the raw material for their production has to be available before such an autocatalytic system could be effective. If

[7] However, dissipation is not only related to absorption, but also has a component due to reflection. The incident solar photon beam is highly collimated. Any dispersion of the solar beam into a greater solid angle is also dissipation (or entropy production). Large cloud albedo on Earth also contributes to entropy production. To understand the relative importance to the global entropy production of Earth and its neighboring planets of the different contributions to entropy production, see Michaelian (2012b)[213].

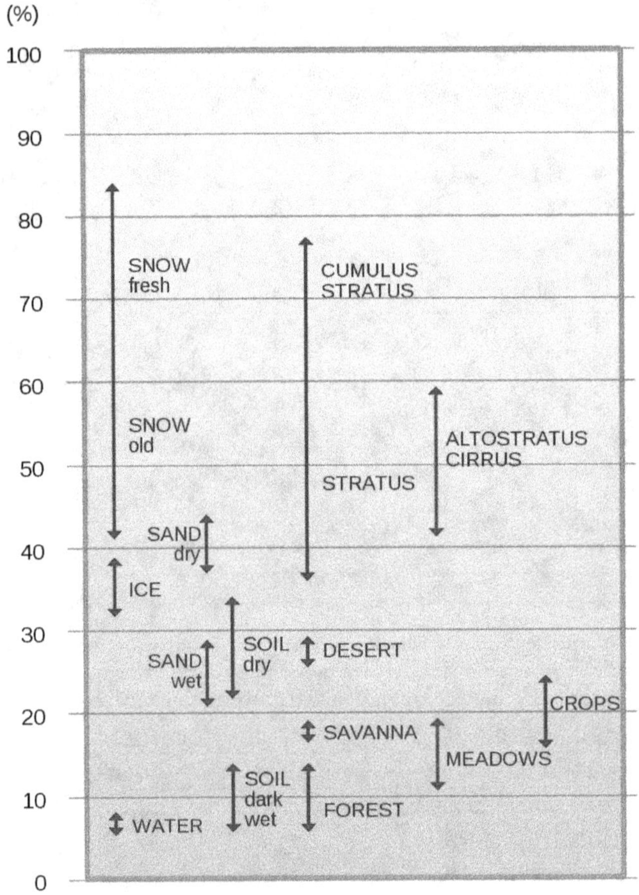

Fig. 16.8. Albedos of different natural surfaces. Pure water has a low albedo for direct incidence, but is transparent to visible light. If there were no cyanobacteria or other organic pigments on the ocean surface, visible light would be absorbed at greater depth meaning there would be a less pronounced temperature profile with depth (see figure 10.4). There would then be less evaporation on the surface and thus a smaller water cycle. Overall, this would lead to less dissipation of the incident high energy solar photons. Natural water with organic pigments has a lower albedo than pure water, especially at large incident angle to the surface normal and for shorter wavelengths. Vegitation cover, especially climax forests, has the lowest albedo of all natural surfaces. Rock surfaces (not listed) have an albedo of about 20-40. Image credit: Wereon. CC BY-SA 2.5.

the far inland and mid ocean surfaces were barren and could not support the production of pigments, highly mobile organisms, that through their excrement and death could spread nutrients ever more outwards in both directions from the nutrient rich ocean shores, would be thermodynamically selected.

The history of multi-cellular animals in the ocean began around 700 Ma, some 100 Ma years before that on land surfaces. However, there is no evidence for fossils from deep water during this period. The first deep sea fossils of multicellular organisms are only about 180 million years old (see figure 16.9) which argues in favor of the idea of multi-cellular animals being invented as nutrient spreaders for the organic pigments on the surface. Animals would have had little such business to perform in the blackness at 3 Km below the ocean surface. Today, those animals found at depths mainly dissipate the chemical potential available in the rain of organic material falling from the surface.

Fig. 16.9. Brittle star Hemieuryale pustulata in life position on cnidarian. One of the oldest known deep sea fossils of multicelular organisms, about 180 million years old. This fossil may have been the anscestor of modern day star fish. The fact that deep sea multicellularism is much more recent than surface multicellularism ($\sim$ 700 Ma) suggests that the most important role played by animals is the role of nutrient transporters for phototrophic cyanobacteria, algea and plants that can carry out surface photon dissipation. Image credit: Ben Thuy, Natural History Museum Luxembourg. Reproduced with permission.

The first colonization of land at approximately 2.6 Ga [393] probably corresponded to cyanobacteria being carried inland from coastal regions during storms. Such a spread of cyanobacteria onto land also provided an increase in the supply of minerals to the coastal waters through bacterial catalysis of rock erosion. Thus a positive feedback cycle leading to greater photon dissipation would have arisen but it would have been limited to near coastal areas. The evolution of insects would have allowed the rapid propagation of nutrients inland to areas barren of the important phosphor and nitrogen needed for the production of pigments. Once land plants became established, the water cycle would have been extended ever further inland through winds driven by pressure gradients caused by vegetation evapotranspiration and condensation, as happens today, for example, in the Amazon river basin (Makarieva and Gorshkov, 2007)[197]. This would have allowed plants to spread ever more inland over the land masses until finally making available another 30% of Earth's surface to more efficient photon dissipation.

Animals play fundamental roles in spreading nutrients and seeds for plants through migration, excretion, and death. The largest animals usually have the greatest range and are most responsible for nutrient and seed dispersion (Doughty et al., 2013)[84].

Top predators are also indispensable to the thermodynamic functioning of an ecosystem. A specific example of this is the re-introduction in 1997 of wolves into Yellowstone park after decades of absence due to extermination by humans. It did not take long to make the surprising realization that the re-introduction of wolves kept the herbivores on the move and thus prevented overgrazing by elk and other herbivores which led to an increase in the robustness and number of willow trees (Ripple and Beschta, 2006)[300]. This, of course, has the effect of reducing the albedo and increasing the net photon dissipation.

Whales and other aquatic mammals of today are descendents of land mammals and their thermodynamic role in the ocean is to spread nutrients, particularly iron, atomic nitrogen and phosphor, to ocean waters far from coastal regions. Fish and mammals (including whales) also facilitate the resurfacing of sedimented nutrients from the depths by feeding on bottom feeders (Roman and McCarthy (2010); Lavery et al., 2010)[301, 183] (see figure 16.10).

Land and aquatic ecosystems thus gradually evolved into ever more complex and photon dissipating systems and this is mirrored today in the succession of ecosystems (Michaelian, 2012a)[212], somewhat analogously to Haeckel's formulation of "ontogeny recapitulates phylogeny" in which the stages of evolution of the species were argued to be mirrored in the individual development of animals from a fetus[8]. The biosphere is evolving as a whole to ever greater photon dissipation rates and this includes the coupling of both biotic and abiotic irreversible processes, such as, for example, the photon dissipation in the leaves of plants and bodies of cyanobacteria coupling to the water cycle (Michaelian, 2012b)[213].

Although the strict Darwinian interpretation strongly refutes it, Gould (2002)[110] accepted the possibility of a "hierarchy" in the action of evolution;

[8] *Recapitulation theory*, as it is known, is now generally discredited.

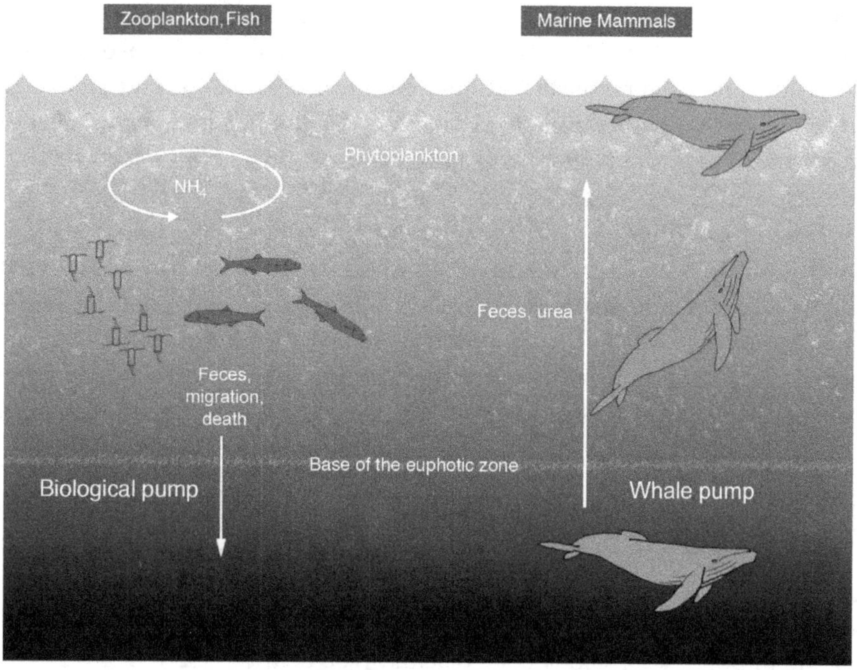

Fig. 16.10. Whales and other marine animals through consumption of bottom dwelling animals and defecation at the surface provide a pump for seeding the ocean surface at great distances from the shores with iron, nitrogen, phosphor and other important nutrients so the photon dissipating cyanobacteria and phytoplankton can thrive. This is the principle thermodynamic reason for the existence, proliferation, and evolution of animals. Image credit: Roman and McCarthy (2010)[301]. Creative Commons, 3.0.

the idea that evolution may act on more than one unit simultaneously, as opposed to only acting upon individual organisms. It is suggested here (and will be detailed in chapter 17) that what appears to be selection at the individual, group, species, clade and ecosystem level is really ultimately thermodynamic selection at the biosphere level, and it is based on the efficacy of global biosphere photon dissipation [212](Michaelian, 2012a).

Finally, as an example of the human contribution to the thermodynamic utility of animals, it has been brought to my attention (Bunge, 2015)[31] that in many parts of the world humans are depleting the limited known phosphate deposits and spreading them on barren lands such as deserts in order to grow crops. This is a good example of how the human animal is fomenting the global photon dissipation process through the deliberate spread of nutrients, although completely unconscious of its underlying thermodynamic significance. Of course, this has to be considered in the historical context of humans destroying the Earth's natural environment. It seems, however, that the human species has finally recognized that it has no alternative and, for the most part, now understands that its survival is dependent on photon dissipating ecosystems and much of the damage is being halted, and in some places even reverted through re-forestation and other conservationist programs. Perhaps even more telling of our role in the fomentation of global photon dissipation, are recent extensive satellite surveys that have determined that anthropic increases to the CO_2 concentration in Earth's atmosphere has actually led to a greening of Earth's surface (Zhu et al., 2006)[429]. Humans mining hydrocarbons stored deep within the Earth and then converting these hydrocarbons into a form (CO_2) that can be used by photosynthesizing organisms that dissipate solar photons represents perhaps the epitome of the thermodynamic function of animals. This presents a new perspective on human-induced global warming which, although certainly affecting negatively the status quo, may represent a natural transition to a greater state of global photon dissipation. This deserves more in-depth studies.

16.7 Life Couples to Other Biotic and Abiotic Irreversible Processes

Taking the opponents of the theory of Gaia (see chapter 1.2.7) seriously – that there can be no natural selection at the level of the biosphere – we are left with no other alternative but to conclude that natural selection in the Darwinian theory of evolution lacks sufficient elements to explain the origin, persistence, and evolution of the biosphere, or even to explain the biotic and abiotic components thereof. Darwinian theory is incomplete and a more fundamental theory of life and evolution is needed.

In order to understand the biosphere and its evolution, it is first necessary to recognize that biological processes in the biosphere are thermodynamically coupled to abiotic processes, and then understand how biosphere evolution is driven by increases in its entropy production through the origin and coupling of its constituent irreversible processes. No irreversible process, or coupling of processes, biotic or abiotic (such as life or the hydrological cycle), will arise unless it produces extra entropy. Entropy production is not incidental to the coupled processes, but rather the fundamental reason of its existence.

A salient characteristic of nature is the search for routes to ever greater global entropy production, often building on pre-existing routes by coupling new irreversible processes to existing ones, and this is made obvious in the well documented evidence for an evolutionary principle of building on inherited structures and processes. Onsager (1931)[256] has shown how diverse irreversible processes can couple in order to remove impediments to greater global entropy production (Morel and Fleck, 1989)[234]. In general, the more complex the dissipative structuring in space and time, i.e. involving many coupled irreversible processes with embedded hierarchal levels of interaction of long spatial and temporal extent (see figure 2.4), the greater the overall entropy production due to the systems interaction with its external environment (Onsager, 1931; Prigogine et al., 1972a,b; Lloyd and Pagels, 1988)[256, 290, 291, 189].

Examples of biology coupling to abiotic irreversible processes are ubiquitous. Pigments in the leaves of plants dissipate photons into heat and this heat provides the latent heat of evaporation of water for driving the water cycle. Over oceans, absorption of sunlight in the pigments of cyanobacteria not only foments the water cycle, but also foments hurricanes as another abiotic process for dissipating free energy. Cyanobacteria fomenting increased evaporation over ocean surfaces can actually originate and steer hurricanes (Gnanadesikan et al., 2010)[108].

The coupling, of course, is also between distinct biotic processes. As an example of how the analysis of the coupling of biotic irreversible processes from within a non-equilibrium thermodynamic framework can lead to a deeper understanding of complex systems such as ecosystems, in the following box we consider the dissipative coupling of populations of different species forming a particular ecosystem. This analysis demonstrates how non-equilibrium thermodynamic principles, in particular, Glansdorff's and Prigogine's general evolutionary criterion (see chapter 3.1), can be used to determine the population dynamics of the ecosystem. Such a thermodynamic analysis of ecosystems based on first principles has obvious advantages over a purely descriptive analysis such as that derived from the Lotka-Volterra equations. The following box can be skipped by those with an adverseness to mathematics. For a more detailed description, see Michaelian (2005)[206].

Non-equilibrium thermodynamics describes the entropy production of a system as a bilinear form of generalized forces, X, multiplied by generalized flows, J, (see chapter 3.1 or Prigogine (1967)[289]), i.e.

$$P \equiv \frac{d_i S}{dt} = \sum_k J_k X_k \tag{16.4}$$

where the index k runs over all irreversible processes operating within the boundaries of the system. These processes can be, amongst many others, flows of mass, heat, or charge, chemical reactions, and photon absorption and dissipation, etc. In ecosystem processes, individual organisms can be considered as units of entropy production and exchange and the forces X_k are then associated with the populations p_k of species k , and the flows J_k with the flows of entropy (Michaelian, 2005)[206].

The flows, in general, can be written as a non-linear sum of the forces (populations),

$$J_k = \sum_{l,m,n,...} a_k + b_{kl}X_l + c_{klm}X_lX_m + d_{klmn}X_lX_mX_n + ... \quad (16.5)$$

giving that the entropy production, Eq. (16.4), is

$$P = \frac{d_iS}{dt} = \sum_{k,l,m,n...} a_kX_k+b_{kl}X_kX_l+c_{klm}X_kX_lX_m+d_{klmn}X_kX_lX_mX_n+... \quad (16.6)$$

where the first term is the one-body entropy production term due to all the individuals of species k, the second term is the two-body species interaction between individuals of species k and l, the third term is the three-body term, etc. Except for social species, the three-body and higher order interactions will be rare and these terms could safely be neglected. The coefficients a_k, b_{kl}, c_{klm}, ... in Eq. (16.6) are terms which measure the strength of the entropy production or entropy flow among interacting individuals of the given species.

According to empirical evidence and Attard's *second law of non-equilibrium thermodynamics* (see chapter 3.1.1), over evolutionary time these constants will evolve to give every greater global entropy production P, subject to the prevailing physical conditions. We have given an explanation of why this happens in terms of microscopic fluctuations in the direction of greater dissipation will tend to be reinforced leading the system most probably to solutions of greater entropy production at a bifurcation (see chapter 3.3).

Over short, human, time scales, during which the ecosystem can be found in a thermodynamic stationary state in which the entropy and other thermodynamic variables are independent of time, $dS/dt = d_iS/dt + d_eS/dt = 0$, these coefficients a_k, b_{kl}, c_{klm}, ... can be considered as constants and Eq. (16.6) along with the general evolutionary criterion of Prigogine $dP/dX_k \leq 0$ (see chapter 3.1) are sufficient to determine the population dynamics of the ecosystem. For a non-linear system like that given in Eq. (16.6), and for systems of three or more variables X_k (populations of species k), cyclic and chaotic attractors can be found (see Michaelian (2005)[206]).

This thermodynamic interpretation of ecosystems provides a much more desirable description than the purely descriptive and *ad hoc* Lotka-Volterra equations since the interaction coefficients $a_k, b_{kl}, c_{klm}, \ldots$ can, in principle, be determined empirically. In the Lotka-Volterra framework, the interaction coefficients can only be determined from a non-unique fit to the overall population dynamics.

In our non-equilibrium thermodynamic view, selection of a particular irreversible process (the terms a_k, $b_{kl}, c_{klm}, \ldots$), or, in other words, the coupling of irreversible processes, is contingent on increasing the global entropy production of Earth in its solar environment (Eq. (16.6) for the whole biosphere). In general, it is not the individual organism's fitness, or even the individuals entropy production, in isolation of all others (this does not even make sense in principle as can be seen from extensive coupling of populations X_k in Eq. (16.6)). However, entropy production is an extensive (additive) quantity and the individual terms $J_k X_k$ are often positive (although not necessarily so if two or more processes occur within the same macroscopic region of space-time; in this case a term may be negative as long as a coupled term is positive and of greater magnitude (see Prigogine, (1967)[289]). Therefore, in this non-equilibrium thermodynamic framework, selection based on increases in global entropy production can occur at any level, including the individual or the biosphere, or at many levels simultaneously as Gould (2002)[110] identified but was unable to determine its nature.

16.8 Biotic Induced Transparency of Earth's Atmosphere

Along with biology, many abiotic physical characteristics of Earth, including the gasses of Earth's atmosphere, have been coevolving over time (see Chapter 17.4). It is probably not a coincidence that Earth's present atmosphere is transparent to the most intense and high energy region of the solar spectrum. Surface based life employing organic pigments is most efficient at absorbing and dissipating this light into heat and it is reasonable to assume that life played an important part in evolving the atmosphere to the transparency it is today despite the geochemical and geophysical forces that also came, and still come, into play. For example, it is known that today's atmosphere is not an atmosphere in chemical equilibrium but is maintained out of equilibrium by life. Without the oxygenic photosynthesis performed by organisms, the oxygen of Earth's atmosphere would rapidly become depleted since oxygen is a very reactive gas which would form oxides with many elements. In this section, I describe some of the well established effects of evolving life on Earth's atmosphere as well as some not so well established speculation for which at least some evidence exists (see Michaelian and Simeonov (2015)[217] for more details).

It is believed that Earth formed in the Hadean at about 4.7 Ga with an original atmosphere rich in hydrogen, helium, carbon dioxide, water and nitrogen, and most probably other more complex molecules built out of photochemical reactions of UV-C light on these primordial gases. Some of the photochemical products being hydrogen cyanide, HCN, aldehydes, and tholins. It is likely however, that Earth's initial atmosphere was lost to space through repeated asteroid and planetesimal bombardment[9], and the new atmosphere that emerged at the end of the late lunar bombardment ($\sim$ 3.9 Ga) was derived from the outgassing of Earth's newly forming mantle, mainly H_2O, CO_2, SO_2 and N_2. The H_2O would have condensed out over time to form oceans, and much of the CO_2 would have been absorbed in the oceans and precipitated out in carbonates forming the great limestone formations similar to those known today. Aldehydes would have formed through UV-B and UV-C photochemical reactions on the primordial molecules. As microscopic dissipative structures, aldehydes in the atmosphere would have absorbed over the UV-B region from $\sim$ 285 to 310 nm (Sagan, 1973)[309].

Lowe and Tice (2004)[193] have suggested a gradual depletion of atmospheric CO_2 through the carbonate–silicate geochemical cycle starting at around 3.2–3.0 Ga through weathering of the newly forming continental crust, including the Kaapvaal and Pilbara cratons. Much of this weathering could have been promoted by bacterial acidic waste secretions, as still happens today on the continents and, in fact, in our teeth! By 2.9 – 2.7 Ga, CH_4 to CO_2 ratios may have become $\sim$1, thereby stimulating the formation of an organic haze that would have given rise to a large visible albedo and reducing surface temperatures to below 60 °C at 2.9 Ga, perhaps allowing oxygenic photosynthetic organisms to thrive and increasing the amount of oxygen and ozone in the atmosphere. Pavlov et al. (2000)[270] showed that 1000 ppmv each of CH_4 and CO_2 (Kharecha et al., 2005)[157] would counteract the faint young Sun sufficiently to keep temperatures above freezing at this time. However, not all of Earth may have remained above freezing, since glacial tillites have been identified in the 2.9 Ga Pongola and Witwatersrand supergroups of South Africa (Young et al., 1998; Crowell, 1999)[421, 66]. Eventual erosion of the continents and tectonic recycling of CO_2 would have allowed the CH_4/CO_2 ratio to reduce again, bringing back a warm greenhouse atmosphere to the late Archean.

Using as a model the results obtained from the Cassini/Huygens mission for exploration of Titan, Trainer et al. (2006)[377] investigated the probable formation of organic haze on an early Earth through photolysis of CH_4 with

[9] It is thought that the moon was formed out of the debris from a collision of a planetesimal the size of Mars with Earth sometime in the early Hadean.

the solar Lyman-line at 121.6 nm in a N_2 and CO_2 atmosphere. In laboratory experiments designed to simulate Earth's atmosphere at the origin of life, in which the CH_4 mixing ratio was held at 0.1 %, and the CO_2 mixing ratio was varied from 0 to 0.5% (suggested to include most reasonable estimates for the early Earth; Pavlov et al., 2000 [270]), they found principally molecules with mass-over-charge ratios (m/z around 39, 41, 43, and 55) indicative of alkane and alkene fragments. The amount of aromatics of 77 and 91 amu decreased with increasing CO_2. The C/O ratio rather than the absolute concentrations of CH_4 and CO_2 was shown to be the factor most correlated with the chemical composition of the products. Aerosol production was seen to be maximum at C/O ratios close to 1, which according to Lowe and Tice (2004)[193] would have occurred at approximately 2.9 Ga. Approximately spherical particles were found in the experiments with average diameters of about 50 nm. Particles of this size, at the estimated photochemical production rates, would have produced an optically thick layer in the UV but a rather thin layer in the visible (Trainer et al., 2006)[377]. However, as observed on Titan, and in laboratory experiments employing an electrical discharge source, these particles readily form fractal aggregates of size $>$ 100 nm, consistent with observations of the atmosphere of Titan, thereby significantly increasing the visible attenuation with respect to the UV (Trainer et al., 2006)[377].

Given the intensity of the Lyman- line (121.6 nm) from hydrogen in the Sun at Earth's upper atmosphere and the probable concentrations of CH_4 at these altitudes, Trainer et al. estimate an aerosol production rate on early Earth of between $1x10^{13}$ and $1x10^{15}$ g yr^{-1}, which alone is comparable to, or greater than, the estimated delivery of prebiotic organics from hydrothermal vents and comet and meteorite impacts combined. The free energy in UV-C surface light during the Archean available for the production of the fundamental biomolecules from primordial gases would have been many orders of magnitude greater than that available through the chemical potential at hydrothermal vents.

From studying the sulfur isotope record, Domagal-Goldman et al. (2008)[82] suggested that a thick organic haze, which blocked UV light in the 170–220 nm region from promoting the photolysis of SO_2 in the lower atmosphere, arose at 3.2 Ga and persisted until 2.7 Ga. Based on these isotope ratios, they suggest that Earth's atmosphere went from a haze-less to thick haze between 3.2 and 2.7 Ga, and then again to a thin haze after 2.7 Ga. The appearance of the haze may be associated with the appearance of methanogens (and anoxygenic photosynthesizers) around 3.2 Ga, which led to a buildup of CH_4, while continent erosion led to a decline in CO_2, and as the ratio of CH_4/CO_2 became close to 1, the organic haze became thicker and spread over the upper atmosphere.

Crowe et al. (2013)[65] suggest that there were appreciable levels of atmospheric oxygen ($3x10^{-4}$ times present levels) about 3 billion years ago, more than 600 million years before the Great Oxidation Event and some 300–400 million years earlier than previous indications for Earth surface oxygenation. The researchers also suggest that the observed levels are about 100 000 times higher than what can be explained by regular abiotic chemical reactions in Earth's atmosphere, and therefore the source of this oxygen was almost certainly biological.

There is evidence of oxygenic photosynthesis by at least 2.78 Ga in the presence of 2-methyl hopanes from O_2-producing cyanobacteria (Brocks et al., 1999)[25] and sterols from O_2-requiring eukaryotes (Summons et al., 1999)[365] in sediments of this age (Brocks et al., 2003)[26]. The buildup of oxygen consumed the CH_4 in the atmosphere, leading to a reduction in the organic haze. The oxygenation of Earth's atmosphere may have begun in earnest at about 2.9 Ga but accelerated at about 2.45 Ga and would have removed most of the CH_4 greenhouse gas from the atmosphere by about 2.2– 2.0 Ga (Rye and Holland, 1998)[308].

Other lines of geochemical evidence suggest that the major oxygenation event occurred in the atmosphere at about 2.2 Ga, with atmospheric O_2 levels rising sharply from $< 1\%$ of present atmospheric levels to about 15% of present atmospheric levels, during a relatively short period from 2.2 to 2.1 Ga (Nisbet and Sleep, 2001; Wiechert, 2002)[249, 406]. Studies of carbon deposition rates and the sulfur isotope record suggest another abrupt rise in atmospheric oxygen occurring at about 0.6 Ga, which probably reached present-day levels (Canfield and Teske, 1996)[37]. Deep ocean environments, on the other hand, are thought to have remained anoxic and highly sulfitic during the long geological period from 2.2 to 0.6 Ga, when atmospheric O_2 was only about 15% of present-day levels (Anbar and Knoll, 2002)[2].

The spawning of wildfires requires an atmospheric oxygen content of at least 13%, and the first evidence of charcoal deposits comes from the Silurian at 420 Ma (Scott and Glasspool, 2006)[330]. Recent results from the analysis of plant material trapped in amber suggest that oxygen levels did not rise to present-day levels of 21% by mass until very recently, remaining at levels of between 10 and 15% by mass from 250 to 30 Ma (Tappert et al., 2013)[372].

The amount of water in the present-day hydrologic cycle, and thus in today's atmosphere, has been predicted to rise by about 3.2% for every 1K increase in surface temperature due to greenhouse warming (Kleidon and Renner, 2013)[161]. Another determination can be made from the saturation pressure of water which increases about 6.5% per degree. With temperatures in the Archean at least 50 °C above those of today, a conservative estimate for the amount of water vapor in the Archean atmosphere would be at least twice as large as today.

Through an analysis of the imprint of "fossil raindrops" from 2.7 Ga discovered in Ventersdorp in the North West Province of South Africa, Som et al.

(2012)[353] concluded that atmospheric pressure in the Archean was probably similar to today's and certainly no more than twice as large as today.

The above uncertain, but best presently available, history of Earth's atmosphere and the better known evolution of the spectral distribution of G-type stars (see chapter 4.1 and Michaelian and Simeonov (2015)[217]) has been taken into account in determining the probable spectral distribution at Earth's surface during certain epochs since the Archean and is given in figures 4.4 and 4.7. From figure 4.7 it is apparent that an important UV-C component penetrated to Earth's surface from before the beginning of life and until at least ~ 2.5 Ga, implying that life evolved below this UV-C component for perhaps 1.35 billion years before oxygen and ozone from oxygenic photosynthesis delegated UV-C dissipation to the upper atmosphere. From figure 4.4 it is obvious that, since the very beginnings of life, there has always been an important flow of low entropy photons in the visible and near UV and therefore it would have been thermodynamically advantageous to evolve pigments which not only absorbed in the UV-C but also in the visible and near UV as, indeed, seems to have been the case (see chapter 4.3). As long as the UV-C component existed, complicated biosynthetic pathways to these pigments were not required. Pigments which absorbed and dissipated in both the UV-C and visible would, in fact, have enjoyed even greater autocatalytic proliferation (see chapter 6) then those that absorbed only in the UV-C.

It is further suggested here that biology evolving the atmosphere of Earth to allow the short wavelength photons to reach the surface would allow not only the photons to be intercepted by the pigments necessarily confined to Earth's surface, but would also allow for the coupling of still more biotic and abiotic irreversible processes, for example the water cycle, winds, and ocean currents, to further red-shift Earth's outgoing energy spectrum and in this way increase the global entropy production of Earth.

17. The Biosphere

17.1 Introduction

This chapter describes how biotic and abiotic irreversible processes become thermodynamically coupled in a hierarchical structuring with feedback between the levels, leading to ever greater overall entropy production of Earth in its solar environment. This global structuring of many interconnected dissipative levels is known as the *biosphere*, here considered as that greater entity composing the processes of life, the lithosphere, atmosphere, and hydrosphere. It is suggested that the biotic component of biosphere arose even before the beginnings of what we commonly consider as life in the form of organic pigments floating on the ocean surface and dissipating UV light into heat. Most of this heat was channeled into latent heat of evaporation, coupling the process of photon dissipation to the primitive water cycle. As described in chapter 3.1 there exists a thermodynamic imperative to couple distinct irreversible processes (whether biotic or abiotic or a mix of both) in order to remove impediments to a greater global entropy production of Earth (Morel and Fleck, 1989)[234].

In 1922 Lotka suggested a principle of natural selection of organisms quite unlike the usual Darwinian notion based on competition for resources. Lotka suggested that natural selection would favor those organisms that augment the available energy channeling through an ecosystem. In Lotka's own words[191],

> "But the species possessing superior energy-capturing and directing devices may accomplish something more than merely to divert to its own advantage energy for which others are competing with it. If sources are presented, capable of supplying available energy in excess of that actually being tapped by the entire system of living organisms, then an opportunity is furnished for suitably constituted organisms to enlarge the total energy flux through the system. Whenever such organisms arise, natural selection will operate to preserve and increase them. The result, in this case, is not a mere diversion of the energy flux through the system of organic nature along a new path, but an increase of the total flux through that system."

Lotka saw the opportunity for speciation through the latching on to new channels of free energy and that this would allow ecosystems to grow and become more robust and favorable to life. One could say that this observation was, in some way, an "energetic" precursor to the theory of Gaia (see chapter 17.4) which suggests that organisms coevolve together, and with their abiotic environment, to increase the suitability of Earth for life. The thermodynamic view emphasized throughout this book suggests that Lotka's new channels for latching on to free energy most often become available through the evolution of new, more efficient, organic pigments which cover ever more of the solar spectrum, as well as the evolution of their associated organismic vehicles which transport them, allowing them and the necessary nutrients to spread into inhospitable regions of Earth's surface for the thermodynamic purpose, not of making Earth more suitable to life, but instead of increasing the solar photon dissipation.

In this view, life has no inherent "interest" in self preservation, but rather is a set of dissipative structures driven by photon dissipation within a biotic-abiotic biosphere which evolves towards ever greater entropy production. Besides the Earth's biosphere, other biospheres, traditionally considered as being abiotic, can provide routes to global entropy production (such as the giant southern vortex on Venus – see figure 0.1) and which route Nature takes towards dissipation, although similarly based on the dissipative structuring of pigments, will depend strongly on the initial conditions of the material and subsequent environmental perturbation but evolution guided by the probabilistic principle of increasing the global entropy production of the planet, or other body, in its stellar environment.

Zotin (1984)[431] and Ulanowicz and Hannon (1987)[383] since the 1980's had already realized that global entropy production was a useful variable for studying the Earth's biosphere. Swenson (1989)[366], and later Schneider and Kay (1994)[323] added valuable insights and empirical data respectively in support of the use of entropy production as a defining variable in ecosystem evolution.

Examples of biotic-abiotic coupling are; biology catalyzing the hydrological cycle (Michaelian, 2009; 2012a; 2012b)[207, 212, 213], and biology catalyzing ocean and wind currents, and the carbon cycle[1]. Such a thermodynamic view of the biosphere provides an explanation of many intriguing biotic-abiotic associations discovered while accessing the Gaia theory (Lovelock, 2005)[192] (see Chapter 17.4) and provides a necessary physical-chemical foundation for Gaia theory.

[1] A most recent example of humans catalysing the carbon cycle in favor of greater entropy production is the steady increase in CO_2 atmospheric concentrations, now surpassing the 400 ppm, since the industrial revolution. It has been recently shown (Zhu et al., 2006)[429] that this has led to a "greening" of Earth's surface, and therefore, greater global photon dissipation.

The thermodynamic view also provides a framework for explaining the observed coevolution of life with its environment, and for the resolution of the paradox of "the evolution of a system of population one – the biosphere" (Swenson, 1991)[367]. One of the earliest criticisms directed against the theory of Gaia coming from prominent biologists and evolutionists was the negation of the ability of the whole Earth to evolve since there was only one Earth and therefore no possibility for competition or the consequent selection necessary for Darwinian evolution. James Lovelock, originator of the theory of Gaia, countered with a particular toy model of how characteristics of the biosphere could indeed evolve based on selection at the individual level and therefore global characteristics of Earth, according to Lovelock, were indeed subject to Darwinian evolution. Lovelock called his model "Daisy World" and it was meant to show how the temperature of a hypothetical planet could be regulated by daisies of only two particular colors, black and white, in order to maintain the temperature at optimal values for their growth. The Daisy World model of Lovelock leads to stability of the planet's temperature, but more importantly leads to optimal photon dissipation rates over the long term, or, in other words, to greater global entropy production. In section 17.4 I give a detailed thermodynamic description of Daisy World. Although its simplicity leads to questions of relevance to the real world, it is a good theoretical example of how thermodynamic selection based on global entropy production acts simultaneously on all levels, including selection at the individual level. However, the Darwinian paradigm of selection only at the individual level is deficient in many respects and some of these deficiencies will be highlighted and addressed in the following sections.

17.2 Hierarchal Dissipation

In a top-down thermodynamic view, organisms should not be seen as individual entities endowed with a metaphysical "will (or drive) to survive" competing for survival against each other and against their environment, but instead as local thermodynamic flows which arise in response to local (on the relevant scale) thermodynamic forces which define their environment. The local thermodynamic potentials providing the forces for the flows are created by other irreversible processes dissipating thermodynamic potentials on a still higher level, and so on up until reaching the highest hierarchal level of the Earth in its solar environment, the biosphere. The moment that a local thermodynamic potential wanes, or becomes depleted, the organism created (irreversible process spawned) at whatever level to dissipate this potential will "suffer", or go extinct.

Organisms (considered as irreversible flow processes in this thermodynamic scheme) on whatever scale, therefore, have higher probability of survival if they are attached to large, robust and stable thermodynamic forces (gradients of generalized chemical potentials). Life, in general, is attached to the greatest thermodynamic potential impressed over the entire Earth; the solar-space photon potential. Since entropy production is given by the product of the thermodynamic flow times thermodynamic force (see equation 3.1), stronger forces, or stronger flows, imply greater entropy production, and, will therefore be thermodynamically selected with greater probability according to Attard's second law of non-equilibrium thermodynamics (Attard, 2008; 2009)[9, 10], which is consistent with the extensive empirical data available listed in chapter 3.2.

Although of lesser potential, there are other thermodynamic forces in the environment such as chemical potentials derived from the degradation (catabolism) of biomolecules originally produced by the photosynthesizers. These weaker forces spawn other dissipative flows such as animals, fungi, heterotrophic bacteria and protists. One specific example of a force based on the gradient of a chemical potential are the fruits of plants which are "designed" for animal consumption and dissipation in order to spread seeds and nutrients needed by the primary photon dissipaters. Another example is the chemical potential in detritus, the rain of the remains of dead surface organisms. Ocean bottom feeders living off detritus, together with their predators, serve to maintain a more homogeneous distribution of the essential nutrients for the photon dissipaters of cyanobacteria or diatoms, particularly at the ocean surface which receives the most sunlight and can become nutrient depleted due to gravitational sedimentation. These other thermodynamic forces based on secondary chemical potentials, although playing an important role in pigment proliferation, are, however, dwarfed in scale by the solar photon potential, and therefore should be considered as secondary or parasitic, but at the same time catalytic, providing the positive feedback essential to maintaining a robust and ever proliferating biosphere.

17.3 Thermodynamic Selection in the Biosphere

All macroscopic irreversible processes, including evolution and those processes required for evolution; replication, mutation, and selection, only find a reason for existence in dissipation. In chapter 3, in particular section 3.2, I have listed the empirical evidence suggesting that there exists a probabilistic universal principle (such as Attard's second law of non-equilibrium thermodynamics or Dewar's maximization of Shannon entropy for the probability distribution of phase-space paths) which implies that nature will most likely be found in those dissipative processes which contribute most to global entropy production. In the case of Earth in its solar environment, this entropy production is mainly photon dissipation. According to such a universal principle, certain processes arising and persisting within the biosphere are selected over others depending on their contribution to global photon dissipation rate. In order to understand how thermodynamic selection works on all scales simultaneously and relate this to the Darwinian paradigm of replication, mutation, selection and evolution at the individual level, and, furthermore, to pinpoint the deficiencies in the Darwinian paradigm, it is necessary to understand how each of these Darwinian concepts, of replication, selection and information coding, and evolution, is associated with global thermodynamic dissipation.

In Chapter 12 I have described a dissipation-replication relation operating through an ultraviolet and temperature assisted replication (UVTAR) mechanism and how thermodynamic selection could have operated on individual RNA and DNA molecules, providing greater reproduction potential to those molecules, or complexes of molecules, which dissipated more photons per unit area per unit time[2]. The most dissipative and thus replicative molecules would be those random RNA and DNA oligos that stereochemically coded for antenna type pigment molecules (such as the aromatic amino acids), or those that stereochemically coded for molecules that made RNA or DNA more amphipathic[3] and thereby less likely to sediment, or those that had some other physical characteristic important for dissipation. RNA or DNA which coded for UV-C absorbing pigments or for molecules helping it adhere to the ocean surface would dissipate more photons into local heat and thus facilitate denaturing in an ever colder sea. This is a description of a mechanism of replication in which the drive behind the replication (in essence, the vitality of life) is derived from non-linear, non-equilibrium thermodynamic principles; the dissipation of the solar photon potential through a photochemical autocatalytic reaction (see chapter 6). All biotic replication, even today, can be described in these terms of fomenting dissipation, albeit today more indirectly (through complex biosynthetic pathways) than at the beginning of life.

Information, in the form of coding sequences for stereochemical affinity to molecules which facilitate photon dissipation, would thus gradually become established within the fortuitous RNA or DNA. Mutation of the oligos through a high energy particle, or other external influence, would have led to an irreversible semi-permanence of the new information within the general population of oligos if this mutation produced a sequence which coded stereochemically for a molecule that increased photon dissipation and thus made replication through such a UVTAR mechanism even more probable. This process of the accumulation of information connected with greater dissipation has been labeled as "evolution" in the biological sciences without awareness of its directed and thermodynamic

[2] Note that this is qute distinct from the conventional, and still contemporary, view of growth and evolution at the molecular scale arising through competition for certain molecular species.

[3] Amphipathic molecules have one end hydrophobic and the other hydrophilic, like the phospholipids, which makes them adhere to the air-water interface (see chapter 15).

character. In the following I will emphasize this thermodynamic character by referring to it as *thermodynamic evolution*.

In this section we will generalize the notions of thermodynamic selection and thermodynamic evolution to cover all hierarchal levels simultaneously, from the molecular level described above, to the level of the biosphere. Within the Darwinian paradigm, although it is understood that the notion of fitness must somehow also be expressed at higher biotic levels, all efforts to derive a mechanism run into problems and paradoxes at these higher biotic levels. For example, how could the fitness of the biosphere be defined in the Darwinian sense? The biosphere is not in competition for survival with any other. This led to the paradox of the evolution of a system of population one, the biosphere (Swenson, 1991)[367].

In the thermodynamic dissipation theory of the origin and evolution of life, there is no problem or paradox in speaking about selection of increases in the entropy production of the biosphere, and indeed, this can occur simultaneously with increases (or even particular decreases) in the entropy production at the individual, or any other, level. Selection occurs simultaneously on all biotic levels and those biosphere changes of greatest probability will be those leading to greater global entropy production of Earth in its solar environment, given Earth's present state. A change in the biosphere at any level may cause the thermodynamic forces at any hierarchal coupled dissipative level (see figure 2.4) to wax or wane, or extinguish, or cause new generalized chemical forces to arise. The thermodynamic flows (molecular complexes, individuals, communities, species, clade, ecosystems, etc. – see chapter 16.7) wax and wane, go extinct, or arise accordingly.

Before detailing thermodynamic selection and evolution at biotic (or coupled biotic-abiotic) levels up to the biosphere, it is first necessary to emphasize some of the characteristics of thermodynamic evolution (replication and selection) at the molecular level, because these characteristics also translate to the higher levels. It is important to understand that each new dissipating level of the biosphere was gradually added to an existing one through a particular invention or symbiosis; each evolutionary event adding to the systems dissipative efficacy. The thermodynamic selection of this growth or evolution of the system will be based, probabilistically, on an improvement in dissipation efficacy. For example, at the oligo molecular level, we saw the possibility of selection based on a dissipation-replication relation through the UVTAR mechanism. At the level of groups or complexes of molecules we saw the possibility of selection based on a dissipation efficacy-replication relation where now information became programmed into the oligos to provide stereochemical attraction of the oligo to the other molecules of the complex, for example the antenna molecule amino acid tryptophan, or to other oligos which formed networks making the group of oligos more amphipathic and less susceptible to sedimentation. We could now take this one step further to the association, also stereochemical, of the dissipative molecular complex with the phospholipids which provided an enclosure for continued replication in colder seas or colder new environments (see chapter 15). The stereochemical information for this association would again become programmed into the DNA either directly, or through some other molecule that had affinity to both RNA or DNA and the phospholipids. These discrete evolutionary events leading to ever greater molecular complexity with ever greater dissipation efficacy can be repeated indefinitely up until reaching the great dissipative complex of the biosphere of today. All the information for defining the biosphere, including its abiotic components (such as the water cycle, the transparency of the atmosphere, sea levels, surface temperature, etc.), has become encoded in the collectivity of all of the genomes of all of the organisms presently living on Earth.

Complex biotic dissipative structures (such as, for example, animals or trees) having evolved through this process of information storage related to dissipation allows them to store information about the generalized chemical potentials existing in their present and past environments. This information endows them with a certain plasticity to "adapt" to different chemical potentials by building on existing structures, or to rapidly return to dissipation of potentials that have returned to the environment after a certain absence, thus allowing them to evolve into ever more universal, efficient, and adaptive dissipative structures.

Another important characteristic for understanding selection and evolution at the higher ecological levels is the *finite time* factor. The UVTAR mechanism mentioned above and described in chapter 12 for molecular replication is not instantaneous; replication occurs over a diurnal cycle, with denaturing and nucleotide activation occurring during the day and extension during the colder overnight periods. This is a finite time process and thus implies that many "external" factors may intervene which could affect (either positively or negatively) the replication process. One example might be the diurnal variation of the ocean surface pH which may intervene to improve or retard the replication process. Any invention (e.g. through random mutation giving rise to a stereochemical coding for another molecule) that would enhance or retard the effect of the diurnal variation of sea surface pH would be thermodynamically selected for, based again on the dissipation efficacy of the new molecular complex through the same familiar dissipation-replication relation. Another external factor could be the diurnal variation of the sea surface salt concentration, etc. The list of external factors that could influence the replication rate would have been very large, precisely because the UVTAR replication mechanism is a finite time process. The finite time factor allowing the complexity and variability of the environment to come into play would thus provide a stimulus for the increasing complexity of the organism and its information content, each new coupling to a distinct molecule increasing slightly the autocatalytic nature of the complex in dissipating the solar spectrum (see chapter 6).

If we now consider this finite time factor at advanced stages of evolution, when the higher biotic levels were arising, it is easy to see that at these higher levels, apart from external abiotic factors affecting the dissipation-replication relation, we could also have biotic factors affecting this relation. This is the feedback referred to earlier. For example, ribosomal RNA (rRNA), initially evolving independently in the same way as DNA, that could build simple enzymes out of amino acids that had the ability to increase the replication rate of DNA, and making its replication less dependent on the diurnal cycle, would also be thermodynamically selected. This could be candidate for a DNA-RNA symbiosis evolutionary event. The list of these inventions through associations, each providing greater dissipation efficacy, is endless and we are now beginning to identify some of them as endosymbiosis events.

Besides biotic and abiotic factors acting over individual molecular complexes, they could also have been acting over groups of such complexes. One example is that of photon dissipation being negatively affected by sedimentation of the complexes due to gravity. RNA or DNA, or their complexes, that sediment to the bottom of the ocean (particularly at night when convection, either atmospheric or oceanic, is low) would become out of the range of the UV-C photons and therefore lost to dissipation and replication. However, it is known that by forming loosely connected networks of double strand DNA, this increases significantly the amphipathic nature (having both hydrophobic and hydrophilic parts) of the system compared to isolated DNA, and the network adheres to the air-water interface (Dai et al., 2013)[69]. The DNA involved would thus remain at the ocean surface where they could receive the greatest UV-C light flux, favoring denaturing during the day. Those DNA that had the most chemical attraction to others (through particular nucleotide sequences) could thus remain at the surface and dissipate more and therefore replicate more efficiently under UVTAR as a group than when separate. Here we perhaps have an early example of "group selection" which, as required for any thermodynamic selection, is again based on an increase in the global dissipation rate.

Besides diurnal variations of the sea surface environment due to Earth's rotation, there would also be seasonal variations due to axial tilt of Earth and its elliptical orbit around the sun. Depending upon the latitude, some regions of the ocean surface may have become too cold in winter to allow the UVTAR mechanism to be effective. In these regions, DNAs with affinity to phospholipids may have become favored since this provided an insulating enclosure where UV-C light could enter and be absorbed on the fundamental molecules, but heat would have remained trapped for some time (due to the low heat conductivity of phospholipids), raising the temperature within the enclosure to a point of optimal UVTAR efficacy. Of course, these phospholipid enclosures would have to be permeable to short RNA or DNA segments and to the fundamental molecules (pigments) of life, but impermeable to longer RNA or DNA segments. The lipid membranes would also need to have the characteristic of growth and binary fission triggered on RNA or DNA replication. All of these are, in fact, characteristics of phospholipids. This facilitation of the UVTAR mechanism may have been the thermodynamic reason for the origin of the first cell-like enclosure (for more details of the emergence of a phospholipid enclosure, see chapter 15).

Through increasing the catalytic efficacy of the complexes in dissipating the solar photon flux, greater replication rates and consequently greater proliferation is conferred (see chapter 6). As the environment changed, or as the complexes spread to new environments, further complexity may have be called for in order to achieve favorable thermodynamic selection. In today's biosphere, thermodynamic selection is acting similarly. However, today, without the potential for direct single photon permutation of covalent bonds (there is no longer UV-C light arriving at Earth's surface), the biosynthetic pathways are necessarily more complex in order to be able to make use of the lesser free energy available in visible photons. Thermodynamic selection and therefore thermodynamic evolution is still, however, based mainly on photon dissipation as it was at the beginnings of life.

An example of this at the biotic level of ecosystems is the re-introduction of wolves into their historical ecosystem in Yellowstone National Park (before uncontrolled hunting led to their extinction). On re-introduction, these top predators began to keep the deer and elk populations always on the move and therefore unable to overgraze, which led to a greening of the park, which, in turn, means greater photon dissipation (see also chapter 16.6). Such a deer and elk dispersion mechanism would also lead to a dispersion of nutrients for the dissipating pigments. The re-introduction of wolves into their natural ecosystem of Yellowstone National Park is synonymous with bringing this system back to its natural climax state, and this state gives greatest photon dissipation. This example shows how thermodynamic selection is operating at the ecosystem level today[4]. It is also still acting at the molecular level today by selecting pigments in the leaves of the trees that are most efficient at dissipating the photon flux.

Still another example of thermodynamic selection acting today at a species societal level is the human-induced increase in atmospheric CO_2 since the indus-

[4] One could certainly argue that the wolves started eating the physically weakest deer and elk, and this is Darwinian selection, but this is missing the global thermodynamic picture which says that the wolves are eating the excessive deer and elk because this increases the global photon dissipation. The ecosystem can be represented by a matrix of many coefficients representing the interaction between individuals of different species. These interaction coefficients evolve such as to optimize the global photon dissipation (see chapter 16.7 and Michaelian (2005)[206]).

trial revolution which has led to a surprisingly important greening of Earth (Zhu et al., 2006)[429], with up to a 50% increase in vegetation in some areas. Although too much atmospheric CO_2 may lead to ecosystem destruction through temperature induced climate change, it is most probable that Nature will indicate to humans, in no uncertain terms, just what the optimal concentration of CO_2 for Earth's atmosphere should be.

Finally, since today's biosphere has both biotic and abiotic components coupled on many different levels, it is relevant to make a few remarks concerning the coupling of biotic and abiotic irreversible process. Both biotic organisms and abiotic processes have the ability to adapt to a changing thermodynamic potential, implying that thermodynamic selection is also a finite time process, as emphasized above with respect to replication. Biotic organisms today adapt through their organismal plasticity (e.g. the ability to migrate, or their ability to survive off different thermodynamic potentials - heterotrophy) or through their genetic apparatus and reproduction (plasticity on the species level). In contrast, abiotic processes have an inherent plasticity, for example, a change in size or direction of a hurricane, in response to a change in the size or direction of a temperature gradient. In our non-equilibrium thermodynamic view of Nature, the processes are not "striving to survive", rather, they are flows responding to changing thermodynamic forces. Darwinian theory speaks of life fighting for survival against an imposed external environment, Gaia theory speaks of "life shaping the environment to its own benefit", but the thermodynamic dissipation theory speaks of life coupling at many different levels to irreversible thermodynamic processes (both biotic and abiotic) which are together evolving to ever greater levels of global entropy production. It is these higher level potentials (in particular the solar photon potential) along with non-equilibrium thermodynamic principles (in particular that of increasing global entropy production) that are the creators, deformers, and selectors of those irreversible dissipating processes that we may call abiotic or biotic organisms.

The thermodynamic functioning of the biosphere and its evolution in time through thermodynamic selection is depicted schematically in figure 17.1. The circles represent the different dissipative processes at different levels in the hierarchy, each occupying a particular area (niche) on the diagram corresponding to sources of free energy. The area within the outermost circle represents the total amount of free energy that could possibly be dissipated given the influx of solar photons and the black-body cosmic radiation of space. The area of the solid discs at the center of each dissipative circle represents the amount of free energy that is actually being dissipated by that process. There is positive feedback from the inner loops towards the outer loops. For example, the water cycle would bring nutrients for more pigment production at the outermost hierarchal level. However, if two circles at a given level overlap the processes must compete for the same free energy niche and the possibilities are that either the two processes couple to form a single bigger dissipating system (that one circle engulfs the other, producing a new level of hierarchy and a larger inner solid disc), or that one of the two processes conquers the free energy available (producing a bigger dissipative circle) leaving the other to wane on a reduced share of the available free energy of the niche (leading to a negative adjustment in the size of its dissipative circle or leading to extinction). In the Darwinian paradigm, these two possibilities are known as symbiosis and competition, respectively, but were only assumed to act at the individual level because selection was assumed to be based on differential reproduction and not on thermodynamic efficacy as suggested here. At the beginning of life only a few pigments were intercepting sunlight and dissipating this light into heat (therefore the small solid inner disc – see figure 17.1). There were few processes and levels involved in dissipation. The albedo of Earth was larger. As time increased, however, more and more dissipative processes (structures) would arise to cover the free energy niches, for example, by feeding off secondary chemical potentials produced as the result of dissipation at a higher level. Given enough time eventually all the available free energy will be dissipated and the whole encircled area will become black. This corresponds to a biosphere of very low albedo, all the incident photons are being absorbed and dissipated into very long wavelength light.

The underlying thermodynamic principles for evolution are the same for both abiotic and biotic irreversible processes. However, there are differences in mechanism which make them appear very distinct, and this has contributed to keeping their analysis separate until now. Although the biotic and abiotic processes of the biosphere are strongly coupled, and it makes little sense to base any conclusions on a model which separates them (for example, cyanobacteria cannot be separated from the origin and evolution of hurricanes – Gnanadesikan et al. 2010[108]), an artificial separation is made here only to emphasize some differences. The main difference between biotic and abiotic irreversible processes appears to be related to the nature of the thermodynamic potential that they arise to dissipate. Biotic organisms primarily dissipate directly the photons arriving from the sun. Covalent interactions between the atoms of the organic molecules provide energy gaps compatible with the quantum energies of visible and UV photons. These molecules, when in water, can thus absorb and dissipate efficiently into heat the incident solar photon spectrum, hence the coupling of life to water and to the water cycle. If not in water, the molecules remain for a longer time in an excited electronic state where they are vulnerable to destruction through photo-reactions, as well as to radiative decay and some other less dissipative means of de-excitation (Middleton et al., 2009)[222]. The non-central forces of covalent interactions gives them directionality properties which allows

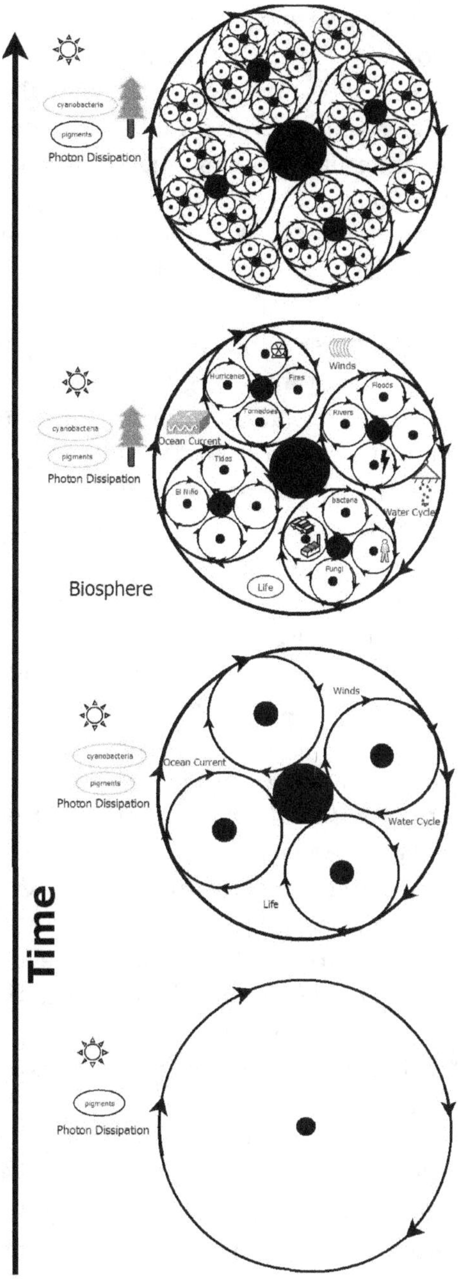

Fig. 17.1. The evolution over time of the dissipation attributed to the biosphere. The circles represent different dissipative processes at different levels in the heirarachy, each occupying a particular area (niche) on the diagram, the size of which corresponds to the free energy available. The area of the solid discs at the center of each dissipative circle represents the amount of free energy actually being dissipated by the process. There is positive feedback from the inner levels towards the outer levels. At the beginning of life only a few pigments were intercepting sunlight and dissipating this light into heat (the albedo of Earth would have been larger). Few processes and levels were involved in dissipation. As time increased, more and more dissipative processes arise, feeding off secondary chemical potentials produced through dissipation at a higher level. See text for more details.

a great, almost inexhaustible, variety of distinct molecular configurations for pigments that absorb from the far UV to the infrared (Gatica, 2011)[101].

This diversity of organic molecular configurations allows for a genetic apparatus of biotic individuals that provides an information storage and retrieval system that accumulates information about the other irreversible processes and thermodynamic potentials in the present and past environments. This endows biotic organisms and species with great plasticity in responding (Darwinists would say "adapting") to changes in the thermodynamic potentials of their environment.

Abiotic processes, on the other hand, tend to dissipate secondary heat gradients in which the relevant quantum energy packets are of much lower energy than those of visible or UV photons. Most of the direct solar photons intercepted and dissipated by abiotic processes are in the infrared (e.g. the absorption of solar infrared light in the atmosphere and oceans gives rise to winds and ocean currents and hurricanes, respectively).

The material of auto-organization corresponding to strictly abiotic processes are usually "glued" together through central and weak hydrogen bonds (e.g. the interaction between water molecules) or through central van der Waals forces. These interactions have no directionality properties, and, therefore, lead to a low diversity at the microscopic molecular level. This precludes a microscopic molecular genetic code and limits the organisms (abiotic irreversible process) ability to accumulate information about the thermodynamic potentials or other irreversible processes operating in their environment. However, information is nevertheless acquired at a more macroscopic level, as, for example, in the distribution of ocean surface water temperatures which "remembers", at least for some time, the history of recent prior hurricanes (and other irreversible heat flow processes) and constrains the birth and evolution of new, future hurricanes.

There is no room in the Darwinian paradigm for the theory of Gaia with its strong coupling of the biotic with the abiotic. Darwinian theory, with its covertly implicit and mystical "will to survive" of the individual is deficient since the "will to survive" was not rendered understandable on physical or chemical law and there is no straight forward and general way to include selection simultaneously at all biosphere levels. The "will to survive" and selection only at the individual level were *ad hoc* assumptions which, although seemingly plausible, in fact led to numerous paradoxes, inconsistencies, and an inherent tautology. Thermodyanmic selection, which associates replication, growth, proliferation and evolution with dissipation makes clear the deficiencies in the Darwinian paradigm and removes the tautology from evolutionary theory. Thermodynamic selection allows the evolutionary program to be extrapolated back to the protoplanetary disc and sheds new light on the origin of life. Furthermore, it allows some predictions for life's future evolution from where we find it today (see chapter 20.1).

It was generally admitted in the biological sciences that selection must be acting simultaneously on different biotic levels, but mechanisms were difficult to come by and, as mentioned above, always ran into problems at the higher biotic or coupled biotic-abiotic levels. In recognizing the insurmountable logical problems of explaining higher level selection leading to evolution from within the Darwinian paradigm, and the need to extend the theory in order to resolve these issues, renowned paleontologist Stephen Jay Gould appealed to a collaboration of the biologists with, of all professions, philosophers. On p. 28 of the "Structure of Evolutionary Theory" Gould writes (Gould, 2002)[110],

> "And yet, the conceptual problems presented by theories based on causes operating at several levels simultaneously, of effects propagated up and down, of properties emerging (or not) at higher levels, of the interaction of random and deterministic processes, and of predictable and contingent influence, have proven to be so complex, and so unfamiliar to people trained in the simpler models of causal flow that have served us well for centuries, that we have had to reach out to our colleagues explicitly trained in rigorous thinking about such issues. Thus we learned, to our humbling benefit, that conceptual muddles do not necessarily resolve themselves "automatically" just because a smart person – namely one of us, trained as a scientist – finally decides to apply some raw, naive brain power to the problem. Professional training in philosophy does provide a set of tools, models and approaches, not to mention a feeling for common dangers and fallacies, that few scientists (or few "smart folks" of any untrained persuasion) are likely to possess by simple good fortune of superior raw brainpower."

Of course, physicists, chemists, and complex system scientists, would probably reproach Gould for not giving them an opportunity for collaboration before calling in the philosophers! It seems that the conceptual difficulties encountered in leaving one's paradigm, in this case in understanding non-equilibrium thermodynamics and autocatalytic chemical reactions, was calculated to be too great for this particular eminent paleontologist, evolutionist and biologist, who preferred to let the philosophers attempt to fix the logical structure of the traditional biological arguments rather than renouncing the cherished paradigm[5].

[5] An example of this laziness in searching for answers beyond one's particular paradigm is the scant intellectual effort expended by the biologists to unravel the "survival of the survivors" tautology in Darwinian theory which was brought to popular attention by the philosopher Karl Popper. The biologists, palentologists and evolutionists became satisfied with Popper's recanting of his criticism of tautology (apparently as a result of his realization of the usefulness of even circular arguments in science) but ignored the fact that Popper continued to call evolution through natural selection a "metaphysical research program". The Darwinian paradigm has become consensus science within the biological sciences community and is ferrosiously defended irrespective of its obvious deficiencies.

It is the author's hope that this book, at the very least, will serve to convince the biologists of the need to look to other well established fields of research and to consider honest collaborations with physicists, chemists, and complex system scientists.

17.4 Gaia and Dissipation

In the early 1970's, James Lovelock was offered a NASA contract to come up with a way to detect life on Mars without the heavy cost of sending equipment to the planet. Lovelock realized that Earth's atmosphere was surprisingly different from that of its neighboring planets due to the large amount of the very reactive species oxygen. On a planet without life, oxygen would quickly react forming oxides with metals and other elements and rapidly settle out of the atmosphere. Life seemed to be maintaining the Earth's atmosphere out of chemical equilibrium. Lovelock reasoned that the observation of an out of equilibrium atmosphere on whatever planet would indicate the probable existence of life. Since the gases of the atmosphere of Mars were in chemical equilibrium, Lovelock concluded that there was no life on Mars.

Studying how life affected Earth's atmosphere led Lovelock to discover many other effects of life on its environment. Shortly after these findings, Lovelock proposed a theory in which the entire Earth was considered to be one great living organism which he named Gaia, in reverence of the ancient Greek goddess of the Earth. His justification for Gaia was based on the fundamental way in which life on Earth interacted extensively and coordinately with its physical environment in such a way as to appear to stabilize the conditions for its own existence. Lovelock thus began to shift his attention away from the individual organism and instead see the complete biosphere as the essential living system. The biosphere had, in fact, many attributes of a living organism and Lovelock compared it to a large tree; the living parts being confined to an outer shell and all the individual live cells of the tree acted together in a coordinated form giving rise to a homeostasis of the giant organism (see figure 17.2).

Empirical evidence for the active role of life in providing a kind of homeostasis for the Earth's biosphere, from control of its temperature, control of the oceans salinity, to stabilization of the atmospheric concentration of oxygen, etc. has come to light through the work of Lovelock and his collaborators in their attempt to verify the theory of Gaia, in particular, through collaborative work with Lynn Marguilis (the biologist wife of the physicist Carl Sagan) and her colleagues. The most liberal proponents of this theory (which, at the beginning, included Lovelock and Marguilis) considered the Earth as a living organism and associated the biosphere with the ultimate unit of Darwinian selection. The more conservative proponents of the theory, however, acknowledged the apparent sta-

Fig. 17.2. James Lovelock, author of the theory of Gaia, compared the Earth to a living tree like these sequoias in northern California. The thin layer of life over Earth represented the outer shell of the tree which was alive. The bark of the tree protects it and allows for the exchange of gases and other nutrients, like our atmosphere does to life on Earth. Most importantly, all the components of the system acted together in a coordinated fashion to achieve homeostasis. Image credit: Tuxyso / Wikimedia Commons / CC-BY-SA-3.0

bilization of the external environment through life and gave credit to Lovelock and collaborators for delineating the evidence for the strong interaction between the biotic and abiotic, but considered the idea of the whole Earth as being a single living organism as merely a metaphor, nothing more.

Gaia theory has gone from being completely ignored by the scientific community, to being despised and aggressively criticized, to being hotly contested and passionately debated from many distinct viewpoints. Perhaps the most copious and strongest criticism of this theory has come from the traditional evolutionary and life scientist, Darwinian purists, who insist that all life falls under Darwinian paradigm with clear rules for reproduction and selection at the individual level. They claim that since the Earth cannot reproduce itself, it simply cannot be subject to the dictates of evolution through natural selection. For example, Steven Jay Gould, during his lifetime distinguished as a "Living Legend" for his work in evolutionary theory, in his last work entitled "The Structure of Evolutionary Theory" (Gould, 2002, p612)[110] states,

> "The earth does not generate children, and did not arise by competitive prowess as the sole survivor among defeated brethren (who must have died or been expelled, I suppose, from the solar system long ago). Therefore, among a plethora of other reasons, the earth cannot be construed as a Darwinian agent or unit of selection."

There are other evolutionists, the most notable being Richard Dawkins, who negate the Gaia theory by going to the other extreme of the very small by suggesting that the gene is the ultimate, and only real, object of selection. Dawkins then denies the Earth a living status since no gene has been found for which the entire Earth could be considered its, what Dawkins calls, "survival machine"[6].

However, notwithstanding much of the valid (as well as unfair and misdirected) criticism against the theory of Gaia, James Lovelock must be credited for generating the awareness among scientists that life is strongly coupled to its physical environment and plays a crucial role in determining and stabilizing the environment. The theory of Gaia has opened up a new window on the phenomena of life. No longer seen as subject to the whims of its external environment, the theory of Gaia has shown us that life plays an integral part in shaping, and continually evolving, its own environment. Lovelock and his collaborators uncovered much empirical data as they went about collecting evidence for the theory, all of which demonstrates the tightly interwoven fabric of life and

---

[6] In 1976 Richard Dawkins published what became a popular book entitled "The Selfish Gene"[71] in which he proposed that the fundamental unit of natural selection was the gene. All of evolution, according to Dawkins, could be reduced to the selfish attitude of genes to want to increase their expression in following generations (Dawkins suggested that the phenotype was merely the "survival machine for the genes"). This theory, which envisions natural selection acting only at the level of the gene, is fundamentaly flawed since it makes the even more outlandish ascertion that a gene has an inherent interest in survival and furthermore denies natural selection acting on higher levels such as organism, species, clade, ecosystem, etc. Dawkin's theory, however, is no longer a theory considered as a serious contender by most evolutionists. In criticism of Dawkins's theory, Gould (2002)[110] makes the point that the gene is merely the vector, and that all physical interactions which take place in deciding on selection happen at the individual and higher levels, and therefore the gene cannot be the ultimate unit of selection. Both of these viewpoints, that of Dawkins and Gould, do not permit evolution at the biosphere level and they are both incorrect since they ignore the fact that life consists of coupled irreversible processes and that Nature's only "interest" is in increasing the global entropy production over time. Life, at any level, is not an organism with an inherent selfish interest in survival, but it is a heirarichal set of dissipative processes which wax and wane and evolve over time in response to local generalized chemical potentials in order to increase the global dissipation of Earth in its solar environment (see figure 17.1).

its abiotic environment. The books by Lovelock on the theory of Gaia[7] make fascinating reading for anyone interested in understanding the view of life as architect, builder, and curator of its abiotic environment.

Much of the data collected by Lovelock and his followers provides invaluable evidence for the present theory concerning the thermodynamic dissipative origin and evolution of life. The most important lesson from Gaia is the demonstration of just how intricately the biotic is connected with the abiotic. It is known from the Onsager's and Prigogine's formulation of irreversible thermodynamics (see chapter 3.1) that distinct abiotic processes with similar symmetry elements may couple whenever their combined effect leads to a greater global entropy production when dissipating the generalized thermodynamic potentials imposed on the system by the external environment. Through photon absorbing and dissipating pigments, life facilitates most of the abiotic irreversible processes that occur on Earth; processes that produce entropy at the expense of the high energy photons arriving from the Sun.

A particularly interesting example of how the data obtained in validating the Gaia theory supports the thermodynamic dissipation theory of life is the sulfur cycle as analyzed by Lovelock (20025)[192]. Sulfur is an essential element needed by all living organisms. Ions of sulfide are eroded out of the rocks through the action of falling rain and carried to the sea by rivers. It has been reasoned for some time that a mechanism must exist for returning sulfur to the land in quantities large enough to maintain life. The mechanism discovered by Frederick Challenger and Margaret Simpson (Challenger and Simpson, 1948)[39] is that much of marine life emits dimethyl sulfide(DMS) into the atmosphere. (It is, in fact, this chemical that gives the sea its characteristic smell, much loved by some.) Lovelock proposed that this is the principle transfer mechanism of sulfur from the sea to the land. Later, meteorologists R. Charlson with Lovelock and others (Charlson et al., 1987)[43] suggested that dimethyl sulfide would oxidize quickly in the atmosphere producing sulfuric acid which would then act as nuclei for the condensation of water drops and thus form clouds. Lovelock then reasoned that because of the high albedo such a cloud dynamics could provide a biotic temperature control mechanism for his auto-regulating planet, Gaia. There is evidence that the greater the sea temperature, the greater the emission of dimethyl sulfide by ocean algae (plankton, in particular *Emiliana huxleyi*). This then would lead to greater cloud formation giving a greater albedo, and thus a reduction of the sea surface temperature.

From the point of view of stabilization of the environment for living organisms, this is a reasonable conjecture; the plankton, by emitting dimethyl sulfide and forming clouds, thereby control their optimal temperature for photosyn-

[7] For example; Lovelock, J. E.: Gaia: Medicine for an ailing planet, 2nd Edn., Gaia Books, New York, 2005.

thesis and reproduction. However, this explanation again implies an implicit (and non-defined) organismal "inherent will to survive and proliferate". There is a different explanation which lends support to the thermodynamic dissipative theory presented here.

As we have mentioned earlier, the evaporation/rain cycle is an important entropy producing process on Earth. Energy used to evaporate the water is consumed at the high temperature of the ocean surface (10 to 30 °C) and then released as infrared energy upon the condensation of water vapor into water droplets at the low temperature of the cloud tops (-15 °C). The rate of evaporation from any body of water not only depends on the temperature of the water and the air above it, but also on the relative humidity of the air above the water surface. An atmosphere at humidity close to saturation (100% relative humidity) does not permit further evaporation. By acting as nuclei of condensation, the sulfur dimetal released by the algae thereby reduce the relative humidity of the atmosphere and thus allow a greater rate of evaporation. The algae are therefore acting as catalysts not only to the dissipation of incoming solar photons by absorbing and dissipating this light into heat within their pigments, but are also acting as catalysts to the other great entropy producing process of the evaporation/rain cycle known as the water cycle. The emission of dimethyl sulfide by the plankton would thereby also lead to greater water and sulfur cycles over land.

Another interesting fact uncovered while pursuing the validation of the theory of Gaia is that life maintains the Earth's water supply in tact. The planet would be dry, very much like our sister planets Venus and Mars if it were not for the process of photosynthesis that freed oxygen molecules from CO_2 allowing them to bind to hydrogen molecules freed by the interaction of hot volcanic basaltic rock with water (Lovelock, 2005)[192]. More than just a catalyst of the dissipative process of the water cycle, life is absolutely necessary for the continued existence of water on Earth.

This strong interaction between the biotic and abiotic is exactly what is expected from a non-equilibrium thermodynamic perspective and we would simply call it a "coupling of irreversible processes" to stabilize and increase global dissipation. Onsager (1931)[256] showed that this would happen as long as the entropy production of the coupled system increases as a result. There are many examples of this type of coupling in the non-equilibrium thermodynamic literature, going back over 100 years, such as, for example, the Sebeck effect in which an electrical current couples to a heat flow, or the Soret effect in which a heat flow couples to a flow of material, both couplings increasing the global entropy production of the system. Here we have seen how photon dissipation in the pigments surrounded by water inside a leaf (or cyanobacteria) couples strongly to evaporation and thus the water cycle (Michaelian, 2012a)[212]. Together, the system as a whole, pigments in plants plus water cycle, is much more efficient at dissipating the solar photon flux. The one great "living" organism of Gaia could therefore be identified as a great non-equilibrium dissipative structure, involving the coupling of both biotic and abiotic irreversible processes, known as the biosphere. The biosphere is "living" off solar photon dissipation, as Boltzmann first understood 130 years ago.

As an example of how photon dissipation is optimized in the biosphere, we reconsider Lovelock's Daisy World, this time from the perspective of the thermodynamic dissipation theory of the origin and evolution of life. Responding to critics who suggested that it was impossible that the biosphere could evolve towards improving the conditions for life on Earth because the biosphere was not in competition with any other, Lovelock presented a simplified but ingenious model of the biosphere, called Daisy World. Lovelock's intention was to show that selection acting on individual organisms could indeed affect the global environment for the benefit of all the organisms. In Daisy World, there only exists daisies, and only two varieties of daisies, one white and the other black. For the sake of the model, daisies were assumed to proliferate best at a particular temperature, neither too hot nor too cold. For the purpose of setting the initial conditions of Daisy World, we assume that originally the genetic allele frequencies for the black and white daisies were equal within the population and that this situation was adapted to the incoming solar light, in the sense of giving a surface temperature optimal for proliferation.

If for some reason the surface solar light would diminish, for example, by a volcanic eruption sending plumes of gas and dust into the upper atmosphere, then the surface temperature of Daisy World would drop and the equal allele frequency would no longer be optimal for proliferation, since now too much light would be reflected by the white daisies to maintain the temperature at an optimal value for proliferation. A mutation which made the white allele dominant would therefore not propagate very well since this would reduce the temperature even further and reduce the reproduction rate of all daisies. A mutation which made the black allele dominant, however, would indeed propagate well as this would increase the surface temperature to bring it back to its optimal value for reproduction. The reverse (increases in the white allele) would, of course, happen if the solar intensity at Daisy World's surface for some other reason increased rather than decreased. In this way, reproductive success at the individual organism level would indeed be able to affect the global characteristics of the biosphere, in this case the temperature, in such a way so as to make the environment optimal for all organisms. Daisy World effectively silenced most of the critics of Lovelock's Gaia theory, although few openly admitted that Lovelock could be right.

The Daisy World model of Lovelock shows how a many variable model can go through adaption at the component level to evolve a global variable (optimal temperature) through a global optimization problem, in this case maximum proliferation of the daisies. However, contrary to popular belief, it has been shown that Nature does not optimize photosynthetic proliferation, but rather evapotranspiration (Wang et al., 2007)[392] which is related directly to photon dissipation. How would the dynamics of Daisy World appear if instead of proliferation being optimized, photon dissipation or entropy production were optimized? Proliferation of black daisies would certainly improve dissipation over proliferation of white daisies (since reflection of light without changing wavelength produces less entropy than absorption and dissipation leading to an increase in wavelength). This would, however, raise the Earth's surface temperature, perhaps to values non-optimal for daisy proliferation. What would be thermodynamically selected in this case? Perhaps a mutation which allowed effective photosynthesis at higher temperatures, or better still (since it means a lower black-body temperature of daisy emission), a mutation that increased the evapotranspiration at the leaves of the daisies to decrease the temperature to lower the daisy black-body emission temperature (and perhaps a more optimal temperature value for photosynthesis). Such a mutation may consist of an increase in the depth of the root system to bring up more water at larger depths, a more porous leaf to allow water to carry away in heat of vaporization the energy left by photon dissipation at the pigments; a means of storing photon energy in regions of little available water and dissipating it at night to avoid undue loss of water in the daisy during the day (a process operative in cactus). Under our thermodynamic dissipation perspective, daisies would evolve towards ever more black alleles in the population and evapotranspiration systems would improve, increasing the size of the secondary dissipative process known as the water cycle, ever lowering the temperature of the black-body emitting biosphere while absorbing and dissipating as many incident photons as possible.

If we now go into the field and observe Nature as it presents itself to us, this black, highly dissipating, daisy world is exactly what we find. Plants are in fact essentially black, absorbing at close to the 90% level over the whole of the UV and visible region of the solar spectrum arriving at Earth's surface (see figure 16.3). We only see plants green because our eyes are most adapted to the green and this is also where there is a small dip (of only a few percent) in the plant absorption spectrum. Plant pigments have evolved over time to cover ever more of the solar spectrum (become ever more black), the root systems of plants have evolved to become increasingly extensive, bringing up water from ever greater depth, evapotranspiration has increased over evolutionary time and the corresponding amount of water in the water cycle has more than doubled over the evolutionary history of life on the continents. Evolution of relatively recent

CAM and C4 photosynthesis, at 32 and 9 million years ago respectively (Osborne and Freckleton, 2009)[263] have opened up new ecological niches for dissipation in water scarce areas. These plants, such as desert cacti, are able to take in CO_2 while reducing water loss to avoid cavitation events in strong sunlight, by either opening their stoma only at night (CAM photosynthesis) or by reducing photo-respiration (C4 photosynthesis). However, this water conserving photosynthesis has not displaced the older, heavily transpiring C3 photosynthesis, which is still relevant for 95% of Earth's biomass.

The real "Daisy World" tells us exactly what Nature is optimizing, not proliferation for life's sake, but photon dissipation for entropy production. Natural selection is actually thermodynamic, at all levels up to the highest level of the biosphere (see chapter 17.3). Selection is not based on reproductive success at the organismal level as depicted in Darwinian theory and in the original Daisy World model. In a real Daisy World we would see black daisies proliferating and from the perspective of the deficient Darwinian paradigm we would say that they are black so as to maintain their temperature at optimal values for reproduction, without realizing that the true variable that is being optimized is photon dissipation.

17.5 Indicator of Ecosystem Health Based on Entropy Production

Ecosystems arise and evolve predominantly through the thermodynamic imperative of dissipating the solar photon flux into heat. Organic pigments coupled to water inside cyanobacteria, algae and plants provide the dissipative structures for this entropy production. Viruses, bacteria, insects and other animals play the role of diversifiers and nutrient and seed dispersers in favor of the proliferation and dispersal of pigments over Earth's entire surface. Since the industrial revolution there has been an enormous negative human impact on the majority of Earth's ecosystems, antagonistic to their nominal supportive role in photon dissipation. Discerning whether or not efforts in reversing the damage are having the desired effect requires an accurate measure of ecosystem health. This section describes an indicator of global ecosystem health based on the entropy production of the ecosystem as a whole which recognizes solar photon dissipation as its ultimate thermodynamic function. Thermodynamic justification for using the "red-edge" as an even simpler remotely sensed indicator of ecosystem health is also given.

With the publication of a seminal paper by Ulanowicz and Hannon (1987)[383] that it was realized that entropy production was an important ecosystem variable which could be used to study the dynamics of ecosystem succession and evolution. Ulanowicz proposed using remote sensing to determine the entropy production as the difference in the integrated entropy spectrum of the photons leaving and entering an ecosystem. The entropy flux was calculated by Ulanowicz at a given wavelength λ to be approximately the energy in the photon flux at that wavelength, $e(\lambda)$, divided by a temperature, $T(\lambda)$, i.e. $S(\lambda) = e(\lambda)/T(\lambda)$. The temperature was determined by assuming the photon flux to be a Bose-Einstein gas in thermal equilibrium giving $T(\lambda) = hc/k\lambda$, where h and k are the Planck and Boltzmann constants respectively and c is the speed of light. Ulanowicz suggested that "mature" ecosystems would have a more red-shifted emitted spectrum and thus greater entropy production. According to Ulanowicz, not only would the emitted spectrum of ecosystems be red shifted with respect to that of areas barren of life, but the albedo (ratio of the reflected to incident light integrated over the visible region of the spectrum) measured over living areas would be lower than over areas barren of life.

Schneider and Kay (1994)[323] considered the proposal of Ulanowicz and applied the thermodynamic formalism to remotely sensed temperature data obtained by Luvall and Holbo (1991)[194]. Given an incident photon spectrum and assuming a black-body spectrum for the emitted radiation, ecosystems measured at a lower temperature would have a more red-shifted emitted black-body spectrum and hence greater entropy production. In this way, Schneider and Kay demonstrated that old growth forest ecosystems had a greater entropy production than new growth forests and, in turn, the latter had a greater entropy production than clear cut areas. A reverse trend was found, as Ulanowicz had predicted, for the albedo, for example, the albedo over old growth forest was measured to be as low as 5% while that over clear cut areas increased to 25% (see also Betts and Ball 1997)[20]. This work showed that it was indeed possible to distinguish between stages of ecosystem succession using thermodynamic principles and employing simple remote sensing temperature measurements. In summary, for a given incident photon flux, older, more established, ecosystems have greater entropy production and thus a lower black-body temperature and this relation between entropy production and the maturity of the ecosystem is now well corroborated.

Wang et al. (2007)[392] have shown that under variation of external conditions, and even under stressful situations, plants optimize transpiration rather than photosynthesis. Transpiration removes the heat of the dissipated photons at the leaf surface by converting it into latent heat of the evaporation of water and thus is directly associated with photon dissipation. Together, photon dissipation and transpiration account for, by far, the greatest free energy dissipation performed by plants (Hernandez Candia 2009, Michaelian 2012b)[129, 213]. If, by extension, it is also true that ecosystems optimize the rate of solar photon dissipation under variation of external conditions, as Ulanowicz proposed and the empirical analysis of Schneider and Kay suggests, and therefore that healthy ecosystems have greater entropy production than unhealthy or stressed ecosystems, then a measure of entropy production should be a reliable indicator of ecosystem health.

The author is not aware of any published data with regard to remote temperature sensing comparing healthy with unhealthy ecosystems. Although using recorded temperature values as a measure of ecosystem health should not be discounted *a priori*, there are, however, a number of complications and problems related with such an approach; 1) ecosystem temperatures are a function of the intensity of the incoming solar radiation 2) comparisons of the temperature must be made over extended periods and therefore prone to atmospheric and seasonal variations, 3) ecosystems do not emit light in a black-body spectrum (Gates, 1980)[100] and therefore an equilibrium temperature is not even a well defined concept for ecosystems. Here, instead, I consider a more accurate determination of the true entropy production of an ecosystem and define this number as the best possible indicator of its present state of health.

The true entropy production due to photon dissipation can be directly obtained from the differences between the incident and emitted entropy flux of the light spectra as Ulanowicz suggested. However, it is not necessary to assume that ecosystems are black-bodies, an approximation in error of between 30 to 40% (Michaelian 2012b)[213] and which, in fact, can be questioned on the grounds that ecosystems are out of equilibrium structures. In the following section I determine an accurate value for the entropy production of an ecosystem using equations for the entropy of a photon flux derived by Planck (1913)[284] (see the appendix), including a contribution for photon scattering without absorption. The analysis takes as input the incident and emitted photon spectra and produces a single number, our "indicator" of ecosystem health, for the entropy production of the area under observation.

In section 17.5.2, I describe an alternative indicator of ecosystem health, still based on total entropy production, but now obtained through a more simple remotely sensed determination of the red-edge; the wavelength at which the absorption of light by plants, algae and cyanobacteria decreases rapidly from very high values (which occurs at wavelengths of around 700 nm). The proposed thermodynamic justification for the association of the red-edge with ecosystem health is that, under nutrient or other physical stresses, photosynthetic organisms would prioritize the production and maintenance of primarily those organic pigments which dissipate the highest energy photons available since this maximizes entropy production under the given restrictive conditions. The entropy production of an ecosystem may thus be directly related to the remotely sensed position of its absorption red-edge (Michaelian 2015)[218]. The red-edge is therefore a simple and reliable indicator of ecosystem health, not requiring full spectrum integration over wavelength and independent of temporal atmospheric or insolation conditions, although, as with a full calculation of entropy production by integrating over wavelength, it has a detectable seasonal variation (Gates 1980)[100] in fact related to nutrient flow restrictions in the manner indicated above.

17.5.1 Entropy Production as an Indicator of Ecosystem Health

Photon dissipation by ecosystems is a coupled process involving various stages. In the first stage, a high energy photon from the Sun is absorbed on an organic pigment molecule of a plant, algae or cyanobacteria. The electronic excitation energy is dissipated through various de-excitation processes, the principal of which is known as internal conversion, into the translational and vibrational modes of the surrounding water molecules, thereby increasing the local temperature of the water. A certain amount of liquid water is thus converted into gas, removing the latent heat of vaporization from the organism. The H_2O gas rises in the atmosphere to a height at which the temperature is low enough for condensation around microscopic particles, leaving part of its heat of condensation to escape into space in the form of many infrared photons. A single high energy photon (visible or ultraviolet) is thus converted into many (20 or more) infrared photons, conserving the total energy but producing entropy in the process; since the initial photon energy has been distributed over the many more degrees of freedom of the numerous infrared photons. Most of the entropy production, about 63% (Kleidon and Lorenz 2005)[160], occurs at the surface of Earth during the first stage of the process where the incident photon is absorbed and dissipated by organic pigments. A further approximately 2.6% can be attributed to the latent heat flux of the ensuing water cycle (Kleidon and Lorenz 2005)[160]. Details of how biology catalyses the hydrological cycle can be found in Michaelian (2012b)[213] and will not be discussed further here except to say that this coupling is important to keep in mind when determining our indicator of ecosystem health based, for example, on remote sensing satellite data which detects light emission from both the ecosystem and the atmosphere.

The entropy production of a specific area of the Earth's surface can be determined by considering the change in the frequency or wavelength distributions of the radiation incident from the Sun, $I_{in}(\nu)$ [Jm^{-2}] or $I_{in}(\lambda)$ [Jm^{-3} s^{-1}], and that radiated by the area, $I_{rad}(\nu)$ or $I_{rad}(\lambda)$, including the change in the directional isotropy of the radiation. Planck (1913)[284] determined that the entropy flux $j(\nu)$[Jm^{-2} K^{-1}] due to a given photon energy flux $I(\nu)$ takes the following form (Wu and Liu 2010)[413] (the derivation is given in the appendix, see equation (A.8)),

$$j(\nu) = \frac{n_0 k \nu^2}{c^2}\left[\left(1+\frac{c^2 I(\nu)}{n_0 h\nu^3}\right)\ln\left(1+\frac{c^2 I(\nu)}{n_0 h\nu^3}\right) - \frac{c^2 I(\nu)}{n_0 h\nu^3}\ln\frac{c^2 I(\nu)}{n_0 h\nu^3}\right] \quad (17.1)$$

where n_0 denotes the polarization state, $n_0 = 1$ or 2 for polarized or unpolarized photons respectively, k is the Boltzmann constant, c is the speed of light, and h is Planck's constant. In terms of wavelength ($\lambda = c/\nu$), the corresponding expression is (Wu et al. 2011)[414] (the derivation is given in the appendix, see equation (A.7)),

$$j(\lambda) = \frac{n_0 kc}{\lambda^4}\left[\left(1+\frac{\lambda^5 I(\lambda)}{n_0 hc^2}\right)\ln\left(1+\frac{\lambda^5 I(\lambda)}{n_0 hc^2}\right) - \frac{\lambda^5 I(\lambda)}{n_0 hc^2}\ln\frac{\lambda^5 I(\lambda)}{n_0 hc^2}\right] \tag{17.2}$$

which has the units [Jm^{-3} K^{-1} s^{-1}]. The total entropy production is thus

$$J = \int_0^\infty d\lambda \int_S (\boldsymbol{j}_{out} - \boldsymbol{j}_{in}) \cdot \hat{n} dS, \tag{17.3}$$

where $\hat{n}$ is the unit normal to the surface S which contains the irreversible processes and the entropy flows, j, have units, for example, of [WK^{-1}m^{-2}].

The radiated part $\boldsymbol{j}_{out}$ is composed of two parts, that due to emission after absorption $\boldsymbol{j}^e_{out}$, and that due to reflection without absorption $\boldsymbol{j}^r_{out}$. For the ecosystem, we may assume isotropic emission into a 2π solid angle and predominantly Lambertian reflection also into a 2π solid angle since scattering from leaves is predominantly diffuse (Gates 1980)[100] and multiple scattering from many leaf surfaces occurs in ecosystems. Therefore, Eq. (17.3) becomes

$$J = \int_0^\infty d\lambda \left[2\pi j_{out}(\lambda) - \int_{\Omega_{in}} j_{in}(\lambda)\cos\theta_{in} d\Omega_{in}\right], \tag{17.4}$$

where θ_{in} is the angle of the incident solar radiation with respect to the normal of the detection surface and Ω_{in} is the solid angle subtended by the Sun as seen from the surface of Earth. For example, if we take the Sun directly overhead ($\theta_{in} = 0$) and the detection surface perpendicular to the zenith, then equation (17.4) can be simplified to give

$$Health = J = \int_0^\infty d\lambda\, [2\pi j_{out}(\lambda) - 0.04 j_{in}(\lambda)], \tag{17.5}$$

where $j_{out}(\lambda)$ is obtained from Eq. (17.2) with $I_{out}(\lambda)$ measured by the detecting spectrometer and $j_{in}(\lambda)$ obtained from Eq. (17.2) with $I_{in}(\lambda)$ the solar spectrum at Earth's surface with the Sun directly overhead. The factor of 0.04 accounts for the solid angle subtended by the Sun at the Earth's surface. The units [SI] of this indicator of ecosystem health (entropy production) are [J K^{-1} m^{-2} s^{-1}].

The distance above the ecosystem at which the spectrometer is flying, and the solid angle of the detector, will determine the extent of the ecosystem considered. Satellite measurements are most global but will include other coupled abiotic dissipative processes as mentioned above, such as the water cycle and ocean and wind currents, which are spawned by the heat generated through photon dissipation in the ecosystem. In a more accurate calculation, one would also have to consider photon dispersion by clouds and the atmosphere.

A few remarks are in order with respect to this measure of ecosystem health based on Eq. (17.5), or more generally (17.4). First, it is an instantaneous measure which will vary throughout the day and is not completely accurate since part of the energy absorbed by the ecosystem during the day is released at night and this radiation is not included in the instantaneous detector measurement. A more accurate measure would integrate Eq. (17.4) over the 24 hour diurnal cycle, but would be significantly more complex to perform. The same applies to the annual cycle. Second, equation (17.5) is more accurate than simple temperature measurements since there is no assumption of thermodynamic equilibrium (a black-body spectrum) and the entropy production is not based on heat flow equations, which can result in up to 40% error in the calculated entropy production (Michaelian 2012b)[213]. Third, by considering the full spectrum of the radiated entropy flow as the emitted plus reflected, $\boldsymbol{j}^{e}_{out}+\boldsymbol{j}^{r}_{out}$, the above calculation also takes into account the entropy production due to the Lambertian scattering of the component which is reflected and referred to as the albedo, which accounts for roughly 8.3% (assuming a wavelength independent albedo) of the total entropy production integrated over the whole of Earth's surface (Michaelian 2012b)[213].

17.5.2 The Red-edge as an Indicator of Ecosystem Health

It is a curious fact that the great majority of phototropic organisms have strong absorption throughout the UV and visible regions of the sun's spectrum but a sudden pronounced drop in absorption at approximately 700 nm. This sudden drop in absorption is known as the "red-edge" (see figure 16.3). Beyond the red-edge almost all light is either reflected or transmitted by the organism until around 1400 nm where the strong absorption bands of water in the organisms become important (see figure 10.2).

The red-edge has been attributed to a gap in the molecular energy levels between the lowest energy vibrational state of the 1st electronic excited state and the highest energy vibrational state of the electronic ground state (Gates 1980)[100]. A second explanation is also given by Gates (1980)[100] suggesting that it may be an evolved characteristic since plant leaves would heat up to beyond optimal temperatures for photosynthesis if the leaves also absorbed the solar energy beyond the red-edge. However, these explanations do not appear convincing, particularly given the fact that photosynthesis is most efficient at wavelengths around the red-edge. The first explanation can now be definitively rejected since there have now been found many deep ocean living bacteria that have strong electronic absorption within the gap beyond the red-edge and that, in fact, use the faint very red light from deep sea hydrothermal vents for efficient photosynthesis (Kiang et al. 2007, Beatty et al. 2005)[158, 16]. Anoxygenic photosynthesis has been discovered using wavelengths as long as 1015-1020 nm (Trissl, 1993, Scheer, 2003)[378, 321]. The second explanation of Gates appears also to be lacking since plants could have equally well evolved to reflect or transmit the UV and blue light and only strongly absorb at a peak centered around 700 nm where, in fact, photosynthesis is most efficient (Kiang et al., 2007)[158].

Here I suggest instead that the red-edge can be explained by the finite size and dead-time (excited state decay time) of present day organic pigments under the premise of the optimization of entropy production in organisms dissipating the solar photon flux. The solar photon flux integrated over the whole spectrum at the Earth's surface at the equator and at midday is of the order of 2×10^{21} photons per square meter per second. This copious flux saturates present day organic pigments given their finite size and finite excited state dead-time. It would thus be most profitable, from the viewpoint of entropy production, to dedicate resources to absorption and dissipation only at those shorter wavelengths where dissipation has the greatest potential for entropy production.

If plants, cyanobacteria and algae have evolved for producing entropy through the dissipation of the solar photon flux, then if these organisms were in some way stressed, for example through nutrient limitation, climate change, or disease, the first pigments to be abandoned with respect to nutrient supply would be those dissipating towards the red since this region has relatively less entropy production potential per unit photon. It therefore follows that healthy organisms or ecosystems will have a red-edge more towards the red while unhealthy organisms or ecosystems would have their red-edge shifted from nominal values towards the blue.

Detailed calculations considering not only the size of the pigments, but also the space needed for the pigment support structures, and the different excited state decay channel lifetimes, are being performed to determine the viability of this postulate; of the red-edge arising from a physical limit to the rate of dissipation in living organisms and, therefore, as to whether the red-edge can be used as a reliable indicator of ecosystem health.

Part V

Beyond Convention

18. Comparison with Prevailing Scenarios for the Origin of Life

Traditional views of the origin of life see it as an extraordinary event, persisting for a kind of inherent self-indulgence, and evolving in the sense of individuals becoming ever more keen in the fight for survival. The present theory sees the origin of life as a non-equilibrium thermodynamic "auto-organization" of material to dissipate an imposed external photon potential. In other words, life is a thermodynamic imperative; a new irreversible process, arising once environmental conditions became appropriate, that coupled to existing abiotic irreversible processes to augment the global entropy production of the Earth in its interaction with its solar environment.

Although prevailing conceptions of the origin of life sometimes, but not always, recognize the need of a free energy source, principally for the metabolism of self perpetuation, such as that derived from generalized chemical potentials, e.g. a proton gradient resulting from an acidic environment (de Duve, 1991)[73] or a thermal gradient of a hydrothermal vent (Wachtershauser, 2006)[389], they fail to acknowledge the existence of a much greater thermodynamic function of life beyond the metabolism of self-perpetuation. Almost no consideration has been given to the fact that life dissipates directly into heat about 99.9% of the free energy arriving in the photons it captures, without making any direct use of this free energy.

The solar photon flux would have been the most intense and extensive free energy source available for dissipation on Earth during the Archean, and still is today (see table 18.1). Chemical and thermal gradient sources of free energy would have been limited in size and extent and a theory of the origin of life based on these sources would still have the very difficult problem of explaining how photon dissipation (including photosynthesis, appearing very shortly after the origin of life (Rosing and Frei, 2004)[302]), could have evolved so quickly from chemical or thermal dissipation. The theory presented here is consistent with continuity in the historical pigment record, suggesting that the visible photon dissipation performed by life of today is linked continuously through time and evolution to the UV-C photon dissipation existing at life's beginnings in the Archean.

Solar radiation, visible (> 300 nm)	2.6×10^6
Solar radiation, UV-C (230 - 280 nm)	3.9×10^4
Lightning	1.9×10^3
Coronal discharge	9.8×10^2
Shock wave from impacts	2.0×10^2
Radioactivity	1.2×10^2
Volcanos	5.4

Table 18.1. Estimated amounts of free energy available from sources existing at the origin of life in the Archean. Units are (kJ $m^{-2}yr^{-1}$). Visible light photons would not have enough energy to break and remake molecular covalent bonds directly. UV-C photons were thus the largest usable free energy source available on early Earth. Solar radiation values are derived from the data given in figures 4.4 and 4.7. Lightning and coronal discharge values were obtained from Chyba and Sagan (1991)[46]. Shock waves, radioactivity, and volcano values were obtained from Deamer and Weber (2010)[74].

On addressing evolution, prevailing theories of the origin of life that do not consider the overwhelming importance of dissipation are subject to the same criticism of circularity of argument that inflicts Darwinian theory. What is natural selection selecting? Since the Darwinian "fitness" of an individual cannot be defined in chemical or physical terms, one is left only with the tautology of "survival of the survivors", irrespective of Popper's recanting[1]. The present theory suggests that natural selection selects biotic-abiotic couplings at all levels simultaneously which lead to ever greater global entropy production, in accordance with; i) Onsager's principle, ii) the reinforcement of a fluctuation in the direction of greater dissipation at a bifurcation (see chapter 3.3), and iii) the overwhelming empirical evidence (see chapter 3.2).

The present theory is consistent with ribosomal RNA genetic tree evidence pointing to a thermophilic or hyperthermophilic origin of life, a result also favorable to the hydrothermal vent proposals of the origin of life (Schwartzman and Lineweaver, 2004)[329]. Any irrefutable evidence indicating that the surface of the Earth was cold at the time of the origin of life would render the ultraviolet and temperature assisted replication (UVTAR, see chapter 12) mechanism of enzymeless replication inviable. Unlike the hydrothermal vent proposal, however, the present theory is reliant on solar photons, not only for forming the fundamental pigments out of the primordial molecules, but for providing the external generalized force driving the irreversible process of the origin and evolution of life. The salient electronic and optical properties of RNA and DNA

[1] Karl Popper (1902-1994) perhaps the most important philosopher of science of the 20th century first brought attention to the circularity in Darwinian theory, the "almost tautological survival of the survivors". Popper called the paradigm a "metaphysical research program". Towards the end of his life, however, Popper recanted on his labeling of the theory as "tautological", but he still referred to it as a "metaphysical research program" (Popper, 1978)[287].

and the other fundamental molecules of life all point to a surface origin with high exposure to UV-C (see chapters 4.3 and 7).

The present theory suggests that life as we know it may be much more universal and abundant than prevailing assumptions of a fortuitous initial event might suggest. However, besides the requirements of liquid water (or some other liquid or gas solvent – see chapter 20), a concentration of appropriate primordial molecules, and an abundant free energy source of UV-C photons, the theory presented here that led to life as we know it also imposes some requirements on the material body's initial conditions and on the evolution of the environmental conditions; an atmospheric window allowing an important UV-C flux to reach the surface, temperatures descending gradually below the melting temperature of RNA or DNA, diurnal temperature cycling, and a water (or other solvent) cycle.

On the other hand, the present theory suggests that not all life is necessarily like our own and that we should be looking for "living" biospheres through a much larger lens. Other "life" in the universe may be based on a coupling of other irreversible processes, such as, for example, other solvent phase cycles (see chapter 20) coupling organic pigment absorption and dissipation to other irreversible processes, such as the giant southern vortex on Venus (see figure 0.1). A more detailed discussion of such biospheres is given in chapter 20.2. However, water is by far the most abundant solvent for organics in the cosmos and has many extraordinary properties that make it well suited to carbon based life, and thus we should expect to find, spread throughout the universe, many extraterrestrial life forms very similar to our own based on a water solvent (see chapter 20).

The present theory avoids the difficulty of the RNA World hypothesis that has made it most vulnerable to criticism; that of requiring *a priori* sufficient RNA information content and reproductive fidelity (Shapiro, 2007)[336] for an initial enzyme assisted replication. In the theory presented here, fundamental molecule (pigment) proliferation is explained through a non-equilibrium thermodynamic principle related to increasing the global entropy production based on the chemistry of an autocatalytic photochemical reaction and does not require a certain base set of information in order to arise. Early replication is instead based on the particular physical environmental conditions of Archean Earth. An indication of the viability of such a proposal is the fact that a temperature cycling mechanism with only one enzyme, polymerase, for amplification of RNA and DNA, known as polymerase chain reaction, has already been shown to function in the laboratory (Mullis, 1990)[239]. A similar, but completely enzymeless, mechanism involving the dissipation of UV-C light by RNA or DNA on the Archean sea surface, coupled with a small diurnal temperature cycling at high ocean temperature, which we have called Ultraviolet and Temperature

Assisted Replication (UVTAR) has, in fact, now been partially demonstrated in the UV-C light induced denaturing of DNA (Michaelian and Santillán, 2014[216] and see chapter 12.2). This could thus provide a plausible, and completely abiotic, mechanism for nucleic acid reproduction based on dissipation (see chapter 12) without requiring a basis set of information for enzyme production.

In contrast to prevailing theories on the origin of life, the present theory does not require the unlikely discovery of an abiotic mechanism that produced an initial high enrichment of chiral enantiomers to explain the homochirality of life today. Instead, the present theory argues that homochirality arose gradually, but rapidly, over time due to a small asymmetry in the environmental conditions that promote the UVTAR mechanism (Michaelian, 2010)[208]. In particular, if indeed life emerged when the sea-surface temperature had cooled to below the denaturing temperature of RNA or DNA, then, since the sea-surface temperature would be greatest in the late afternoon, the absorption of the slightly right-handed circularly polarized atmospheric light (Angel et al., 1972)[3], or submarine light (Wolstencroft, 2004)[409], of the afternoon, could have contributed to a slight advantage for reproduction of RNA or DNA D-enantiomer oligos. Double strand oligos of the L-enantiomer would absorb less well the slightly right-handed circularly polarized light of the afternoon, and thus could not raise local water temperatures as often for denaturation, thereby curtailing their replication and subsequent evolution (Michaelian, 2010)[208] (see chapter 14).

19. Paradigms in Need of Reform

The thermodynamic dissipation theory of the origin and evolution of life presented in this book, with its foundations in physical and chemical law, brings into question many of the contemporary paradigms concerning life and this will open the door to severe criticism of the theory. However, breaking with paradigms becomes an inevitable strategy when an impasse is reached and no further progress can be foreseen within the paradigm of consensus. General acceptance of any new paradigm takes time, sometimes generations, and extraordinary evidence is required before an old paradigm is forsaken for a new. Max Planck (1858 –1947), founder of quantum theory, had a starker view of the history of science asserting that "A new scientific truth does not triumph by convincing its opponents and making them see the light, but rather because its opponents eventually die, and a new generation grows up that is familiar with it." Irrespective of this possibility of opening the flood gates to criticism, in the following sections of this chapter I briefly present the contemporary paradigms that are in need of reform if the thermodynamic dissipation theory of the origin and evolution of life is correct and describe how the new theory offers new perspectives on old problems and paradoxes common to these conventional paradigms. Evidence in support of the theory has been given in detail throughout this book and in the following only a brief recordatory reference will be made to the evidence.

19.1 Entropy and Information

In 1948 Claude Shannon in his seminal paper "A Mathematical Theory of Communication" (Shannon, 1948)[335] determined the general limits on information transfer in a given communications channel. Looking for a function that would measure how much information could be contained in a Markovian[1] succession of n events with probabilities $p_1, p_2, p_3, ...p_n$, Shannon wrote (p. 10),

> "We have represented a discrete information source as a Markoff process. Can we define a quantity which will measure, in some sense, how much

[1] A Markovian process is the name given to a random system that changes states according to a transition rule that only depends on the current state. No memory of the history except knowledge of the current state is required.

information is "produced" by such a process, or better, at what rate information is produced?"

Shannon realized that such a function H would have to satisfy three basic criterion;

1. H should be a positive and continuous function of the p_i.
2. If all the p_i are equal, $p_i = p$, then H should be maximal, and also a monotonic increasing function of the number of events n. With equally likely events there is more choice, or uncertainty, when there are more possible events.
3. If a choice can be broken down into two successive choices, the original H should be the weighted sum of the individual values of H.

Shannon then showed that the only unique function which could satisfy the above three criterion was,

$$H = -k \sum_{i=1,n} p_i \log p_i \tag{19.1}$$

where k is some positive constant.

Shannon's equation (19.1) was already known as the Gibbs equation in statistical mechanics and thermodynamics at the time, named after one of the founders of thermodynamic theory, Josiah Willard Gibbs (1839-1903). It gives the entropy of a macroscopic state of a system having n consistent microstates, each with a probability of p_i ($i = 1, n$), and where k was the Boltzmann constant k_B.

That the Shannon equation for the information content of a succession of events had the same form as the Gibb's equation for the entropy of a macroscopic state was at first not considered as more than a coincidence until Jaynes re-interpreted the Gibb's thermodynamic entropy to be a measure of the Shannon information that was lacking and needed in order to completely specify the microstate, given the measured values of the observables which defined the macrostate of the thermodynamic system. This lacking information, is, of course, extremely large for thermodynamic systems which have on the order of 10^{23} particles. Such lacking information is much larger than anything that could be imagined in a string of Markov events defining a flow of information as Shannon had originally planned his metric.

Jaynes identified entropy with a measure of the lack of information that one had about the system. The greater the lack of information, the greater the entropy. Jaynes apparently converted a physical objective thermodynamic variable (entropy) into a subjective observer dependent variable (lack of information).

Are the two concepts completely different, not even casually related, or is there a deep underlying connection between the two?

When Shannon made the proposal for the limits on information transmission, he asked what he should call his new variable. It was the idea of John von Neumann to give it the name "entropy" because in his words "You should call it entropy, for two reasons. In the first place your uncertainty function has been used in statistical mechanics under that name, so it already has a name. In the second place, and more important, nobody knows what entropy really is, so in a debate you will always have the advantage". Von Neumann's prediction has come true with Shannon's acceptance of the use of the word "entropy" for his concept of lack of information. Shannon's entropy certainly displays the same mathematical form as thermodynamic entropy, but are the two the same, and is the Shannon information theory and his entropy more general than physical theory and physical entropy?

The view which most intimately describes Nature is inevitably the one which is more fundamental, based on the fewest laws, axioms, or suppositions. To the present author, a set of symbols arranged in a specific order, or "information", has no physical relevance without the presence of an interpreter, selector, or some kind of decoder. The existence of such a decoder is already not very fundamental, while a definition of entropy as a measure of the dispersion of conserved thermodynamic variables over microscopic degrees of freedom is certainly more fundamental, being something with precise meaning that can be measured unambiguously. On the other hand, when such a decoder is present in Nature, it anyway can be shown that the fundamental utility of the information is for sustaining or fomenting thermodynamic dissipation.

Not all, however, would be in agreement with my interpretation. Hubert Yockey (1995)[419] saw evolution as information, and the origin of life for Yockey is the origin of information. The important difference between the living and the non-living for Yockey was the information contained in the genetic program known as the genome and the efficacy of the program was related to the fact that "the life message is digital, linear, and segregated" (Yockey, 2005)[420]. However, Yockey saw no way to bootstrap such a complicated program, the genome, from nonliving material. The origin of life, according to Yockey, must therefore be taken as an axiom, that cannot be derived or proved from other facts. Such a situation left Yockey open to interpretations of being in favor of Intelligent Design, a proposition he spent much effort negating and, in fact, pointing out the fallacy in the Intelligent Design arguments.

Any description of life which has nothing to say on its origin is, however, incomplete and, if after sufficient reflection, nothing appears forthcoming, it would be most prudent to relegate the theory to the dust bin of the history of science.

19.2 The Struggle for Survival

Like any irreversible process, life at whatever level necessarily "feeds" off an external generalized chemical potential produced at a higher level of dissipation[2]. This external potential defines, to a large extent, the external environment of a living organism. More accurately stated, life develops as a spontaneous organization of material in response to an applied external potential coming from its environment. Under this thermodynamic view based on physical and chemical law, life is not acting out of self-interest with a view to survival, but rather is a dissipative flow that waxes and wanes in response to the local forces of the local external environmental, those forces that are driving it to becoming. persisting and proliferating.

Within this environment, however, there are other physical factors, for example, temperature, which affect the ability of the material to organize into dissipative structures. If the external potential or these other physical factors change, for whatever external or internal reason (e.g. through feedback mechanisms), life may "adapt"[3], through its inherent plasticity (see remarks made in chapter 17.3), and continue dissipating the potential in the new, perturbed environment. During this process of "adaption", individuals may die or species go extinct or morph gradually into others. It is the imposed external potential that "selects" the individual or species, but this should not be looked at as an organismal struggle against its environment for survival, but more as an adaption of the dissipating process to the new constraints of the environment, such that the system as a whole, environment plus organism, can continue dissipating the solar photon potential. Survival, non-survival, and morphing are all adaptive responses in this thermodynamic view.

The vitality of life, for example, its reproduction, proliferation, mobility, "adaptation" to new environments, and evolution, all are responses favored by the dissipation of an external potential. Without access to an external potential, life, after using up any internally stored chemical potential, would come to a literal dead end and go extinct. It is a well documented fact that the great extinction events in the historical record of life on Earth occurred after the solar

---

[2] When speaking about "levels" in this thermodynamic sense, there is no casual relationship assumed with the levels of the conventional ecological pyrimid. At the highest thermodynamic level is the dissipation of solar photons into heat by pigments in water and the production of chemical potential in the covalent structuring of organic elements. These latter two give rise to the water cycle and herbivores respectively in the dissipative chain, and so on until the free energy content becomes minimal and energy of high entropy is emitted into space. Feedback between the different thermodynamic levels (for example, the water cycle fomenting the greening of the planet and thus photon dissipation) is the source of selection on various simultaneous levels.

[3] The word "adapt" in this thermodynamic sense means any change of the system+environment which increases the overall dissipation rate.

photon potential at Earth's surface waned, for example after a large meteoritic impact such as the Chicxulub event[4].

19.3 Life has no Purpose

Perhaps one of the assertions that can most perturb a modern biologist and provoke them into a feisty defense of Darwinian theory is the suggestion that life must have a purpose. The predictable negative reaction to this suggestion by those charged with defending the Darwinian paradigm can be understood and forgiven since they have been incessantly molested by the most vocal detractors of evolutionary theory since Darwin's day. These unrelenting detractors are those who have little stomach for a careful and systematic study of Nature and simply do not comprehend science or the scientific method. Instead, they prefer to allow themselves be programed to defend a set of naive beliefs about the natural world handed down to them, along with a set of instructions on how to live their lives, by previous generations of like-minded persons. Afterall, there is a great prize offered for this intellectual laziness, a promise of an eternal and happy afterlife.

However understandable the biologists defense of the Darwinian paradigm, their conditioned negative and reactionary response to the proposition of the purposefulness of life, leaves them in the very difficult position of having to defend the notion that life is somehow a phenomena very different from the other natural phenomena and they are forced, either reluctantly or carelessly, to accept the notion that life obeys a different set of chemical and physical laws than those that prevail for all other natural phenomena. After careful study, however, we find an underlying reason for becoming and evolving for all natural phenomena; the dissipation of some generalized chemical potential.

It is rather disingenuous to suggest that the great evolutionary program which has led to organisms of astounding diversity and abilities, that have even learned how to explore and understand their cosmos, was all driven by random variation and natural selection operating through an undefinable, and even unfathomable, "fitness function", without any natural utility to it all. Life does indeed have a purpose, and that is to dissipate the solar photon potential and other secondary potentials in its environment. It is thermodynamic law, which is ultimately and intimately related to the structure of space and time which gives purpose to life (see chapter 3.5).

Understanding the purpose of life in this sense makes it much more palatable to our intuition derived from our experience in the natural world. It also suggests

[4] The Chicxulub event was a major impact of an astroid measuring approximately 10 km in diameter with Earth 65 million years ago and sending up a large amount of dust that blocked out the sun for an estimated 12 month period, sufficient to kill off the dinosaurs and many other species of large animals and plants.

that some predictability exists as to the future of life, and even that of human life. Finally, understanding the purpose provides us with a new tool from which to study life.

19.4 RNA World

The "RNA World" is a popular hypothesis among scientists that suggests that the first steps of life were entirely based on the molecule RNA and that the rest of life was built up around this molecule through improving its "fitness" based primarily on replication efficacy, eventually giving way to a DNA, protein and lipid world falling under the dictates of the Darwinian paradigm. The main reasons behind accepting the idea of a beginning of life based on RNA are the following;

1. RNA is a fundamental molecule of life, found today in all three great domains of life; eukaryote, bacteria, archea.
2. RNA is central to the translation process of converting stored information in DNA into proteins. Ribosomal RNA (rRNA), which forms part of this translation mechanism, is one of the most conserved information sequences in life, its very slow mutational rate allowing its use in constructing genetic trees.
3. RNA today can act as different enzymes (called ribozymes) consisting of short segments of RNA with catalytic functions, such as ribosomal RNA (rRNA) which perform the catalytic function of forming the peptide bonds between the amino acids to produce proteins. Other ribozymes also exist that can perform self-cleavage (the hammerhead ribozyme) and RNA polymerase ribozyme that can synthesize a short RNA strand on a RNA template single strand.
4. RNA is able to store information in its linear sequence of nucleotides, exactly like DNA.
5. Many of the enzymatic cofactors in life's catalytic processes of today use RNA (or DNA) nucleotides or derivatives of these, examples are ATP, NADH, Acetyl-CoA, etc. suggesting a pre-protein world of RNA and nucleic acid based enzymes.
6. Amino acids can conjugate to the 3'-end of RNA, providing more complex enzymes involving RNA and proteins (like the ribosomes).
7. In the present biosynthesis of the nucleotides, RNA ribonucleotide production precedes the DNA deoxyribonucleotide production. The deoxyribonucleotides are produced subsequently from the ribonucleotides by removing the 2'-hydroxyl group.

A similar number of arguments, however, suggest that there is still no inherent overwhelming reason to presume that an RNA world preceded a DNA world. In fact, from the perspective of the thermodynamic dissipation theory of the origin and evolution of life, both of these fundamental molecules have similar dissipative characteristics, making them equally probable at the origin of life. The arguments against an exclusively RNA World are;

1. Both RNA and DNA have similar electronic and optical properties, making both of them excellent dissipaters of UV-C light, and so, from a purely photon dissipative perspective, neither should be *a priori* favored at life's very beginnings in the primordial soup of organic pigments.
2. RNA is less stable than DNA, particularly at high temperature. At 60 °C, hydrolysis of the phosphodiester bonds joining ribonucleotides becomes very damaging to RNA (except for RNA strands which are purine rich, see chapter 9). The phosphodiester bonds and the ester bond between tRNAs and amino acids (useful, for example, for the evolution of ribosomal protein synthesis) are both more stable at pH between 4 and 5. RNA also has a denaturing temperature of only about 45 °C compared to short DNA of about 85 °C. Given all the evidence for a hot early Earth with temperatures around 80 °C and ocean pH values of about 6 at the origin of life, it would be difficult to conceive of a non-enzymatic template extension of RNA at these high temperatures and acidic pH values.
3. Although Mg^{2+} is as important for stabilizing RNA secondary and tertiary structure as it is for DNA, a high Mg^{2+} concentration can catalyze RNA degradation, which has been identified as a particular problem in the case of RNA template copying (Szostak, 2012)[371].
4. DNAzymes, DNA based enzymes with catalytic activity, have also been evolved in the laboratory, for example DNAzymes that can splice RNA (Gysbers et al., 2015)[121], meaning that catalytic activity is not only confined to RNA (ribozymes) as once thought.
5. Viruses, one of the life forms considered to be most primitive, are based both on RNA and DNA, but predominantly DNA.
6. Catalytic activity begins in RNA only at some minimum ribonucleotide polymer length. Therefore, if the replication of the first molecules of life was based on the catalytic activity of RNA, and not on environmental factors as is suggested in this book, then one is left with the difficulty of explaining how this particular chance polymerization came about in the first place (less likely the greater the RNA length).
7. Single strand RNA (messenger mRNA) can be synthesized over a double strand DNA (Zubay, 1962)[432].
8. With both RNA and DNA available in the original organic soup, there would have been a much greater variation in the optical, electronic, and physical

properties of the original molecular complexes which could have enhanced the rate of evolution of dissipation through, for example, the formation of hybrid duplexes consisting of one strand DNA and the complimentary RNA (DR duplexes – see chapter 9).

Bernhardt (2012)[19] has written an article with the revealing title "The RNA world hypothesis: the worst theory of the early evolution of life (except for all the others)". Hubert Yockey (1995)[419] has gone even further in his criticism of the RNA World hypothesis, even denying that there is evidence for a primordial organic soup and labeling self-organization as "dialectical materialism". Borrowing a metaphor from Jonathan Swift, Yockey suggested that current origin of life research, based on the RNA World hypothesis, "floats improbably in mid-air like the roof of a house built by an architect of the Grand Academy of Lagado". This savant had contrived a method of building houses by beginning at the roof and working downwards. "The architect pointed out that among the advantages of this procedure," Yockey notes, "was that once the roof was in place [before the walls or foundation] the rest of the construction could proceed quickly and without interruption by weather." The "roof" of the theory of life that Yockey refers to is the suggested organic soup hypothesis along with the RNA World hypothesis based almost exclusively on the catalytic activities of RNA, but lacking a foundation and walls. In this sense, one must agree with Yockey, without the explicit recognition of the dissipation of a generalized chemical potential as the foundation of life, the roof, whether viewed from the RNA World or information theory hypothesis, is unstable, unable to float freely. (The relation between information content and dissipation, which arises naturally within our non-equilibrium thermodynamic paradigm, is discussed in chapter 19.1).

Given the similar photon dissipation efficacy of RNA and DNA and the similar probability of photochemical formation under a UV-C light potential, it is probably wiser to assume that RNA and DNA with their molecular antenna complexes were first evolving simultaneously but separately; DNA perhaps in warmer equatorial climates or nearer to the ocean surface, and RNA in colder polar climates or at greater ocean depths. Eventually the two came to form a symbiosis which allowed even greater photon dissipation than the two entities floating separately. Life's history is full of symbiosis events giving rise to more complex systems and this is just a reflection of Onsager's description of the coupling of irreversible thermodynamic processes if this gives greater global entropy production (see chapter 6.3.1).

19.5 Metabolism or Replication First

The complicated nature of reproduction in living organisms today, with all the specialized enzymes needed and application under very strict time coordination, has fostered the belief that reproduction could not have arisen before the existence of at least some type of primitive metabolism. There is an apparent dichotomy here that has led to a debate as to whether metabolism or replication came first. Prevalent origin of life theories are usually characterized as belonging to either metabolism first or replication first categories. However, this dichotomy is really a false dichotomy, indeed, replication could not exist without metabolism.

Metabolism usually refers to those complex metabolic processes necessary for maintenance and reproduction of today's organisms. However, at the origin of life the metabolism related to maintenance and reproduction would have necessarily been much simpler. Under the thermodynamic dissipation theory of the origin of life, the dissipation of an absorbed photon into heat would have been the primordial metabolism carried out by life. The locally deposited heat would have helped in the reproduction process, as has been outlined in chapter 12. In fact, even today, all organismic metabolism is ultimately based on the dissipation of visible light into heat. The UVTAR mechanism, therefore, is one in which both metabolism and reproduction occur simultaneously. Our theory brings to light a very strong relation between dissipation and replication. Dissipation (some form of metabolism) must always accompany any irreversible process, and replication is certainly an irreversible process.

19.6 The Last Universal Common Ancestor

It is commonly believed that all life must have arisen out of a single Last Universal Common Ancestor (LUCA) that gave rise to the three great domains of life. This common ancestor is thought to have lived between 3.5 and 3.8 Ga. The reason for believing in LUCA is that all life today, across the three domains, share many similarities. In particular; all life is based on the molecules RNA and DNA, on the same 22 amino acids, and on almost identical genetic translation mechanisms (the machinery that converts information stored in DNA into proteins) involving ribosomal RNA. For example, so similar are the translation mechanisms that a tree could produce the necessary enzymes and proteins for a human if supplied with the human genetic code. The fact that entirely new life forms do not appear from time to time, nor in the entire historical paleontological record, at first sight seems to indicate that the origin of life was indeed a special event and reinforces the idea of the LUCA.

From the viewpoint of the thermodynamic dissipation theory of the origin and evolution of life, the uniqueness of RNA and DNA in all life most probably has to do with its exceptional electronic de-excitation properties of these molecules due to having a conical intersection (see chapter 7) which foments photon dissipation. There is no other known molecule, biotic or abiotic (besides the precursor molecules of the individual nucleobases – see chapter 6.1), that can dissipate its electronic excitation energy into heat within a picosecond (10^{-12} seconds). RNA and DNA offered themselves as superb acceptor quencher molecules to other donor UV-C pigment molecules charged with electronic excitation energy through photon absorption. This characteristic of RNA and DNA has been been looked at as a protection mechanism against destruction of these molecules by the same UV-C photons that were necessary for their production, however, it must be emphasized that the ultimate driver of the molecular structuring and other associative processes, as well as molecular autocatalytic proliferation is photon dissipation, and this is affected through the autocatalytic photochemical reaction in which the molecule, or complex of molecules, acts to dissipate the same photon potential that created them (see chapter 6). This is much more than simply photoprotectionism.

At the end of the Archean, once UV-C light was blocked by life produced oxygen and ozone from arriving at the surface of Earth, around 2.8 Ga, corresponding to the first appearance of oxygen photosynthesis (see Michaelian and Simeonov, 2015[217] – see also figures 4.4 and 4.7), in order to extend dissipation into the more intense visible light region, it was necessary for life to invent new mechanisms of proliferation no longer based on UV-C light, and this required building up a certain complexity of biosynthetic pathways leading to the fundamental molecules which were now being used in new ways. Dissipation became ever more important at the longer wavelengths of higher intensity. Once the biosynthetic pathways had been established and UV-C light completely vanished from the surface, it would have been very difficult, perhaps impossible, for completely new dissipative organisms to form out of the primordial precursor molecules without the aid of this high energy light. This appears to be an adequate explanation for why new forms of life do not appear in the historical fossil record.

It is known that endosymbiotic events were common and important drivers of evolution throughout the history of life. For example, the mitochondria or the chloroplasts, once free living bacteria, became engulfed within the eukaryote cell to become their organelles. It is thus probable that shortly after the origin of life and before $\sim$ 2.8 Ga, at which point biosynthetic and energy producing pathways necessarily had to become more complex because of lack of surface UV-C light, there may have been many different dissipative organisms proliferating in parallel, based on either DNA, RNA or a hybrid of the two, as the acceptor molecule for energy de-excitation. It is most likely that an endosymbiotic event occurred early between an organism (or simple molecular complex) based on

DNA and one based on RNA which allowed greater dissipation and which may have been the last universal common ancestor. After the extinction of UV-C light at about 2.8 Ga, only these symbiotic organisms survived because RNA and DNA made the transition into dedicated information storing and transfer molecules that were necessary for bootstrapping the biosynthetic pathways for producing the fundamental molecules and newer pigments which now became ever more effective in dissipating visible light.

19.7 The Great Oxygenation Event

For about a billion years of life's early history there was only trace amounts of oxygen in Earth's atmosphere and this permitted intense UV-C light to arrive at its surface, a component that we have argued as being essential to the early synthesis of the fundamental molecules and their proliferation. The discovery of oxygenic photosynthesis in which water became the electron donor, thus freeing oxygen from its hydrogen chaperons, marked an important new direction in life. Oxygen, with its great oxidation capacity, was a deadly poison to most of the simple organismal biosynthetic pathways that had developed up to that time. There is little doubt that many lineages of pigment producing bacteria went extinct or had to change habitat at the initial stage of this great event. From the perspective of the thermodynamic dissipation theory for the origin and evolution of life, what could have been gained by such an event?

Many researchers have argued for the greater efficiency of carbon fixation in oxygenic photosynthesis as compared to that in anoxygenic photosynthesis. They suppose that Darwinian selection would therefore have favored oxygenic photosynthesis. However, given that the 99.9% of free energy available in the incident photons in either oxygenic of anoxygenic photosynthesis is not used in carbon fixation, this increase in efficiency seems a rather unpersuasive argument to advance for such an important paradigm change as the great oxygenation event. It is also argued that oxygen would have helped to shield the ever more complex biosynthetic pathways from destruction by UV-C light, however, it is much more reasonable to assume that any biosynthetic pathways that had evolved up to that time would have necessarily been robust to UV-C, and, in fact, would have been completely adapted to dissipating this region of the spectrum. The important question to be answered is therefore: Is there an advantage related with increased dissipation that could have been gained by the great oxygenation event?

Since both oxygen O_2 and ozone O_3 (which is produced by UV-C light interacting on O_2) are strong absorbers over a wide region of the UV (see figure 19.1), and are resistant to destruction by these wavelengths, they can be considered as UV pigments. As we have seen in chapter 4.3 many pigments which absorb in the UV-C also absorb in the visible (see for example figures 4.9, 4.12, and 4.15). These oxygen "pigments" emitted by the oxygenic photosynthetic organisms into the atmosphere in order to dissipate in this region of the solar spectrum could thereby have consumed UV light, freeing up the organisms on the surface to dedicate resources to dissipating in the more intense visible region of the spectrum.

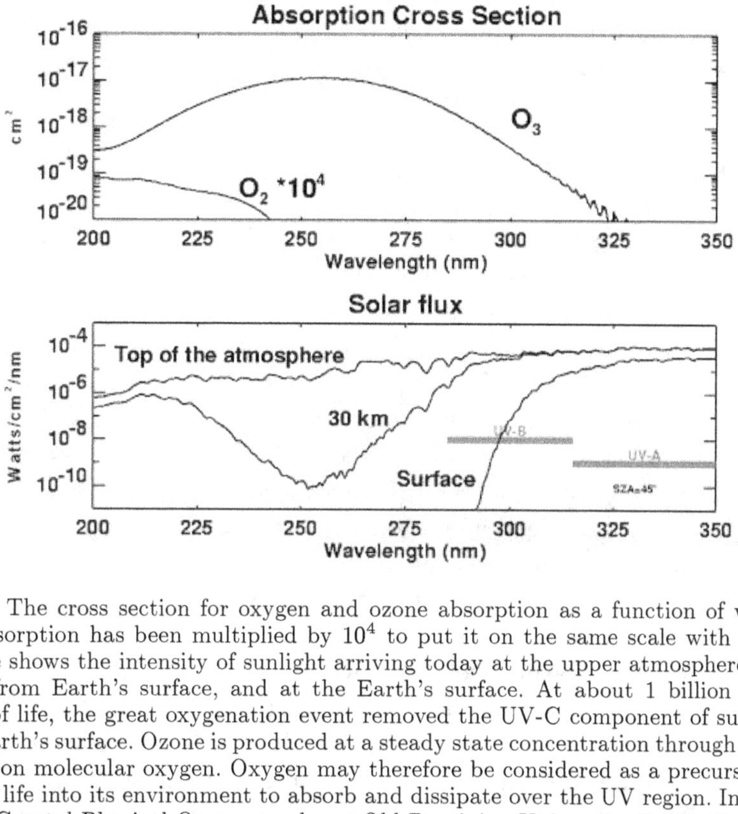

Fig. 19.1. The cross section for oxygen and ozone absorption as a function of wavelength. Oxygen absorption has been multiplied by 10^4 to put it on the same scale with ozone. The lower figure shows the intensity of sunlight arriving today at the upper atmosphere (100 km), at 30 km from Earth's surface, and at the Earth's surface. At about 1 billion years after the origin of life, the great oxygenation event removed the UV-C component of sunlight from reaching Earth's surface. Ozone is produced at a steady state concentration through UV-C light interacting on molecular oxygen. Oxygen may therefore be considered as a precursor pigment released by life into its environment to absorb and dissipate over the UV region. Image credit: Center for Coastal Physical Oceanography at Old Dominion University, Public Domain.

Ozone, O_3, in Earth's upper atmosphere absorbs strongly in the UV-C, and, in fact, reaches a steady state concentration at a steady state concentration of O_2 under the influence of the flux of solar UV-C light, indicating that this production of O_3 is yet another example of microscopic dissipative structuring (see chapter 3.4).

The development of oxygenic photosynthesis may thus have been thermodynamically driven for dissipation rather than as traditionally seen as a Darwinian advantage for more efficient metabolism. It is becoming increasingly clear that many organisms of today exude pigments into their environment, ostensibly to foment the dissipation of the solar photon potential, since this behavior would seem to be wasteful, having little Darwinian utility to the organism itself. Other examples of this behavior, besides oxygen emission, include cyanobacteria and other phytoplankton exuding chromophoric dissolved organic matter (CDOM) into the sea-surface microlayer. Such organic matter includes mycosporine-like amino acids (MAAs), metal-free porphyrins, and aromatic amino acid. An article describing this phenomena of the secretion by organisms of pigments into their environment to foment photon dissipation, is under preparation (Simeonov and Michaelian, 2016)[?]. This finding provides strong evidence in favor of the thermodynamic dissipation theory of the origin and evolution of life.

19.8 Evolution through Natural Selection

The term "natural selection" was coined by Darwin and defined by him as "a principle by which each slight variation [of a trait], if useful, is preserved."(Darwin, 1859)[70]. In more modern terms, natural selection can be defined as the process by which biological organisms are selected based on differential reproductive "fitness" of their phenotypes (physical and behavioral characteristics). Fitness, in this modern Darwinian sense, is determined not only in terms of the differential survival rates in a particular environment but in terms of differential reproductive success of an individual, a process which Darwin separated from natural selection and termed "sexual selection". What appeared to be important in determining overall fitness, was the accumulated number of reproduction events during the lifetime of an organism. Any characteristic that enhanced this trait would, accordingly be naturally selected. According to the Darwinian paradigm, natural selection will invariably operate on any system that displays three characteristics; 1) reproduction, 2) variability, 3) heritability. There is, however, also the possibility of lateral (horizontal) gene transfer in which useful mutations can be passed from one organism to another in the same generation, i.e. not necessarily heritable.

An example showing that some kind of natural selection is indeed a force for change in the biological world is the rather recent discovery of bacteria acquiring resistance to our existing arsenal of antibiotics. Humans are now facing the critical problem of not having sufficiently effective anti-bacterial medicines to cope with the new "superbugs" that have evolved, ostensibly through this process of natural selection. This appears to be incontrovertible evidence in favor of the theory of evolution through natural selection.

There is no doubt that evolution through some form of "natural selection" is indeed occurring but the Darwinian paradigm, in which selection is determined by the "fitness" of an individual, or even the "average fitness" of a population of individuals, is neither an accurate nor a useful description of the evolutive process and is insufficient to betray the underlying physical-chemical mechanism of evolution. The reasoning behind my criticism is the following: As the data collected for validating the theory of Gaia has made very clear (see chapter 17.4), organisms are not adapting to a static physical environment, but are coevolving with the environment towards thermodynamic states of ever greater entropy production. The term "fitness" of an individual, or even of a population of individuals, as employed in the Darwinian paradigm, and considered as being fundamental to its explanatory powers, therefore has no clearly defined meaning. We can, however, construe a meaning to the term "fitness" if we apply it to the rate of dissipation of a well defined external potential, for example, to the biosphere as a whole if we specify it in terms of the rate of dissipation of the solar photon potential. In our thermodynamic view, individual organisms are local thermodynamic flows which arise in response to local thermodynamic forces produced by other irreversible processes occurring in the biosphere at a higher level in the dissipative hierarchy, and the organisms "spontaneously" created through dissipative structuring of material to dissipate these forces will wax and wane, or deform, according to the intensity of the local force itself. The organism, or a population of organisms, is the response of Nature to a local thermodynamic force and not an entity with its own ascribed mystical "fitness" or "will to survive and propagate". For example, herbivores dissipate the chemical potential stored in the organic material of photosynthetic organisms that dissipate the solar photon potential. Selection is thermodynamic and operating simultaneously on all dissipative levels (see figure 2.4), or equivalently on various ecological levels (genes, individuals, populations, species, clades, ecosystems ... up to the biosphere) such as to produce a biosphere of ever greater entropy production.

Going back to the example of the apparent adaptation of bacteria to antibiotics, I can now give a more complete and global description of how this adaptive bacterial process is actually working in benefit of the entropy production of the biosphere. As repeatedly stated, the bacteria (or whatever biotic organism at any ecosystem level) has no inherent self-interest in surviving. But why then should Nature bother with foiling the efforts of humans to eliminate dangerous bacteria? The reason is simply that bacteria (and viruses) play a very important biological role in fomenting the global entropy production of the biosphere. By infecting their human or animal hosts, they are fomenting a number of behavioral traits which ultimately increases the global photon dissipation rate. First, particularly important during the days before antibiotics, an epidemic, like the black plague, would cause humans to flee from the cities towards the country and thus spread over larger areas, dispersing nutrients (by means of excretion and death) necessary for plants and cyanobacteria to multiply and thereby perform the fundamental photon dissipation of life. By causing deadly health epidemics in highly populated areas, they are also reducing the size of the animal population to that which is "sustainable" with respect to maximum photon dissipation in that particular area of animal concentration. In none of the epidemics, either before or after the invention of antibiotics, have viral or bacterial infections completely eliminated the local human population, and this is also true for epidemic infections for the other animals. Bacterial and viral infections of animals is a local control mechanism which secures greater entropy production at the global level.

How then could humans finally eliminate the threat of death by bacteria or viruses? We could always develop more and more sophisticated anti-bacterial or anti-viral drugs, but as long as there exists a human chemical potential for fomenting bacterial epidemics (not only in terms of a human host for reproduction, but human behavioral habits such as congregating in small areas allowing efficient contagious dynamics), bacteria will be deformed by the external conditions to adapt to the local thermodynamic potential. Only by spreading and reducing the human population enough such that bacterial or viral infections would not have any beneficial effect on the global entropy production rates would we expect that such epidemics would disappear, being no longer relevant to global entropy production. Indeed, the reproductive success of viruses or bacteria depend on the population density of its animal host and, for a given animal population, the global photon dissipation is inversely related to the excess animal population density with respect to its optimal density.

19.9 Evolution has No Direction

A notion akin to the supposition that life has no purpose is that evolution has no direction. Darwin, however, did see a direction in evolution, deducing that the mechanism of natural selection would lead to organisms ever more adapted to their environment until leading to what he suggested would be "perfectly" adapted organisms. However, ignoring variations in the external conditions, for example astronomical effects on the climate, this dynamics would lead to an eventual stationary state in which evolution would come to a halt when all species achieved perfect adaptation to their particular niche in the static biosphere. However, based on the observation of the relative constancy of species background extinction rates over large times, once taking into account the high correlation of familial origination and extinction rates (Gilinsky, 1994)[104], we see no evidence of this arriving at a perfectly adapted state. Even taking into account external variations such as the astronomical perturbations, one would still expect to find a stable cycling of optimally adapted species since the genes preserve, and an organism can call on, a record of previous adaptations. It may be argued that the external conditions over Earth have been steadily changing (for example, the slow increase in the solar output, see chapter 4.1), however, it would be very difficult to explain abrupt changes in species composition in terms of incremental changes in the solar flux.

If evolution was instead directed by increasing entropy production through innovation, rather than by adaptation based on an ill defined "fitness", then we would expect abrupt changes when new pigments or dissipative complexes (such as oxygen and ozone – see section 19.7) were conceived by Nature, and then, we could say, judging from the present albedo of Earth and the position of the red-edge in the absorption of the solar spectrum by life (see chapter 19.15), that there is, and always has been room for improvement in the photon dissipation efficiency of Earth's biosphere. In this framework, evolution will continue almost indefinitely, reducing ever more the albedo of Earth and the black-body temperature at which Earth emits radiation into space until this radiation reaches an equilibrium at the black-body temperature of cosmic background radiation at 2.7 K (or at the relevant black-body temperature of the Universe at any time in the future).

19.10 The Ecological Pyramid

The cyanobacteria and plants provide the chemical potential to support the herbivores, and the herbivores support the predators, and so on up until the top predator sitting at the apex of the ecological pyramid, a place usually reserved for humans, or so suggests conventional wisdom. The ecological pyramid is based on the relation of free energy flow between species and our highness at the apex of the pyramid is only a metaphor for our own prejudices regarding our perceived importance among all the predators in the scheme of things. From a thermodynamic point of view, however, humans and other top predators do very little of the global thermodynamic dissipation. Their contribution to the dissipative biosphere is in their ability to distribute nutrients and seeds for, and to pollenize, the photon dissipating (photosynthetic) organisms so that they may grow vigorously and dissipate photons over all of Earth's surface.

The concept of the ecological pyramid for describing the organization of living systems is also inadequate in that it ignores the importance of non-biotic processes interacting strongly with the biotic. For example, the water cycle is strongly coupled at the microscopic level to the photon dissipation due to organic pigments in water and, at the macroscopic level, to the distribution over dry land of one of the most important nutrients for the photon dissipating species, water. The nitrogen, sulfur, and carbon cycles, the water cycle, winds, ocean currents, and hurricanes are other coupled biotic-abiotic processes which are not accounted for in the scheme of the traditional ecological pyramid.

Rather than as an ecological pyramid, the biosphere is arranged as a hierarchy of dissipative processes with positive feedback among the different processes, within a great dissipative cycle based on photon dissipation in plants and cyanobacteria, such as that shown in figure 2.4. The highest dissipative level, i.e. the one performing the greatest thermodynamic work, is the interaction of solar photons with the organic pigments, and thus the organic pigments and their supporting structures take precedence in the thermodynamically ordained biotic evolutionary scheme over all the other processes. Animals only have thermodynamic relevance in how they help the pigments to proliferate and disperse over ever more of Earth's surface. It may be that the evolution of intelligence has led (or is leading) to a new mechanism of dissipation which has the potential to rival that of the biosphere or, in fact, to be even greater, for example through terra forming of other planets within the cosmos. If that were the case, and it may be very likely, then at least somewhere in the cosmos intelligent animals may really be players at the top of their biosphere's dissipative circle. This possibility will be analyzed in more detail in chapter 20.

19.11 Hydrothermal Vent Origin of Life

One of the most intensely studied ideas concerning the origin of life is that of life arising at the bottom of oceans near to hydrothermal vents (see figure 19.2), first proposed in 1981 by Corliss, Baross and Hoffman[59]. Perhaps the most important reason that hydrothermal vents are considered by some as being the true cradle of the origin of life is that hyperthermophiles (with optimal growth temperatures above 80 °C and surviving temperatures of up to 121 °C) from both the Archea and Bacteria domains are found to be at the very base of phylogenetic tree obtained through ss rRNA genome analysis (see figure 19.3)(Stetter, 2006)[358]. Hyperthermophiles obtain their free energy for reproduction through a number of different mechanisms. Many species of hyperthermophiles are chemolithoautotrophic, obtaining free energy from inorganic redox reactions (chemolithotrophic), followed by either anaerobic and aerobic respiration, using CO_2 as the only source of carbon needed to build-up organic cell material. Others are heterotrophic, being able to obtain energy from either the decay of organic material or chemolithotrophy, depending on the environmental conditions (Stetter, 2006)[358], as well as through photosynthesis using the faint red light given off by these hydrothermal vents (Yurkov et al., 1999)[424].

The hypothesis is that the hydrothermal vents provide the required nutrients, and numerous free energy sources; either a chemical potential, proton (pH) potential, or a heat gradient, for driving the primitive metabolism of early life. The similarity is pointed out between the chemistry of the H_2–CO_2 redox couple that is present in hydrothermal systems and the core energy metabolic reactions

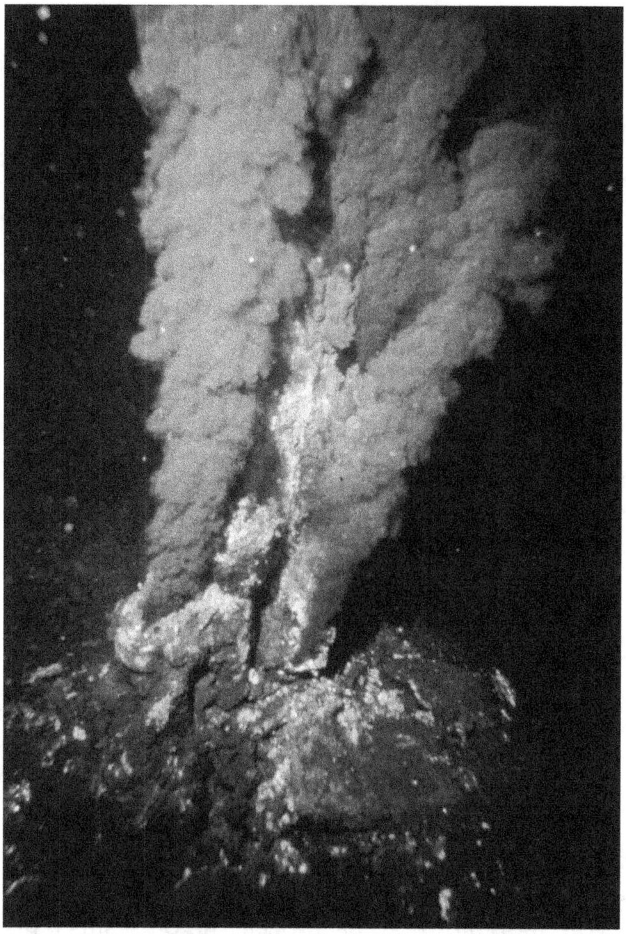

Fig. 19.2. Many prebiotic chemists believe that life began at hydrothermal vents at the bottom of ocean floors like the one pictured here. They argue that the temperature and pH gradients, along with chemical potentials and catalysts, could have driven the primitive metabolism of early life. However, the salient UV-C absorbing and dissipating properties of the fundamental molecules of life argue against this idea and instead suggest that life originated on the surface of Earth exposed to UV-C light. This light would have provided a much more potent source of free energy necessary to maintain a dissipative process like life. The surface waters are also rich in required nutrients such as magnesium ions and lipids. Image credit: P. Rona (NOAA Photo Library), Public domain.

of some modern prokaryotic autotrophs (Martin et al., 2008)[201]. The detection of thermophilic archea and bacteria thriving in these environments was taken as further support for the hydrothermal vent origin of life hypothesis.

Supporters of the hydrothermal vent origin of life also suggest an advantage of ocean water protecting life from the supposedly dangerous UV-C light that was arriving at the surface of Earth during the Archean.

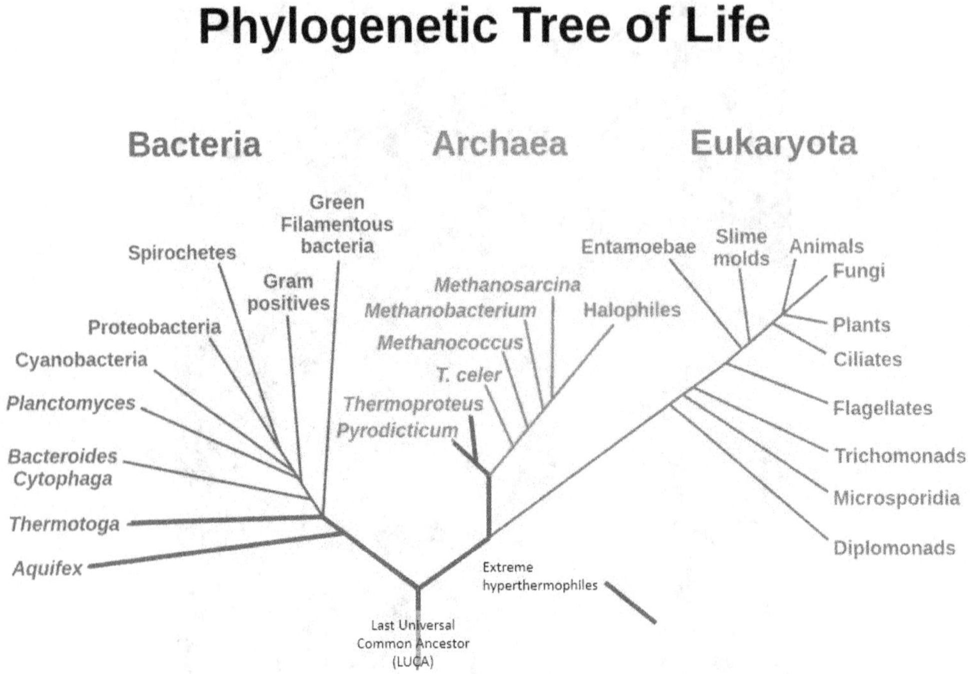

Fig. 19.3. Phylogenetic tree obtained from ssRNA analysis of organisms. The thick lines at the base of the tree all represent hyperthermophilic lineages (optimal growth temperatures > 80 °C). Image credit: John D. Croft, CC BY-SA 3.0

Although early life may have colonized hydrothermal vents very early in its history, or hyperthermophiles may have been the only surviving organisms after intense asteroid bombardment of Earth occurring around 3.9 Ga that boiled ocean waters, the proposal that life originated at hydrothermal vents has a number of problems not easily brushed aside. For example Stanley Miller (who first produced many of the fundamental molecules of life through electric discharges on prevalent atmospheric gases, see chapter 1.2.3) suggested that because of their very high flow temperatures (350-450 °C), hydrothermal vents were not places of production of the fundamental organic molecules of life, but rather places of destruction of these. He argued that sea water containing organic molecules was being subducted by plate tectonics and that all water would eventually pass, about every 10 million years, through these hydrothermal vents, thereby destroying these molecules. Furthermore, there has been very little demonstration of any substantial amount of abiogenic production of the fundamental molecules with experiments performed at the physical and chemical conditions of hydrothermal vents. Miller and Bada (1988)[225], in fact, calculated that the molecular destruction would be orders of magnitude greater than any possible production of organic molecules at these sites.

Notwithstanding the lively and productive debate between hydrothermal vent origin of life scientists and surface origin scientists, our thermodynamic dissipative view of life provides another angle from which to criticize the hydrothermal hypothesis; based solely on thermodynamic probabilities it would be extremely improbable that life originated at hydrothermal vents. As we have made clear above, metabolism, photosynthesis and other biochemical processes necessary to maintain life are a very small, almost negligible, portion of the total dissipation that life performs today. Life, as a dissipative structure is "interested" in dissipation, not in maintaining itself for some kind of metaphysical pleasure to be alive. At the very isolated and dispersed locations of hydrothermal vents at the bottom of oceans, there is very little free energy to dissipate. In fact, the ratio of free energy arriving at the surface in the form of solar photons to the free energy available at hydrothermal vents today is about $10^6 : 1$ and this number would not have been much different during the Archean (see table 18.1), so from a non-equilibrium thermodynamic viewpoint, neglecting for simplicity the necessity of a viable mechanism, this ratio would also be the approximate probability in favor of life originating on the surface rather than at a hydrothermal vent.

Perhaps most important, however, is the fact that the fundamental molecules of life have salient characteristics that argue for a surface origin under intense UV-C light. In section 16.2 I have described how Nature has apparently selected particular isomers and tautomers of the nucleic acid bases who's electronic de-excitation times are extraordinarily rapid, even at the price of a weaker bound RNA or DNA molecule. This is also the case for the methyl bases and other rare bases of transfer RNA (tRNA) (Guo et al., 2013)[120]. I have argued that this is in line with non-equilibrium thermodynamic selection which will select those configurations which are more efficient at dissipating the solar photon potential, and I have given a description of how non-equilibrium thermodynamic selection would have been made operative through a mechanism that relates replication with dissipation (see chapter 12). Others have suggested that rapid non-radiative de-excitation is merely an evolved protection mechanism against photochemical reactions that could destroy the nucleic acids. Those who advocate a hydrothermal bottom of the ocean origin of life where UV-C light would have been non-existent, however, are left in the very difficult position of having to explain this salient feature of native RNA and DNA as simply *ad hoc* or an irrelevant coincidence, or a much later development. The extremely efficient dissipative nature of only the native tautomers of the nucleobases of all present life, even that existing at hydrothermal vents, in my view, provides a very strong argument against a deep ocean origin of life.

There are also many other vestiges remaining in the fundamental molecules of life pointing to a UV-C environment at, or very near, the beginnings of life which would also have to be considered as mere accidents or coincidences in a deep sea hydrothermal vent scenario for the origin of life. For example, the fact that many fundamental molecules of life absorb in this wavelength region (see figure 4.7 and table 4.1). There would be no reason for RNA, DNA, and other fundamental molecules absorbing and dissipating rapidly in the UV-C region if this light was non-existent at the bottom of the ocean. Further evidence along this line of reasoning comes from Kim et al. (1992)[159] who have shown that a particular tryptophan residue of the *Escherichia coli* photolyase, Trp-277, can alone bind to thymine dimers and split them using UV-C light at 280 nm with a quantum efficiency of 0.56. Besides the two known chromophores of *Escherichia coli* photolyase operating in the visible, this chromophore of Trp-277 operating in the UV-C is most probably a vestige indicating that *Escherichia coli* photolyase is a modern version of a more primitive predecessor photolyase operating in the UV-C at 280 nm.

Although hydrothermal vent proponents could counter that these salient features of RNA and DNA may have evolved at a later time when life ventured to the surface and became exposed to UV-C light, since there are no known vestiges of different RNA or DNA (e.g. non-Watson Crick pairing) in organisms living at the ocean depths, it is much more plausible that Nature built upon what it already had rather than performing re-engineering feats from the bottom up (such feats are almost never seen in the history of life). It is therefore most likely that life originated on the surface dissipating UV-C light and gradually evolved pigments for dissipating into the visible and infrared, eventually learning how to dissipate and use for its metabolism the free energy available in the faint infrared light given off by surface magma, perhaps augmenting this small and highly variable supply of free energy with a chemolithic process, before colonizing the hydrothermal vents at the bottom of the oceans. There are, in fact, organisms known today that carry out photosynthesis with the faint infrared light given off by the hydrothermal vents and which alternatively can use lithochemical redox reactions for obtaining their free energy (Yurkov et al., 1999; Beatty, et al., 2005)[424, 16].

19.12 Cold Origin of Life

At the other end of the thermal spectrum are those who believe that life started out on ice, at cold temperatures. The reason for believing in this is the greatly enhanced stability of the fundamental molecules at low temperatures, allowing them to build up to concentrations thought to be necessary for producing a sufficiently thick organic soup such that a "vitality" might spring forth. Levy and

Miller in 1998[187], published a very influential article describing experimental determinations of the stability of the bases of RNA to hydrolysis at different temperatures. Their results showed that at 100 °C, the optimal growth temperatures of many present day hyperthermophiles, the half-lives of the bases are too short to allow for the "adequate" accumulation of these compounds in a organic soup scenario. Their experiments gave half-lives for A and G $\sim$ 1 yr; U = 12 yr; C = 19 days at 100 °C. They concluded that, "unless the origin of life took place extremely rapidly (<100 yr), a high-temperature origin of life may be possible, but it cannot involve adenine, uracil, guanine, or cytosine". The rates of hydrolysis at 100 °C also suggested to them that an ocean-boiling asteroid impact would have "reset the prebiotic clock, requiring prebiotic synthetic processes to begin again". However, at 0 °C, A, U, G, and T appeared to be sufficiently stable (half-lives $\sim 10^6$ yr) to consider as most plausible a low-temperature origin of life scenario. The lack of stability of cytosine at 0 °C ($t_{1/2}$ = 17,000 yr), however, suggested to them that the G-C base pair may not have been used in the first genetic material unless life arose quickly ($<10^6$yr) after a sterilization event.

Miller has also pointed to a further advantage of a cold origin of life, a concentration mechanism for molecules known as "eutectic concentration" in which, as water freezes into its crystalline structure, its organic impurities tend to concentrate in remaining pockets of liquid water. As the pockets of liquid water become smaller, the distance between molecules becomes smaller and the concentration increases. In some cases, this reduction of distance between molecules more than compensates for the lowering of the reaction rates at the colder temperatures. Miller himself carried out experiments with vials of cyanide and ammonia kept on dry ice (-78.5 °C) for 25 years, finding a small amount of nucleobases as one of the products of the experiment. Monnard et al. (2003)[231] have also shown how, over a 38 day period, activated nucleobases of RNA with Mg^{+2} and Pb^{+2} ions as catalysts can polymerize by forming phosphodiester bonds in eutectic concentration experiments at -38 °C, producing strands of up to 30 bases long.

In making their stability analysis however, Levy and Miller did not take into account the fact that polymerization of the nucleotides into strands significantly reduces the probabilities of hydrolysis. Saladino et al. (2005; 2006a)[312, 314] have shown that polymerization prevents decomposition of the nucleotides (see section 8.2). Values of pH somewhat acidic, as would be expected for a high CO_2 concentration in the Archean atmosphere, also provide for greater stability of the nucleotides, as, in fact, Levy and Miller's own work showed.

However, the negatives of a cold origin of life, notwithstanding the advantage of having molecules of greater stability, is obviously the fact that all chemical reactions or enzymatic activity, no matter how simple or primitive, will be dramatically reduced at low temperature (Shimizu et al., 2007)[337]. The rate of a chemical reaction, in fact, slows exponentially with the lowering of temperature and it would take enzymes much more complex and sophisticated than those of today to denature RNA and DNA at below freezing temperatures.

Another problem confronting a low temperature origin of life scenario is the binary fission of lipid bilayers of a protocell at low temperatures (see chapter 15). It would therefore be difficult to imagine a viable replication system operating only through the environmental conditions and it would be necessary to assume much more unlikely (at these temperatures) metabolic processes that could promote replication. Furthermore, isoprene lipids have been detected in rock in the Isua district of Greenland, dated at 3.8 Ga (Jürgen and Huag, 1986)[145], and these are of the glycerol-ether lipid composition which have very high transition temperatures, consistent with a high surface temperature at the origin of life.

A cold origin of life is also not consistent with atmospheric models predicting surface temperatures in the Archean of between 85 and 110 °C (Kasting and Ackerman, 1986; Kasting, 1993)[150, 151] and also not consistent with geochemical evidence in the form of $^{18}O/^{16}O$ ratios found in cherts of the Barberton greenstone belt of South Africa indicating that the Earth's surface temperature was around 80 °C at 3.8 Ga (Knauth, 1992)[163] (Knauth and Lowe, 2003)[165], falling to 70±15 °C during the 3.5–3.2 Ga era (Lowe and Tice, 2004)[193]. A cold origin of life is furthermore in direct contradiction to the evidence from the oxygen isotope content of zicron crystals suggesting that liquid water was running on the surface during the Archean (Mojzsis et al., 2001)[230], and with the evidence from ribosomal RNA, proteins, and other metabolic RNA, that the last universal common ancestor was hyperthermophilic (see section 19.6).

Finally, concentrations of the fundamental molecules in the primordial soup depend not only on their disintegration rate, but also on their production rate. The possibility of non-equilibrium autocatalytic formation of the fundamental molecules was not considered in the determination of the production rates obtained by Levy and Miller. This type of proliferation is what, in fact, determines the vitality of life (see chapter 6.3).

It is therefore much more likely that life arose at a high temperature and that there existed stabilization mechanisms for the fundamental molecules which protected them from hydrolysis, for example, their association in polymers, or their association with other molecules, during, or very soon after their production.

19.13 Pigments provide Photoprotection

In 1973 Carl Sagan published a most important and influential article regarding the origin of life in which he suggested a high probability for a UV-C window in Earth's Archean atmosphere (Carl Sagan, 1973)[309]. In the same article,

Sagan also suggested that incipient life would somehow have to be protected from this "dangerous" light. With very little existent data at that time on the photochemical properties of RNA or DNA, Sagan's conclusion made good sense to the prebiotic chemistry and origin of life communities since it was known that even a small amount of UV-C light could instantly kill most present day microscopic organisms.

The ultraviolet protection issue thus grew very rapidly into a defining prerequisite for any model or scenario contemplating the origin of life and even today permeates almost all of biological thinking on the origin of life. Contemporary proponents of the hydrothermal vent origin of life hypothesis consider the inherent ultraviolet protection provided by the overlying water to be one of its strong points. Over the years, in fact, new pigments have been continually discovered, absorbing from the ultraviolet to the infrared, and one after the other they are assigned, without much consideration or analysis, to either a photo-protective function, or to an antenna function for increasing photosynthetic rates.

However, this protectionist outlook, based on the prejudice of our knowledge of ultraviolet damage to today's evolved complex life, does not seem to be a solid narrative and it is more likely a red herring which has led the biochemistry and origin of life communities astray. For example, most of the fundamental molecules (those found in all three domains of life) have important absorption in the UV-C but have excited state decay times as long as nanoseconds (DNA and RNA have decay times of picoseconds). These fundamental molecules could therefore not possibly have acted as protective agents for DNA and RNA, because they, themselves, would have been much more prone to destruction by photochemical reactions[5]. Instead, it seems more correct, but still not the whole story, to argue that DNA and RNA acted as protective pigments for these fundamental molecules through electronic excited state resonant energy transfer. In fact, the whole story is only told when read from within the non-equilibrium thermodynamic framework in which evolution was driven towards dissipating ever more of the solar spectrum and where new pigments would thus appear and evolve only as a response to increasing the wavelength region of photon capture and thereby increase the global photon dissipation rate. Antenna pigments of the Archean (now known as the fundamental molecules of life) were playing a similar role as antenna pigments of today, increasing the global photon dissipation rate, not protecting life from this light.

Supporting this proposition is the fact that the pigments in a particular leaf of any contemporary plant absorb strongly from the UV-C to the red edge at about 700 nm (see figure 16.3). There is only a small dip in absorption at around 550 nm corresponding to the green, and this is precisely where animal eyes have

[5] The longer the time a molecule stays in an electronic exicted state, the more probable a chemical reaction will occur, changing the nature of the molecule (see chapter 6 for a description of some of the types of reactions allowed in an electronic excited state).

evolved to be most sensitive, surely to maximize functionality within the forest environment. Are all these accessory pigments really needed for protection of the photosynthetic apparatus? Would it not be more logical, if indeed they were meant for protection, that their absorption would be limited to only those wavelengths at which the photosynthetic apparatus was vulnerable?

Zvezdanovic and Markovic (2008)[434] have done a careful study of the bleaching of chlorophyll *a* and *b* (the opening of the cyclic tetrapyrrole ring, see figure 4.10) which compromises photosynthetic function under UV-A, UV-B, and UV-C at high light intensity (~ 12 W/m^2, almost three times the UV-C intensity predicted at the beginning of life). Comparing in vitro the bleaching of pure chlorophyll with that of chlorophyll extracted from spinach leaves which contains a large component of the so called "protective pigments" carotenoids (carotenes and xanthophylls, see figure 4.11), they found that in none of the three UV energy regions, did the carotenoids have any significant effect in protecting the chlorophyll from bleaching over a large range of chlorophyll concentrations. Another interesting fact was that the chlorophyll bleaching by UV-C light for the pure chlorophyll was significantly reduced when the chlorophyll concentrations were large and this they attributed to the hydrophobicity induced chlorophyll aggregate formation at high concentrations and the consequent formation of an extremely rapid non-radiative decay channel for the complex (similar to a conical intersection). Furthermore, H-aggregates, which are arrangements of strongly coupled monomers with the transition moments of the monomers perpendicular to the line of centers (face-to-face) with dipolar coupling between the monomers, leads to a blue shift of the Soret absorption band (Czikklely et al.,1970; Nuesch and Gratzel, 1995)[68, 251]. This would have relevance to the dissipative utility of chlorophyll aggregates during the Archean, allowing photosynthesis in the visible to follow quickly on the heals of UV-C photon dissipation. Such chlorophyll aggregates, or stacks, are still found today in the chloroplasts of plants and bacteria.

Anthocyanins are flavonoids which absorb strongly over the UV-B and UV-C and also in the visible (they are pink, purple, and red pigments, see figure 4.12) and are also commonly assumed to be protective plant pigments. However, Stapleton and Walbot (1994)[357] have found that in field trials of corn plants induced with a defective gene for producing flavonoids, there was no measurable increase in the amount of DNA damage in the form of cyclobutane pyrimidine dimers (CPDs – the most common form of light-induced DNA damage, a result of UV-induced dimerization of adjacent pyrimidines on the same strand of DNA) with respect to the controls with the intact gene for flavonoid production. Furthermore, Gould and Quinn (1999)[111] have found that the location of the majority of the anthocyanins in leaves, for a wide range of species, was in the mesophyll and lower epidermis rather than in the upper epidermis where they would have been more effective as protective screens.

More and more evidence is therefore suggesting that the accessory pigments, and even chlorophyll itself, have little to do with photosynthesis or with protection of the photosynthetic apparatus. It is thus becoming apparent that the accessory pigments exist to capture and dissipate as much of the entire high energy solar spectrum as possible because this is the ultimate thermodynamic function of life. The finite size of the pigments and their finite decay time puts limits on their ability to cope with the large flux of sunlight. It is probable that the red-edge, at which photon absorption abruptly ends in both cyanobacteria and plants (see figure 16.3), is being pushed ever more towards the infrared as new more efficient pigments are invented by Nature (see chapter 19.15).

19.14 Photosynthesis is Optimized in Nature

Only approximately 0.1% of the free energy in sunlight captured by the leaves of a plant is used in the fixation of carbon through the process of photosynthesis (Gates, 1980)[100]. Experiments performed in the 1930's had shown that under intense light conditions, for every chlorophyll molecule that absorbed a photon and participated in the photosynthetic process, there were about 300 other chlorophylls in the plant that simply dissipated their photon excitation energy directly into heat (Mohr and Schopfer, 2015)[229]. This represents an extremely poor efficiency for a photosynthetic system that has had the opportunity to evolve for at least 3,500 million years considering that humans have developed systems capable of converting up to 40% of the free energy in sunlight into usable electrical energy within only 40 years of technological innovation.

Through a set of controlled experiments, expressing transpiration as a function of leaf temperature, CO_2 flux (as a surrogate for stomal resistance), and sensible heat flux, Wang et al. (2007)[392] have shown that plants optimize transpiration, not the photosynthetic rate, under varying the external conditions. Since the transpiration rate is directly related to the photon dissipation rate in pigments within the leaf, their finding is equivalent to suggesting that photon dissipation is optimized in plants.

Still other evidence that nature does not optimize photosynthesis is the fact that higher plants and algae have a large number of photosystem II reaction centers ($\sim$ 32% of the total) that do not contribute to photosynthetic electron transport at all (Chylla and Whitmarsh, 1989)[47], but merely dissipate the captured photon energy directly into heat. Why plants devote resources to the synthesis of reaction centers that apparently do not contribute to carbon fixation has been a perplexing problem for the biologist working from within the Darwinian paradigm, which insists that natural selection should have terminated with such waste, but it makes perfect sense from the thermodynamic view of life as the evolution of photon dissipation efficacy.

19.15 The Red-Edge; a Result of a Lack of Molecular Excited States

The traditional explanation of the pronounced drop in the absorption spectrum at the red-edge at $\sim$ 700 nm of living organisms (see figure 16.3) attributes it to the physical properties of the organic pigments, in particular that the region between 700 and 1000 nm is a region of the spectrum corresponding to few possible molecular energy states, lying between the lowest energy vibrational state superimposed on electronic excited state and the highest energy of the molecular vibrational states superimposed on the electronic ground state (Gates, 1980)[100]. However, given the extraordinary number of possible molecular configurations related to the directional properties of the carbon covalent bonding between organic elements, this explanation is not very convincing. In fact, new pigments have recently been found absorbing very well beyond the red edge. Many deep ocean living bacteria have strong absorption beyond the red-edge and, in fact, use the very faint red light from deep sea hydrothermal vents for efficient photosynthesis (Yurkov et al., 1999; Beatty et al. 2005; Kiang et al. 2007)[424, 16, 158]. Anoxygenic photosynthesis has been discovered using wavelengths as long as 1,015–1,020 nm (Trissl 1993; Scheer 2003)[378, 321].

A second explanation for the red-edge, also given by Gates (1980)[100], is that it may be an evolved characteristic since the plant leaf would heat up to beyond optimal temperatures for photosynthesis if it absorbed the solar energy beyond the red-edge (which amounts to approximately one half of the total energy received by the leaf). This second explanation of Gates also appears to be contrived since plants could have equally well evolved to reflect the UV and blue light with only a strong absorption peak centered around 700 nm where, in fact, photosynthesis is most efficient (Kiang et al. 2007)[158]. Energy absorbed in the blue Soret band of chlorophyll is anyway dissipated directly to the Q band in the red before it can be used in the photosynthetic process. Furthermore, efficient photosynthesis is known to be operating at temperatures of up to 75 °C in the green non-sulfur bacteria Chloroflexus and Synechococcus (Meeks and Castenholz, 1971; Nes and Nes, 1980)[203, 246].

Instead, the red-edge is most probably related to the fact that the most thermodynamically important photon interaction with material has to do with dissipation, and not with photosynthesis. The photon flux from the Sun today is copious, on the order of 10^{22} photons $m^{-2}s^{-1}$midday at the equator, and even today, after billions of years of evolution, the contemporary pigment complexes can only handle on the order of 26% of this flux due to the finite size and finite excited state life-times of the pigments (Michaelian, 2015)[218]. From the viewpoint of dissipation, it would thus make thermodynamic sense to dedicate space, nutrients, and other limited resources to pigments that absorb and dissipate those photons of short wavelength since these give the greatest entropy production per photon dissipated. The wavelength at which biotic life can no longer cope with dissipation due to pigment finite size and finite excited state lifetimes, and to the copious incident photon flux, is probably what defines the red-edge. Photons of wavelength larger than the red-edge are simply reflected or transmitted by the organism. As described in chapter 17.5.1, reflection of a narrow incident beam into a larger solid angle, even without a change of wavelength, also produces entropy and would thus be thermodynamically selected over transmission. Reflection is, in fact, stronger in most plants than transmission (Gates, 1980)[100].

Detection of a "red-edge" on extrasolar planets may be an indication of a life as we know it based on molecular pigments (Seager et al., 2005)[331], however, the position of the "red-edge" in wavelength would not necessarily be the same as that of vegetation on Earth since, from the perspective of dissipation, the position of this edge would depend on the stellar spectrum and integrated photon intensity, and the time that "vegetation" has had to evolve efficient pigments of a dissipative system. More concerning the remote sensing of extraterrestrial life will be given in chapter 20.3.

These facts argue for the notion that photon dissipation in plants, cyanobacteria and algae takes precedence over photosynthesis, and that the red-edge is an artifact of dissipation with little to do with optimizing photosynthesis, which is, in fact, most efficient at long wavelengths around those of the red-edge.

19.16 Panspermia

In the early 1970's Fred Hoyle and Chandra Wickramasinghe (see figure 5.1) realized that they could obtain better fits to interstellar gas and dust absorption spectra assuming organic material instead of the then commonly suspected inorganic silicon grains. Their good fits of the interstellar emission spectra to freeze dried bacteria (figure 5.2) suggested to them that the universe may be teaming with bacterial life. Based on this evidence, they developed their theory of Panspermia which postulated that the origin of life was probably a very special event which occurred perhaps only once somewhere in the universe at some early epoch and was transported from planet to planet throughout the universe.

Although the core of the theory of panspermia has relatively few followers, there is no doubt that, just as with Jim Lovelock's Gaia theory, the data that has been collected by proponents of the theory has been truly exceptional in changing our conception of the universe and the possibility for life within it.

This data includes the uncanny fit of interstellar absorption spectra and emission spectra to that of freeze dried bacteria (see figure 5.2) and the finding of microorganisms and viruses in Earth's upper atmosphere, which, according to Hoyle and Wickramasinghe came from extraterrestrial cosmic dust. Hoyle and Wickramasinghe have even suggested that newly discovered viruses negatively affecting the human population may be recent interstellar arrivals (Hoyle et al., 1986)[134].

Finding many of the amino acids used by life inside the Murray and Murchison meteorites was another event that provided followers of panspermia a renewed cause for enthusiasm. The findings of various martian meteorites on Earth (material derived from a large meteoritic impact on Mars sending material flying out towards Earth), also speaks for the possibility of life from a given planet contaminating another. For a time it was even suspected that the material from the martian meteorite ALH84001 found in Antartica (see figure 19.4) contained fossils of ancient nanobacterial[6] life on mars. More careful studies, however, have concluded that rather than being biological fossils, these features are most probably self-propagating mineral-fetuin complexes of abiotic origin (Raoult et al. 2008)[297].

Other evidence in favor of the theory of panspermia comes from experiments demonstrating the survival of bacterial spores exposed for months to the harsh space environment of the exterior of orbiting spacecraft (subjected to ultraviolet, x-rays, gamma rays, and even energetic protons). Once brought back to the laboratory and placed on a nutritive medium, these spores produce vigorous bacterial colonies. On exposure to extreme environments with little water, bacteria produce a protective sheathing forming a spore which is extremely robust and have been known to remain viable, eventually "come to life", even thousands of years after their formation when the environmental conditions become conducive. More surprising, although still controversial, are data suggesting bacterial spores extracted from ice cores drilled on Antartica have survived perhaps up to 8 million years trapped and protected in ice (Bidle et al., 2007)[21]. There is even a suggestion that viable 250 million year old bacterial spores have been revived from salt crystals extracted from deep under ground (Vreeland et al., 2000) [387]. If these reports were true, inter-galactic panspermia could certainly be viable. However, an analysis of the genomes of these particular bacteria suggest that they are recent and therefore, most probably, contaminants (Graur and Pupko, 2001)[117].

[6] Nanobacteria are hypothesized life forms of nanometers (10^{-9} m) in size. The smallest size of confirmed bacteria appears to be 200 nm. The evidence for nanobacteria fossils on the meteorite is alternatively explicable through non-biotic percipitation mechanisms, suggesting that morphology alone is not suficcient to identify life. The existence of nanobacteria is controversial and so far they have not been uneqiuvocally identified in any source.

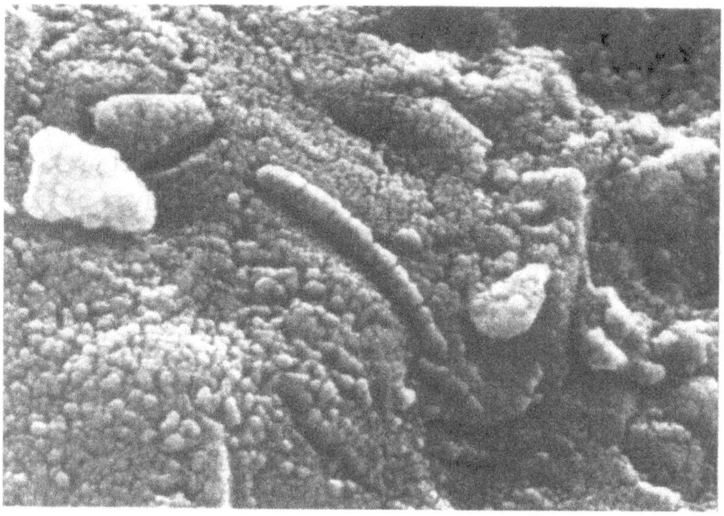

Fig. 19.4. Structures found on the Allan Hills (ALH84001) meteorite found in the Antartic and comming from Mars. The strucutres were originally identified as nanobacteria but more careful analysis suggest that they are probably of abiotic origin. Image credit: NASA, Public domain.

An even more exotic, but perhaps equally plausible, theory is that of directed Panspermia also proposed by Hoyle and Wickramasinghe. In this theory, intelligent beings have decided to seed the universe with life and they would do that most efficiently by sending bacterial spores on random trajectories throughout the universe. A very entertaining account of the development of the theory of panspermia, of the professional hardships that had to be endured by Hoyle and Wickramasinghe for their very unorthodox ideas, and the latest evidence in their favor, is given in the book "A Journey with Fred Hoyle" by Chandra Wickramasinghe (2013)[405].

Although a panspermia origin of life on Earth is a theory that still cannot be ruled out, it does not solve the mystery of the origin of life, it only relocates it to another place and earlier time. Granted, life could gain perhaps an extra 10 Ga (10^{10} years) of evolution, and an immensely greater probability for its origin (perhaps 400 billion to one – the number of stars in our galaxy) if it were truly a fortuitous event. However, there would be no, or very little, utility for such a process of panspermia if life springs forth naturally through spontaneous material organization into pigments and supporting structures in response to the dissipation of a stellar UV-C photon flux, as this book proposes.

Perhaps a local panspermia, in which bacteria are transferred in rock launched into space by an asteroid collision with a planet rich in life, known as lithopanspermia, restricted to a particular solar system, may provide a route to colonization of harsher planet environments once the diversity of life on the originating planet had become great enough, for example the evolution of the extremophiles on Earth colonizing Mars, the clouds of Venus, or the ice covered oceans of Europa. Whether such a local lithopanspermia has occurred in our own solar system will surely be discovered within the next 50 years or so. However, perhaps we do not even have to wait as long since, as will be argued in the following chapter, from the dissipative thermodynamic perspective, we have already discovered extraterrestrial life in the early stages of development on other planets of our solar system which were produced by mechanisms that do not require panspermia.

Finally, the extraordinary overlap of the absorption spectrum of the fundamental molecules of life with the probable incident photon spectra in the UV-C at Earth's surface during the Archean (see figure 4.7), and the phylogenetic analysis showing that the last universal ancestor was either thermophilic or hyperthermophilic, suggests that life had a very intimate relation with the early physical conditions prevalent at Earth's surface from its very beginnings and that, therefore, the fundamental molecules of life, and therefore Earth life, most probably originated, not on other planets in the cosmos, but here on Earth itself.

Part VI

The Dissipative universe

20. Dissipative Life throughout the Universe

In this book I have presented an explanation for the origin of life in which the "self-organization" of material into pigment molecules with conical intersections, or into pigment molecules with affinity to those molecules with a conical intersection, occurs naturally and spontaneously under an externally imposed UV-C photon potential. The general process is known as microscopic dissipative structuring and was first described in detail by Ilya Prigogine (see box on page 59 and chapter 3.4). This microscopic self-organization of material occurs as a non-equilibrium thermodynamic response to dissipate the imposed generalized chemical potential, just as the self-organization of the material of the atmosphere occurs in the form of a hurricane to dissipate the externally imposed temperature gradient. In this view, life arose as a photochemical, autocatalytic process in which the catalysts for the dissipation of the external photon potential were the photochemically produced pigments themselves. Life evolved outwards from these fundamental pigment molecules to an ever greater complexity in order to dissipate ever more efficiently the solar spectrum, to dissipate over ever more of Earth's surface, and later to dissipate other chemical potentials in its environment (for example those existing at hydrothermal vents or those chemical potentials produced by phototrophic life).

It is apparent that this evolutionary program based on photon dissipation has been operating on Earth for at least 3.85 Ga under universal laws of physics and chemistry and probably even earlier in the form of pigment production in the Hadean and, most likely, even earlier in the protoplanetary disk around our star. Such a photon dissipation program should, therefore, also be operating almost everywhere in the universe. Not only on other bodies similar to Earth, but, in fact, anywhere in the cosmos where high energy photons interact with organic elements within a liquid solvent (or even a gas) environment providing a thermal bath to bring the pigment molecules rapidly back to the ground state so they can be ready once again to absorb a new photon.

Organic pigments have indeed been found throughout the cosmos as detailed in chapter 5. For example, organic molecules have been found in significant quantities on the surfaces of inner orbit comets (see section 5.4). It is probable that these molecules were not formed in the atmospheres of red giant stars and later collected by these comets, as suggested by Hoyle and Wickramasinghe (1978)[135], but rather that they formed on the comets themselves as the comet passed sufficiently close to the Sun to experience intense UV-C light and to cause regions of liquid water to form within the ice at the surface of the comet. As on Earth, on the comet surface the proliferation of these molecules to concentrations much beyond their expected equilibrium concentrations would be a non-equilibrium thermodynamic imperative, a result of their catalytic activity in dissipating high energy photons from the Sun (see chapter 6).

If indeed the thermodynamic dissipation theory accurately characterizes the essential physical and chemical features of the process of the origin of life, an interesting question is, "What are the essential conditions required for life similar to our own (i.e. organic element based and dissipating from the UV-C to the red) to arise elsewhere in the universe?". From our non-equilibrium thermodynamic perspective, the essential conditions would be:

1. A second or later generation star that would have a solar system formed out of a gas and dust cloud containing elements heavier than hydrogen and helium, in particular carbon, produced by nucleosynthesis in earlier generation stars.
2. A star of enough mass giving it a surface photon spectrum sufficiently strong in the UV-C or UV-B region, light which has enough energy per photon to break and form covalent bonds, but not enough to ionize and thereby destroy the molecule or make it vulnerable to photochemical reactions. Stars of too great a mass have a very high energy spectrum and a short a lifetime and therefore life similar to what we know may have begun on the planets of such a star who's atmosphere was adequate to shield the molecules from the very high energy photons, but the star would not last long enough to support a lengthy evolutionary program as has occurred on Earth. From the perspective of the thermodynamic dissipation theory of the origin of life, star types with the necessary characteristics for life similar to ours are thus limited to the K, G, F and A -types (see figure 20.1). This amounts to about 23% of main sequence stars.
3. Some kind of atmospheric protection (e.g. N_2 or CO_2) against dissacociation from the very high energy ionizing light ($<$ 220 nm).
4. A water, or other liquid organic, or gas solvent environment providing the required microscopic degrees of freedom for effective dissipation of electronic excitation energy of the pigment into the higher entropy vibrational and rotational energy of the solvent molecules. Water (or some other solvent for organics) would also provide routes for photochemical reactions (Ferris and Orgel (1966)[96]) and would provide a generally stable temperature and pH environment.

5. Surface or atmospheric temperatures in the range where the particular organic solvent is in a liquid or gas state so that it could act as an effective dissipating agent.

Throughout the universe, life based on pigments derived from carbon covalent organic chemistry and water will most probably be the most common due to the ubiquity of carbon, water, and stars of the appropriate photon spectrum mentioned in point 2. above. There is now sufficient observational evidence to indicate the ubiquity of planets orbiting around second or later generation stars of mass > 0.5 solar masses emitting a relevant UV-C to UV-B component (see figure 20.1). Microscopic dissipative structuring would occur with high probability at UV-C wavelengths through photochemical routes to carbon based pigment molecules which absorb and dissipate at UV-C wavelengths, this due to the great myriad of possible molecular configurations available for carbon atoms with four unpaired electrons in the outer shell, providing four directional covalent bonds per atom. Such an immense assortment of possible molecular configurations would also imply a similarly large assortment of electronic and stereochemical configurations for the molecules, leading, through dissipative structuring, to an almost infinite variety of molecular pigments and their complexes.

Class	Temperature	Apparent color	Mass (solar masses)	Radius (solar radii)	Luminosity (solar luminosity)	Approximate main-sequence life span (years)	Hydrogen lines	% of all Main Sequence Stars
O	30,000–60,000 K	blue	64	16	1,400,000	~10 million	Weak	~0.00003%
B	10,000–30,000 K	blue white	18	7	20,000	~100 million	Medium	0.13%
A	7,500–10,000 K	white	3.1	2	40	~1 billion	Strong	0.6%
F	6,000–7,500 K	white	1.7	1.4	6	~5 billion	Medium	3%
G	5,000–6,000 K	yellowish white	1.1	1.1	1.2	~10 billion	Weak	7.6%
K	3,500–5,000 K	yellow orange	0.8	0.9	0.4	~50 billion	Very weak	12.1%
M	2,000–3,500 K	orange red	0.4	0.5	0.04	~100 billion	Very weak	76.45%

Fig. 20.1. Classification of stellar types based on their surface temperatures and some related properties. From the viewpoint of the thermodynamic dissipation theory of the origin of life, the essential requirement for life similar to Earth's is that the star emit sufficiently in the UV-C to UV-B region, is of second generation or younger, and has a long enough life to permit the development of an evolutionary program. This includes star types K, G, F and A. Image credit: Spectral Types, Wikipedia, Creative Commons.

Although water is by far the most common solvent for organic molecules in the universe, other solvents may play a similar dissipative role at the incipient stages of life. Liquid methane and methanol and ethane and ethanol on Titan's surface may be examples of such solvents for organics that may be involved in an active carbon based dissipative "biosphere" (see chapter 5.4). Methanol provides a similar environment as water for rapidly dissipating photon-induced RNA and DNA electronic excitation energy (Middleton et al., 2009)[222]. Such a solvent may play the role of water in a life similar to ours, also based on RNA, DNA and fundamental pigments, but at physical and chemical conditions very different from Earth's. In fact, rivers, lakes and clouds of liquid methane have been observed on Titan, demonstrating temporal changes indicative of an active methane (and other hydrocarbon) cycle (see figure 5.17). It may thus be too restrictive to confine the "habitable zone" around a star, as commonly done, to that based solely on the temperature-phase relationship of water.

The ultra-fast dissipation characteristics of the nucleotides do not seem to depend critically on the organic solvent (Middleton et al., 2009)[222]. Furthermore, it has been shown that the rapid de-excitation does not require a liquid solvent at all since sub-picosecond lifetimes are also measured for the electronically excited nucleobases seeded within a carrier gas such as argon (Canuel et al., 2005)[38]). Therefore, sufficient conditions for an evolutionary program based on photon dissipation in a given region of the cosmos may thus be simply; elements capable of covalent interactions, an incident photon flux with photon energies similar to the strength of the bonds of the covalent interactions, and a solvent, or carrier gas, to help dissipate the molecular electronic excitation energy.

Based on the ubiquity of infrared emission bands (IEB) detected in nebula and star forming regions (see chapter 5), there is little doubt concerning the existence of organic pigments in interstellar space, particularly in the form of polycyclic aromatic hydrocarbons (PAH's) (see figure 20.2) or mixed aromatic/aliphatic organic nanoparticles (MAONs – see figure 5.8). These polycyclic hydrocarbons absorb and dissipate strongly in the UV-C and also appear to be formed in regions where these wavelengths are intense, suggesting that they were formed as microscopic dissipative structures (see chapter 3.4). These molecules are very robust and resistant to further transformation under photochemical reactions, generally due to an inherent conical intersection. Calculations suggest that up to 20% of all the carbon in the cosmos may be tied up in these molecules (Dwek et al., 1997)[90]. The cosmic ubiquity of these UV-C absorbing organic molecules in space can be explained through the same photochemical autocatalytic proliferation mechanism described in chapter 6 and analyzed in detail in Michaelian (2013)[214] and Michaelian and Simeonov (2016)[220].

There must also exist dissipative systems common throughout the universe very different from what we commonly consider as life. In fact, from the most general definition of life as a dissipative structure, we could say that this very different life exists even here, on our own planet. The water cycle, hurricanes, ocean and wind currents are all dissipative structures and they are all based on the heat generated by photon dissipation in organic pigments. These processes are therefore different in form, but similar in principle and in molecular foundation, to the organic life that we are familiar with. The coupling of pigment UV-

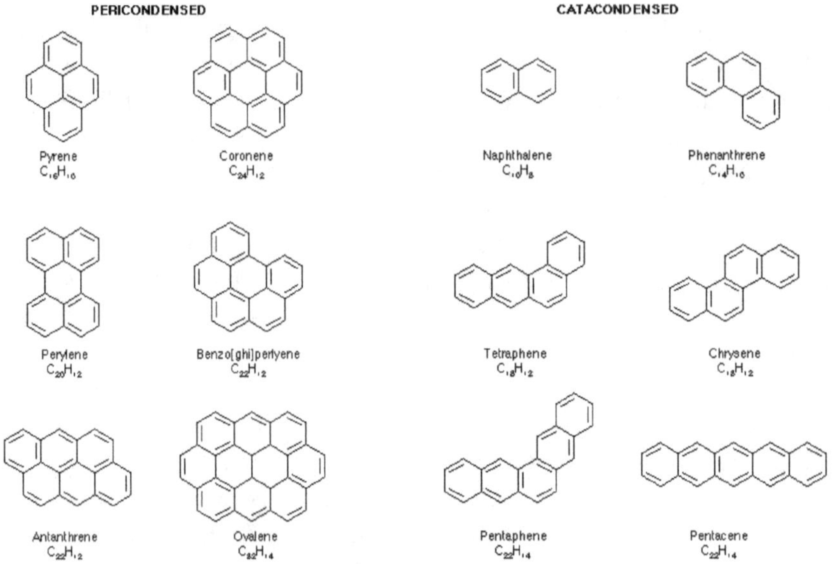

Fig. 20.2. Some common polyaromatic hydrocarbons (PAH's) found in space. These molecules have large UV-C absorption cross sections and were probably made by similar wavelength light interacting with the precursor intergalactic gas, suggesting that they are microscopic dissipative structures. Image credit: NASA, Public Domain.

visible photon dissipation to these abiotic dissipative structures, which dissipate infrared wavelengths (temperature gradients) is in accordance with Onsager's principle of the coupling of irreversible processes (see chapter 6.3.1) which will occur on the observance of certain symmetry requirements[1] and if the coupling increases the overall global dissipation rate.

Finally, life as we know it also has the potential to seed photon dissipation in regions of the universe distant from its origin, for example, large meteoritic impacts that transport material with living organisms inside, in the form of spores, from one planet to another (lithopanspermia), or intelligent beings establishing themselves and terra forming the other planets of their solar system, and even those of neighboring solar systems.

20.1 Life Similar to Ours

Pigments found throughout the cosmos are generally organic compounds because only carbon has the capacity to produce the rich chemistry necessary to create the large variety of molecular configurations and associated electronic

[1] Curie's principle states that macroscopic causes must have fewer symmetry elements than the effects they produce. For example, a directed heat flow could give rise to chemical reactions, but chemical reactions cannot give rise to a directed heat flow.

excited states for dissipating over a large spectral region. Conjugation of the covalent bonding provides for chromophores within the UV and visible regions (see chapter 16.5). From this dissipative viewpoint, it makes sense then that most extraterrestrial life based on photon dissipation would involve conjugated carbon aromatics.

The evolution of life very similar to Earth's, based on RNA and DNA, other pigments, and water, assuming the validity of the present dissipation theory of its origin, would require particular initial conditions for the planet. For example, an intense UV-C or UV-B photon flux, and liquid water at temperatures that gradually dropped below the denaturing temperature of first DNA and then RNA (see chapter 12), and this would most probably happen on a planet at a similar distance from a similar star as our Sun.

Within a volume of a radius of 100 light years (ly)[2] centered on our Sun, there are 512 stars of the same spectral type as our Sun (see table 20.1). This implies an areal (planer) density of G-type stars of approximately 1.63×10^{-2} G stars/ly^2. Since our galaxy has a diameter of approximately 41,000 ly, and assuming a similar areal density of G-type stars over the whole galactic disk[3], this gives the result that our galaxy alone should contain at least 86 million stars similar to our Sun. Since the formation of planets around a star is now known to be common and, in fact, a normal consequence of the stellar evolutionary process, and since the majority of stars in our galaxy are older than our Sun, it can be expected that up to 86 million stars in our galaxy have planets with life at a similar stage or more advanced in terms of the evolutionary dissipation process than our own.

There are other considerations which may change this number, perhaps by an order of magnitude, for example, it is known that a large planet like Jupiter attracts comets and asteroids through its strong gravity, thereby avoiding heavy bombardment of the inner planets that could periodically reset the evolutionary clock of life. Similarly, in galactic regions of high stellar density and of older stars, such as in the core of the galaxy, the likelihood of ionizing radiation from a nearby supernova explosion resetting the clock of life also becomes an important consideration.

On planets orbiting these 86 million stars of our galaxy, life similar to ours, based on RNA and DNA, almost certainly would have developed. The stage of development would depend largely on the age of the star, but also on the planet's

---

[2] A "light year" (ly) is the distance light travels in one year, which equals approximately 9.5×10^{12} Km.

[3] The density of stars is much greater in the galactic core, but most of the stars located there are of population II stars which are redder and of lower luminosity than the hotter blue stars in the spiral arms. Many red stars may not have sufficient emission in the UV-C to break and reform covalent bonds necessary for dissipative structuring of the fundamental pigments. For the purpose of calculation, we therefore assume an equal areal density of G-type stars over the whole galactic disk.

Light years from Sun	Number
0-10	2
10-20	5
20-30	11
30-40	12
40-50	34
50-60	51
60-70	57
70-80	88
80-90	109
90-100	143
Total within 100 ly	512

Table 20.1. G-type stars within differnt distance bins up to 100 light-years from the Sun.

initial conditions and subsequent subjection to perturbations (for example, the two examples of catastrophic events mentioned above). Using Earth's history as a guide, epochs, consisting of different photon dissipative stages of development of life similar to ours on a particular star, would be, in chronological order;

1. The establishment of a thick organic pigment soup on primordial ocean surfaces with the pigments derived through microscopic dissipative structuring proliferated through autocatalytic photochemical reactions and restricted to absorption at UV-C and UV-B wavelengths.
2. The coupling of the heat of photon dissipation due to the pigments to non-local primitive solvent cycles and solvent and atmospheric currents.
3. The evolution of the organic pigments with absorption and dissipation from the UV-C to the UV-B, UV-A and visible regions of the stellar spectrum.
4. The formation of complexes of organic pigments (such as the aromatic amino acids) together with molecules having conical intersections (such as RNA and DNA) acting as acceptor quencher molecules, giving rise to the first virus-like systems.
5. Primitive cells involving phospholipid enclosures allowing the spread of the pigments into more hostile regions while internal cell conditions such as temperature, salinity and pH, could be adequately controlled for reproduction through photon dissipation in a UVTAR-type mechanism.
6. Alteration of the planets atmosphere (for example, through the invention of oxygen photosynthesis) to allow the most intense region of the stellar spectrum to arrive at the surface where it could be most efficiently dissipated by the pigments and their supporting structures. Autocatalytic proliferation would now be increasingly based on longer wavelength light. Relegation of

the UV-C photon dissipation to oxygen and ozone in the upper atmosphere. RNA and DNA and some of the other fundamental molecules of life now begin to take on new functions, such as information storage, still pertaining to dissipation but not to direct absorption in the UV-C.

7. Complex proteins and enzymes, now vulnerable to destruction by UV-C light but protected by ozone and oxygen, capable of catalyzing reproduction using only visible light. The beginnings of extensive visible light dissipation.
8. Complex cells, the result of numerous symbiosis events, each giving rise to greater entropy production (photon dissipation) efficacy per unit biomass, equivalent to the first eukaryote cells.
9. Mobile multicellular organisms capable of nutrient transport over finite distances for benefit of the photon dissipating organisms.
10. Colonalization of land by the dissipative pigments through nutrient dispersal performed by small animals.
11. Large sea and land animals capable of transporting nutrients over great distances, either far out to sea (mid-ocean) or far in-land (away from ocean shores).
12. The development of a dominant species with high individual intelligence and collective behavior required for building societies, perhaps initially thriving off the dissipation of fossilized chemical potentials. The development of efficient solar cells. Exponential increases over time in energy dissipation attributed to this species.
13. Terra-like forming of stellar system planets by the intelligent species. Photon dissipation attributed to the intelligent species begins to outstrip that performed by local "natural" biospheres.
14. Capturing and dissipation of the majority of stellar photons through construction by dominant species of giant sails of synthetic polyaromatic hydrocarbons leading to Dark Stars.
15. Migration of intelligent life and its photon dissipation technology to the rest of the galaxy.
16. Manipulation of the energy content of black holes, perhaps even allowing manipulation of hyper-dimensional space-time (Kaku, 1994)[147].

Each of the 16 stages represents an important increase in global dissipation. Earth's dissipation development is now leaving stage 12 and entering stage 13. With only one planet readily available for study at this stage, our Earth, it is difficult to predict average time periods spent by a particular world in each of the stages, and the duration of each stage would probably vary greatly depending on the initial conditions and the geological and perturbational history of each planet. A more detailed analysis of the major evolutionary events for Earth associated with the first 12 stages enumerated above are given in figure 20.3.

3.85 - 3.2 Ga	3.2 Ga - 2.7 Ga	2.7 - 2.45 Ga	2.45 - 2.2 Ga	2.2 Ga - present
~ 475 Wm^{-2} total 230 - 290 nm (UV-C) ~ 5.5 ± 2 Wm^{-2} 290 - 320 nm (UV-B) ~ 0 Wm^{-2} 320 - 700 nm (UV-A, VIS) ~ 327 Wm^{-2} - Thermophile but habitable condition ~50-70°C. - Molecular UV-C dissipation; pigment complexes of nucleic acids, amino acids, flavins, porphyrins etc. - Non-enzymatic ultraviolet and temperature assisted replication (UVTAR). - Lipid vesicles with membrane-bound antenna pigments. - LUCA branching into Bacteria and Archaea. - Ancient stromatolites of protocyanobacterial anoxygenic photosynthesizers.	~ 439 Wm^{-2} total - Thick upper atmosphere organic haze, increasing albedo at all wavelengths; cooling of Earth's surface locally to glacier temperatures ~ 0-20°C. - Molecular UV-C, UV-B, UV-A and visible dissipation; MAAs, scytonemin; visible dissipating pigments: (bacterio)chlorophylls in Protocyanobacteria, bacteriorhodopsin in Archaea, rising in importance. - Branching of oxygenic cyanobacteria within the protocyanobacterial lineage; gradual O_2 accumulation in ocean.	- Organic haze depleted by declining CH_4/ CO_2 ratio, increasing the intensity between 320 -700 nm. - Non-glacial temperatures > 20°C. - Ozone, from oxygenic photosynthesis, began to reduce UV-C intensity at the surface. - Geologically rapid increase in atmospheric oxygen with the beginning of the Paleoproterozoic. - Probable appearance of first eukaryotes. - Cyanobacterial phytoplankton in the ocean. - Further diversification and proliferation of visible-absorbing pigments: carotenes, xanthophylls, phycobilins.	~ 672 Wm^{-2} total - Ozone extinguishes surface UV-C light. - Oxygen Catastrophe, Huronian glaciation and snowball Earth; temperatures below freezing. - Aerobic respiration supersedes anaerobic. - Diversification of cyanobacteria; cyanobacterial multicellular colonies. - Dominance of modern visible-absorbing pigments: phycobilins, chlorophylls, carotenoids.	~ 865 Wm^{-2} total - 9% in the UV-A and UV-B. - Atmospheric O_2 levels rising sharply from < 1% of present atmospheric levels to ~ 15% of present atmospheric levels; oxygen today at 21%. - Mean surface temperatures > 20°C. - Endosymbiotic events give rise to mitochondria, chloroplasts in modern eukaryotes. - Evolution of multicellular eukaryotes: algae, fungi, plants, animals. - Novel pigments in land plants: flavonoids, anthocyanins, betalains.
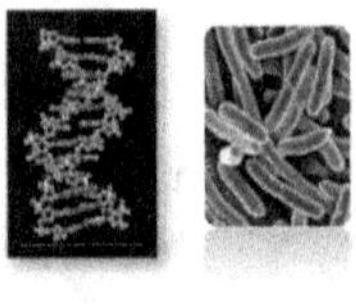 	 	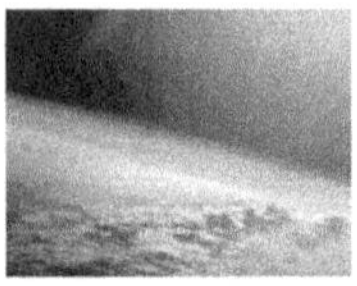 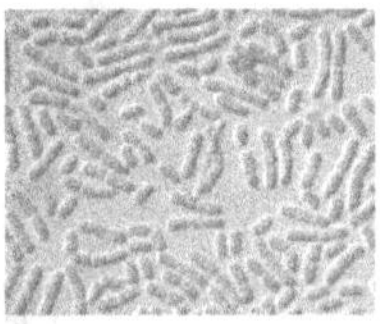	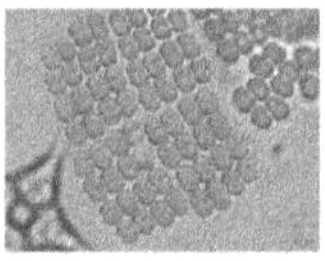 	 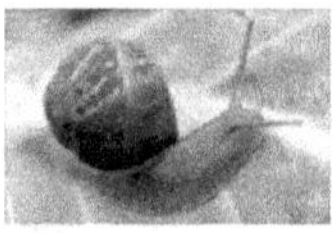

Fig. 20.3. Timeline demonstrating the correlation between the prevailing surface solar spectrum (see figure 4.3) and the appearance and distribution of organic pigments and their complexes. Major dissipation milestones in the evolution of life on Earth are also noted in the figure. Image credit: Michaelian and Simeonov (2015)[217]. Creative Commons.

Finally, the dependence of Earth's biological evolution on its initial conditions, and also on external perturbations, suggests that it was probably not inevitable that life as we know it based on pigment photon dissipation appeared on Earth. For example, the elastic dispersion into a greater volume of an initially collimated photon beam arriving from the Sun also produces a significant amount of entropy, contributing, in fact, almost one half of the entropy production on Venus (Michaelian, 2012b)[213]. It may be, for example, that during ice-ages, the majority of the entropy production is shifted from photon dissipation by plants and cyanobacteria to photon elastic dispersion from ice, or, more probably, a combination of these two forms of dissipation which becomes competitive with photon dissipation by organic pigments alone under the prevailing external perturbation giving rise to cold climate conditions.

In the following section I discuss the possibility of other, distinctly different, forms of dissipative "life" arising on planets or stellar systems different from ours.

20.2 Life Different from Ours

Life as we know it is a dissipative process based on the autocatalytic formation of organic pigment molecules and the later evolution of these into more complex biotic and coupled biotic-abiotic systems, each evolutionary stage representing an increase in the global photon dissipation of the planet. The foundation of this process is the formation and proliferation of organic pigments under an imposed photon potential. As described in chapter 5, this part of the process of life has been occurring naturally on most of the planets of our solar system, on comets, and throughout interstellar space, wherever there exist the organic elements and UV photons with enough energy to break and make covalent bonds. The foundations of life (the building block pigments) are therefore dissipative in origin and nature and are available everywhere in the universe.

The particular nature of the pigments, their physical and optical properties, depends on the available primordial molecules and on the prevailing UV spectrum of the host star. On Earth the pigments which "self-organized" under the UV-C potential were initially the nucleobases and the other fundamental molecules of life formed principally out of HCN, but gradually diversified into near UV and visible light pigments and their support structures. The biosphere of Earth today still incorporates these pigments at its foundations and the heat of photon dissipation is still coupled to abiotic irreversible processes such as the water cycle and wind and ocean currents.

An interesting question is : What kind of biospheres may exist in the other regions of the cosmos where UV light interacts with material but where physical conditions are much different from those on Earth? Some indication of what types of biospheres might be expected can be obtained from observation of other planets in our own solar system where evidence for the first stages of dissipative life (see section 20.1), for example, pigment accumulation, has already been documented (see chapter 5.4).

On Venus the dissipating pigments are sulfuric acid (H_2SO_4) molecules forming clouds reflecting strongly in the visible and absorbing strongly in the far UV-C (<230 nm, Burkholder et al., 2000)[33] and near infrared (> 700 nm, Palmer and Williams, 1975)[264], and perhaps some ferric chloride ($FeCl_3$) absorbing in the near UV (Krasnopolsky, 2016)[170] which gives rise to the structure seen in these clouds in the UV. Venus may be a case in which the microscopic self-structuring of *inorganic* pigments occurs in preference to the structuring of organic pigments due to the prevailing high temperatures on the surface (464 °C) and lower atmosphere which would rapidly disintegrate Earth-type organics. The photon energy dissipated in the clouds by these inorganic pigments is driving the great southern and northern vortices (see figure 0.1) which, together, form the Venusian dissipative biosphere.

On Mars the pigments may be more Earth-like. For example, Pershin (1998; 2000)[278, 279] has postulated the existence of porphyrins and hopanoids (a carotenoid) absorbing in the UV and visible as detected by their fluorescence emission spectra obtained from the Utopia Planitia region of Mars using the Hubble space telescope. Pershin has suggested that these may be the remains of ancient pigments from cyanobacteria-like organisms that once thrived in the sediments during the early Hesperian period (roughly corresponding to the Archean era on Earth starting at 3.7 billion years ago when large regions of Mars were thought to have been covered by liquid water). The survival of such pigments near the surface under arid and harsh UV conditions for such a long period of time is questionable, however. It may be more reasonable to assume that these pigments are recent self-organized microscopic dissipative structures formed under UV-C light within a local environment of temporal liquid water (stage 1 of the evolutionary dissipative program of life as presented in section 20.1). The heat from the dissipation of sunlight in these pigments, as well as in other inorganic pigments, appears to give rise to the strong winds provoking great dust storms that can encircle the whole planet, providing further thermal dissipation and leading to a quite different kind of dissipative "biosphere" but one still founded on organic (and inorganic) pigments.

On Titan it has been revealed by the Cassini–Huygens mission of 2000 that the atmosphere contains haze-like nano-sized solid particles that are most probably the result of the condensation of organics. Their is supporting data in that methane and nitrogen molecules excited by UV photons react to form polymeric hydrogenated carbon-nitride compounds, called tholins, that give the distinctive thick layer of orange-brown haze in Titan's lower stratosphere (Waite, 2007; Nguyen et al., 2007)[390, 247].

Cassini RADAR observations found that these organic nanoparticles condense on surface sand grains that are blown by winds into longitudinal dark-colored dunes on the surface, leading to important UV and visible surface dissipation. Some of these organic nanoparticles also appear to be dissolved in the numerous lakes and rivers of liquid methane and ethane in Titan's polar regions, which exhibit active liquid-gas phase cycling (see Fig. 5.17), similar to the water cycle on Earth, although at a much lower temperature (-179.5 °C) (Atreyaa et al., 2006) [8]. In this biosphere, organic pigment molecules are acting as catalysts to the methane cycle, analogously to the way Earth-based pigments act as catalysts to the terrestrial water cycle, by dissipating the incident solar photons into heat which is then utilized in the heat of evaporation of the solvent (Michaelian, 2012b)[213].

Undoubtedly, other biospheres founded on organic or inorganic pigments will be discovered on other planets and moons in the near future with increasing frequency as China and India enter the space age. But, what other dissipative "life" not based on organic pigments and very different from our own may exist? In the previous section I have given an example of the Earth in a snowball regime in which the entropy production is primarily due to photon dispersion and not photon absorption and dissipation. We would have a hard time calling this life since snow and ice do not appear to be based on organic pigments (the fundamental molecules) and there does not, at first sight, appear to be much of an evolutionary history related with snowball biospheres. However, the dissipative nature of such a system, and its "self-organization" into a dissipative biosphere (in which photon dispersion produces entropy), cannot be denied. In fact, a closer look at the case of Earth reveals that organic pigments were probably involved in the formation of the snowball state through the catalysis of the water cycle which led to the accumulation of snow and ice, and this may also be true on other snowball worlds but with different pigments and a different solvent.

20.3 The Search for Extraterrestrial Life

A design for an efficient search for extraterrestrial life can only be contemplated once the nature of life as dissipative structuring is understood. We now understand how UV-C or UV-B light would be needed to remake covalent bonds in order to produce these robust microscopic dissipative structures which act as pigments that we call the "fundamental molecules of life". This would rule out almost all stars that are not in the categories K, G, F or A -types, leaving only about 23% of stars in our galaxy for a more detailed search (see figure 20.1). An efficient search design would consider life's stages of dissipative evolution as outlined in section 20.1, and how these stages are affected by the initial conditions and subsequent perturbations of the relevant planet or other material body. For example, if we wanted to discover extraterrestrial life at its very initial stage, that of the autocatalytic formation of organic (or inorganic) pigments, then I believe that we could emphatically state that we have already found it on numerous planets and moons of our solar system, for example on Venus, Mars and on the Saturnian satellites Titan, Iapetus, Phoebe and Hyperion (see chapter 5.4). If we were looking for foreign life at stage 2, of the irreversible coupling of the photon dissipation provided by these pigments to other non-local and even global abiotic irreversible processes such as solvent cycles or atmospheric or oceanic currents, then we could again emphatically state that we have also already found it, for example in the methane/ethane cycle catalyzed by organic pigments on Titan, or in the winds that sweep across Mars driven by photon dissipation in the porphyrin-like pigments (see chapter 5.4).

If we were looking for life at stage three in which pigments have evolved towards dissipating light in the UV-B, UV-A and visible regions, then here again the evidence seems to be in favor of already having detected extraterrestrial life at this stage, for example, in the tentative evidence for inorganic pigments in the clouds of Venus or the likely detection of porphyrin-like molecules on the surface of Mars (Pershin, 1998; 2000)[278, 279].

If we wanted to look for extraterrestrial life at stages beyond stage 3, for which we, at this time, do not have any empirical evidence, for example, if we wanted to discover life at stage 4 consisting of antenna pigments in complexes with conical intersections, then we would have to look closer, with more specialized instruments on spacecraft on planetary intercept missions, or develop remote sensing techniques suitable for the task. In the following I give some possibilities for remote sensing which could detect extraterrestrial life at the stages beyond stage 3 of the scheme for the evolution of planetary photon dissipation efficacy.

To make the initial searches for extraterrestrial dissipative life more familiar and therefore reliable, we should constrain our first searches to carbon based organic life with a history similar to our own. As suggested in the previous sections, we should then narrow our search to G-type stars like our sun, of which there are about 500 such stars within our solar neighborhood of 100 light years (see table 20.1). A remote sensing technique for the detection of life would require the ability to measure the intensity distribution over wavelength of the light emitted and reflected by a particular planet. The greater the red-shift of the emitted light with respect to the incident light, the greater would be the stage of evolutionary dissipation. The technology to measure this is indeed already available, certainly for planets and moons of our own solar system and recently for planets of other stellar systems which have now been detected directly by the infrared light they give off (Encrenaz, 2014)[93]. This technology will receive a major boost when the James Webb telescope, designed to be operative in the infrared, becomes available for science in 2019 (see figure 20.4).

Fig. 20.4. The James E. Webb space telescope designed to detect long-wavelength (orange-red) visible light, through the near-infrared and into the mid-infrared (0.6 to 27 μm). One of the missions of the telescope is to detect extrasolar planets. Measurement of the infrared light emitted by a planet and a full spectrum of its star will permit a determination of the planets global entropy production and thereby its stage of thermodynamic evolution in the sense of development of photon dissipation. The large silver sails beneath the telescope are solar heat deflectors to allow the telescope to be maintained below 50 K for optimal performance of its infrared detectors. The telescope is scheduled to be launched in October 2018 into a solar orbit that will follow the Earth at the Earth-Sun L2 Lagrange point (at a distance of 1.5 million kilometers, about 5 times the distance of the moon, where the sails of the telescope will be able to block the light from the Sun, Earth, and Moon simultaneously. At the L2 Lagrange point, the telescope will be able to maintain its position with respect to the Earth and Moon at the larger orbit due to a balancing of centrifugal force with the gravitaional force of the three bodies; Sun, Earth, and Moon). Image credit: NASA/JWST, Public Domain.

Given the many indications presented throughout this book of how organisms, and the coupling of organisms to other biotic or abiotic processes, become more complex over time to increase global photon dissipation, if we wanted to find life similar to ours at an advanced stage, then we should obviously look for those planets of stellar systems that have the greatest stellar photon dissipation. This could be done by determining the incident and emitted photon spectrum and then calculating the average entropy production per square meter of the planet through the use of Planck's formula for the entropy of an arbitrary photon beam as described in detail in the Appendix. Normalizing the entropy production by a $1/R^2$ factor to take into account the differences in incident photon intensity due to different distances R of the planet from its star, should give a relative measure of the entropy production of each planet, an indication of how far advanced the thermodynamic evolution of photon dissipation may be on the respective planet.

Another possibility for remote sensing, requiring much greater spatial resolution but perhaps allowing us to discover mobile life at stage five in which animals may be transporting nutrients required for pigment manufacture over large distances, would be to calculate the variations of entropy production over the surface of the planet. A small spread in the areal entropy production (a spatially homogeneous emitted spectrum) for a highly dissipating planet could be a useful indication of animal dispersion of the nutrients required for pigment production.

Yet another technique of remote sensing would be to detect the equivalent of a "red-edge" in the reflected spectrum of the planet. The theoretical basis for this proposition was given in chapter 19.15. The edge detected would not necessarily be in the red region of the spectrum but would depend on the intensity of the incident flux at the surface of the planet and the stage of dissipative development of the pigments. For a given incident stellar intensity, the further the red-edge is displaced towards the infrared, the greater would be the thermodynamic dissipative development of the planet. It should be emphasized, however, that resolution limits imply that remote sensing of planets at large distance is global and hence coupled abiotic processes (such as a solvent cycle and related clouds) must be considered that will make the "red-edge" difficult to detect. For example, we presently have difficulty detecting the red-edge of Earth using satellites in Earth orbit (Seager et al., 2005)[331].

20.4 A Human Niche in a Dissipative universe

Although to some it may come as a relief to learn that life, and human life in particular, indeed has a meaning and a purpose, to others dissipation will certainly not be the meaning that they were expecting. There is no immortality offered in this thermodynamic perspective, and no divine place or spiritual truth to be encountered in our cosmic searches, other than that, of course, of truly understanding the reasons behind the beautiful tapestry woven by the physical and chemical laws of our universe. For many millions of years, since our emergence as a species, humans have unwittingly performed the same thermodynamic work as that of all the other animals; spreading nutrients, seeds, and performing pollination in favor of the photon dissipating pigments. The drama associated with this unconscious thermodynamic work is what has made our lives interesting and worth living, but all the particular drama appears to have no ulterior or transcendental meaning beyond that which would foment photon dissipation or to what could be considered as contributing to thermodynamic noise.

However, our future may indeed be more interesting. Nature has endowed humans with a special thermodynamic niche, an intelligence niche, which allows us to construct a new meaning to our lives oriented around dissipation. We even appear to have gained, through the evolution of intelligence and the subsequent formation of technological society, the luxury of choosing the particular meaning to our lives and changing it, almost as we see fit. Whatever direction we choose, however, this new meaning will have to be ultimately related with thermodynamic dissipation if we are to maintain our stability within a dissipative cosmos and thus our survivability as a species. It seems that we are choosing correctly to explore the other planets and, undoubtedly, somewhat in the near future, even other solar systems. This will require a tremendous dissipation of free energy simply for providing the accelerations required to be imparted to our space craft. Terra forming other planets, converting them from simple stage one, two or three planets of evolutionary dissipation (see section 20.1) into Earth like planets with the possibility to evolve through all 16 stages of dissipative evolution, would further our contribution to the cosmic dissipation.

Michio Kaku (1994)[147] and Carl Sagan before him (Sagan, 1985)[310] have speculated that if humans can avoid self-destruction over the next million years or so, then our technology would eventually allow us to arrive at stage 16, the point at which we would acquire the ability to manipulate hyperspace and time through concentrations of astronomical amounts of free energy, allowing rapid travel to other parts of the universe through "worm holes" created by synthetic black holes, or perhaps even time travel to the past, or to parallel universes.

All of this is, of course, wild speculation, but something along these lines is consistent with the historical evidence of the ever increasing dissipative efficacy of humans. Indeed, the most thermodynamically important revelation, a product of recent human evolution, is that human consumption of free energy (dissipation attributed to humans) has been increasing exponentially ever since the establishment of societies, beginning perhaps 10,000 years ago. Although the biosphere as a whole still greatly outstrips the dissipation attributed directly to humans, our contribution is becoming an ever more relevant fraction of this dissipation. With the Terra forming of planets at stage 13 of the evolution of planetary dissipation (see section 20.1), the dissipation attributed to humans will have superseded that due to Earth's biosphere.

Although we apparently have the liberty to now choose our direction of evolution, there does not seem to be any alternative available for escaping from our thermodynamic duty if humans are to remain over time viable as a species. For example, if we all collectively decide today that we are no longer going to advance technologically and decide instead to dedicate human life to its pure enjoyment at the technological level that we now have, as appears to be the philosophy of the Amish religious sect [4], then we will be unprepared for the natural disasters that will inevitably befall us. For example, even if we avoid nuclear war and destruction by global warming, deep ice ages will come every 26,000 years or so, a close super nova explosion will engulf us with deadly radiation every 100,000 years or so, large meteoritic impacts causing a dust winter (like the one that killed off the dinosaurs 65 million years ago) will happen every 20 million years or so, and so on. Without the technology to deal with these certain threats, we will certainly extinguish as a species and the dissipation clock will be reset to some more primitive epoch.

Science and technology are increasing at an exponential rate commensurate with human dissipation and this appears to be completely natural and unstoppable for a species with enough intelligence to become aware of the delicate extinction thresholds (both internal and external) they could otherwise unconsciously arrive at. Whatever increase in technology brings with it a concomitant increase in free energy dissipation. Greater free energy dissipation, or, in other words, greater thermodynamic stability for the human species, is dependent on human technology advancing to the point of being able to handle all internal or external threats. There is simply no escape from this and we should probably consider ourselves lucky to have survived so far given the lack of scientific literacy of most of Earth's population.

[4] Contrary to popular belief, the Amish religious sect does not, in fact, shun technology, but instead believe in the judicious use of it in order to follow as close as possible the guidelines for living as promoted in the Bible.

21. Summary

The theory presented in this book is based on the framework of non-equilibrium thermodynamics. Employing this framework, we have taken Boltzmann's extraordinary insight into the struggle for organismal existence in terms of entropy production to its ultimate consequences regarding the origin of life. Irrespective of preconceived notions about living systems being very far from equilibrium, and therefore intractable under existing non-equilibrium thermodynamic frameworks, we have seen that the formalism borrowed from the Onsager-Prigogine school of Classical Irreversible Thermodynamics in the non-linear regime has been sufficient to understand the vitality of life, from the self-organization and proliferation of pigments to the evolution of the biosphere. It is true that it has been necessary to augment this framework with a probabilistic principle favoring transitions over time towards stationary states of ever larger entropy production. I have given an indication of how such a principle might be formally derived from the fact that for non-linear systems at an unstable bifurcation point, fluctuations in the direction of the solution leading to greater dissipation of the external potential will be reinforced most and therefore such fluctuations will be more successful in taking the system down the road to that solution (see figure 3.7 in chapter 3.3). Often this solution of greater entropy production corresponds to a more complex self-organized system with coupling to other irreversible processes. There is ample empirical evidence to indicate that this thermodynamic selection, or tendency to greater dissipation through self-organization, occurs as a response to more effectively dissipate the imposed generalized chemical potential. Such thermodynamic selection acts on all heirarchal levels of self-organization from the microscopic fundamental molecules up to the biosphere.

The physical, chemical, optical, and electronic properties of RNA and DNA and the other fundamental molecules of life indicate that these molecules originated as self-organized microscopic dissipative structures that proliferated over the ocean surface to absorb and dissipate the prevailing Archean solar UV-C photon flux. Their production and proliferation to far beyond concentrations expected under near equilibrium conditions can be understood in terms of non-linear, non-equilibrium thermodynamic principles directing the autocatalytic photochemical reactions which form these pigments in such a manner that they catalyze the dissipation of the same thermodynamic potential (the solar photon flux) that produced them.

Many salient properties of RNA and DNA, such as their wide absorption spectrum in the UV-C and extremely rapid non-radiative decay characteristics, their homochirality, their sharp denaturing curve at high temperatures close to the surface temperature of Earth at the origin of life, their chemical affinity to other UV-C absorbing molecules designated as being fundamental because they occur in all three domains of life, all indicate that these first molecules of life were driven to formation, replication and complexification through photon dissipation. The similarity of these properties for both RNA and DNA suggests further that they coexisted, simultaneously, since their very beginnings and only later formed a symbiosis to improve photon dissipation rates.

I have suggested an autocatalytic photochemical reaction starting from a common primordial molecule, HCN, first studied by Ferris and Orgel (1966)[96], that could have given rise to the nucleobases and their proliferation over the oceans of the Hadean and Archean epochs when Earth's surface was exposed to intense UV-C light. I have further suggested a similar autocatalytic photochemical process, referred to as UltraViolet and Temperature Assisted Replication (UVTAR), leading to the polymerization of nucleotides and giving rise to particular RNA or DNA oligos that dissipated strongly the Archean UV-C flux. We have recently obtained experimental evidence of the viability of the first part of this proposed mechanism, that of UV-C induced denaturing of DNA.

Natural selection in both living and non-living non-equilibrium systems is not based on survivability (an affirmation which, in fact, is tautological) but is determined probabilistically at unstable bifurcation points, favoring those fluctuations out of many which happen to be in a direction towards greater dissipation of the external generalized chemical potential. An abiotic example of this is how, at an unstable bifurcation point of a temperature gradient over a liquid system heated from below and in the regime of conduction, a fluctuation of a microscopic heated volume element in the direction of heat flow will experience an upward buoyant force and it will become amplified into an organized structure known as a Bénard cell (see chapter 3.1). An explicit biological example of this is a microscopic fluctuation that changes a DNA sequence (codon) such that it gives rise to a chemical affinity at the site for an antenna molecule, for example, for the amino acid tryptophan. Such a fluctuation (or mutation) would lead to greater local heat being deposited under an Archean UV-C flux and therefore greater probability of photon-induced denaturing during daylight hours and thus the possibility for extension overnight. This leads to a dissipation-replication relation, through the UVTAR autocatalytic photochemical mechanism, which effectively selects and amplifies the particular DNA+antenna complex. This thermodynamic selection is operative on all biological levels up to the biosphere, at which level there exists a coupling of life to abiotic irreversible processes such as the water cycle. At this highest level, those fluctuations at any level that lead to greater global entropy production are reinforced and selected .

Complexification of life would thus have been driven by increases in photon dissipation efficacy, first through complexing hydrophobic molecules with RNA and DNA to keep them at the sea surface, and then complexing with antenna UV-C pigment molecules (the fundamental molecules of life), making use of resonant energy transfer and the extremely rapid dissipative characteristics of RNA and DNA afforded by their inherent conical intersections. In fact, any innovation that improved photon dissipation, such as stereochemical coding for antenna molecules or coding for hydrophobic molecules, would have been thermodynamically selected since replication (proliferation) is tied to dissipation through the autocatalytic photochemical reactions, an example of which is the UVTAR mechanism.

Photon dissipation continues to this day to be the most important thermodynamic work performed by life, but now employing pigments operating in the near UV and visible regions of the solar spectrum. The production of pigments today to far beyond their expected equilibrium concentrations is similarly driven, but now indirectly, through complex biosynthetic pathways, employing autocatalytic photochemical reactions in the near UV and visible.

Evolution has been, and still is, primarily concerned with increasing the rate of photon dissipation in the biosphere. This has been achieved by; evolving pigments with extraordinarily rapid times for conversion of electronic excitation energy into ground state vibrational energy, evolving pigments of larger cross section for photon absorption and of increased spectral range, quenching radiative decay channels (e.g. fluorescence and phosphorescence), and proliferating these pigments, with the aid of mobile animals, over ever more of Earth's surface.

The implicit but unjustified assumption of an organismal "will to survive" in the Darwinian theory can now be replaced with an explicit and physically grounded "will to produce entropy" (colloquially speaking) in the thermodynamic dissipation theory of the origin and evolution of life. We should no longer speak about individual selection (although this retains an approximate meaning when irreversible coupling between processes is minimal) but rather about hierarchal thermodynamic selection starting with the fundamental molecules (RNA or DNA, aromatic amino acids, cofactors, and the other UV-C pigments) up to the biosphere, involving both its biotic and abiotic components. Random variations occur spontaneously at all levels, including at the individual organismal, lower pigment molecular level, or even the microscopic level, and these are "selected" (a better word would be "proliferated") based on how good a catalytic agent they are in fomenting the entropy production by dissipating the locally imposed generalized chemical potential given both local and global constraints (nutrients, water, photon potential, etc.). What is optimized is not individual "fitness", which, in fact, cannot even vaguely be defined without incurring in circular argumentation, but rather "entropy production" which can, and, in fact, has been, accurately measured (see Michaelian, 2012b[213] and chapter 2).

Because the fundamental laws of Nature appear to be universal, such a photon dissipation program should also be operating on any material body at any place in the cosmos where there exists atoms heavier than hydrogen and helium capable of covalent bonding and where there is a UV-C or UV-B light flux from a local star. The universe should be teaming with life similar, and not so similar, to our own. For life similar to our own, we could expect to find it at different stages of dissipation development; from the first stage of pigment production and proliferation, through complexification into greater dissipative structures, and up to the stage of an intelligent species terra forming other planets, to even, perhaps, the manipulation of space-time worm holes (Kaku, 1994)[147].

Accepting this thermodynamic view of life and this categorization into the different stages of dissipative evolution, it can be emphatically stated that we have already found evidence of extraterrestrial life (and even evidence of extraterrestrial biospheres) existing in the first few stages of dissipation development; for example in the proliferation of organic pigments on Titan dissipating solar photons and fomenting the methane rain cycle, or in the proliferation of inorganic pigments in the clouds of Venus dissipating solar photons and fomenting the great polar vortices. In fact, the first dissipation stage of life, that of pigment proliferation for fomenting the dissipation of photons from the local star, appears to have occurred throughout much of relatively empty space of the cosmos as identified by the infrared emission spectra attributed to the ubiquitous polyaromatic hydrocarbons (PAHs) and the mixed aromatic/aliphatic organic nanoparticles (MAONs) detected in the interstellar gas and dust clouds. This self-organized pigment material is undoubtedly contributing to some of the baryonic dark matter of the universe.

In summary, the most important physical facts and empirical evidences in favor of the thermodynamic dissipation theory of the origin and evolution of life are;

1. All irreversible processes, life included, require the dissipation of a generalized chemical potential for their origin, persistence, proliferation, and evolution.
2. The solar photon potential is, and always has been, the largest source of free energy available for dissipation at the surface of Earth and life today is almost exclusively occupied with the dissipation this potential.

3. Today, organic pigments absorb over the entire solar spectrum at Earth's surface, from the UV-C to the red-edge at 700 nm, filling in regions of the solar spectrum where the atmosphere and water do not absorb.
4. Many fundamental molecules of life absorb and dissipate strongly in the UV-C. Many of these have a conical intersection which allows them to dissipate the photon-induced electronic excitation energy in sub-picosecond times.
5. There are many salient characteristics of RNA and DNA, such as their Watson-Crick bonding between nucleotides, leading to strong absorption and rapid dissipation of UV-C light, which have been thermodynamically selected even at the price of a weaker bound RNA or DNA molecule.
6. Many replication related characteristics of RNA and DNA are commensurate with the external environmental conditions of the early Archean, such as their sharp denaturing temperature at close to the temperature of the ocean surface at the beginning of life. Their robustness to damage by UV-C photons indicates their exposure to this light from an early stage.
7. There are known routes to the formation of the nucleotides through the application of UV-C light to HCN in an aqueous solution at temperatures close to those of the origin of life. It appears that these systems arrive at stationary states in which a particular isomerization of an intermediate molecule is favored under the UV-C light flux, and this is precisely the isomerization needed in order to form the nucleotide. The process also appears to be very UV-C light dissipation intensive and therefore the whole formation process may be considered as microscopic dissipative structuring.
8. Non-equilibrium thermodynamic principles applied to autocatalytic photochemical reactions occurring under the imposition of an external photon potential can provide an explanation for the proliferation of these fundamental pigment molecules to concentrations many order of magnitude greater than could be expected under near equilibrium conditions.
9. Many fundamental molecules (common to all three domains of life) absorb strongly in the UV-C and have chemical affinity to RNA and DNA (in some cases, such as the aromatic amino acids, even affinity to their codons or anticodons) and can transfer their electronic excitation energy to RNA and DNA through resonant energy transfer mechanisms. The complex of DNA + Fundamental Molecule forming a greater dissipating system than the components summed but acting separately.
10. Fundametal molecules which could have increased photon dissipation by forming complexes with RNA or DNA, such as the aromatic amino acids acting as antenna molecules and the hydrophobic amino acids acting as buoyancy floats exposing nucleic acid to greater ocean surface UV-C light, are known to have chemical affinity to RNA and DNA, and even to their codons or anticodons.

11. Microscopic dissipative structuring under an external generalized chemical potential is a known and well studied phenomena within the non-equilibrium thermodynamic community and there is experimental evidence that at least some of the fundamental molecules, for example the nucleotides, can be dissipatively structured under a UV-C photon potential.
12. There is ample evidence for the ubiquity of organic pigments throughout the cosmos, particularly in star forming regions where there exists a strong component of UV-C light from early stars, suggesting that microscopic dissipative structuring occurs even within a cold and diffuse space environment.
13. An enzymeless replication mechanism appears to require the involvement of the dissipation of UV-C light in the production of activated nucleotides, the creation of phosphodiester bonds needed in extension, and in the enzymeless denaturing of double helix RNA, DNA, and hybrid duplexes.
14. There is evidence in today's pigments and other fundamental molecules of a more ancient UV-C absorbing base structure on top of which new structures have been built. An example being the amino acid tryptophan which has been demonstrated to have the capability to revert the cyclobutane pyrimidine dimers formed on DNA using either visible or UV-C light (see figure 16.2).
15. Increasing the number of conjugated bonds in a molecular system is a simple way for the absorption to be pushed towards longer wavelengths while aromaticity is a process pushing absorption towards shorter wavelengths. Both these mechanisms have been amply employed by life.
16. There is empirical evidence that not only the metabolic rate per individual, but the metabolic rate per unit biomass has been increasing over the evolutive history of life on Earth (Zotin, 1984)[431]. This speaks for the thermodynamic selection of increases in entropy production over time.
17. There is a strong coupling of biotic dissipative processes with abiotic dissipative processes, such as the dissipation of photons in organic pigments with water cycle and ocean and wind currents. This again speaks for the thermodynamic selection of increases in entropy production over time.
18. It is becoming increasingly clear that many organisms, even those of today, exude pigments into their environment, ostensibly to foment the dissipation of the solar photon potential. This traditional perspective sees this behavior as wasteful, having no apparent Darwinian utility to the organism itself. One example of this behavior is oxygen emission in oxygenic photosynthesis, but others include cyanobacteria and other phytoplankton exuding chromophoric dissolved organic matter (CDOM) into the sea-surface microlayer (Simeonov and Michaelian, 2017)[344].
19. It is well documented that mobile animals are indispensable for spreading scarce nutrients such as phosphor and nitrogen to photon dissipating or-

ganisms over land and over the ocean surface and are also indispensable in recycling to the ocean surface those nutrients that sediment to the ocean floors.

20. The non-equilibrium thermodynamic perspective on selection and evolution is superior to the Darwinian paradigm in that it removes the tautology in the "survival of the survivors" and does not require the supposition of a metaphysical "will to survive" for the organism.
21. Simultaneous thermodynamic selection, promoting greater global entropy production, on all hierarchal levels of the biosphere removes the problems of selection at higher (or lower) levels than the organismal level in the Darwinian paradigm and resolves the paradox of the evolution of a system of population one (the biosphere).
22. There is ample empirical evidence to indicate that the evolution of abiotic processes on Earth, and even in the entire cosmos, is governed by optimizing global entropy production and it would be mysterious indeed if life were not also treatable under this apparently universal imperative.

Dissipation is much more fundamental than its potential to produce and evolve life. In fact, all processes occurring in the universe, from the nuclear reactions occurring in stars to the gravitational collapse of gas and material into galaxies and black holes, are processes unequivocally associated with dissipation. The evolution of the universe is completely consistent with dissipation, from its initiation in a very ordered state to its eventual heat death, or its death as an irregular and chaotic state at its collapse but without an inversion of the arrow of time during contraction (Hawking and Penrose, 1996)[126]. The force of gravity has been suggested to be not a fundamental force at all, but rather to be emergent, to arise from dissipation (Padmanabhan, 2010)[266]. In fact, the approximate equations of Newton and even the exact gravitational field equations of Einstein can be derived by maximizing dissipation given the values of the fundamental constants, a result alluded to by the fact that the laws of black hole dynamics have an uncanny resemblance to the laws of thermodynamics (Padmanabhan, 2010)[266]. Dissipation is also occurring with universally conserved quantities other than energy; such as, momentum, angular momentum, charge, Noether charge, etc. and much work remains to be done to uncover the dissipative-evolutionary relations here, and to delineate their effect on the observable universe.

22. Epilogue

If humans can survive for the next few hundred years despite themselves (without self-annihilation during the present critical stage at which our technology is much further advanced than our global intelligence warrants), then we are destined to play an ever increasing role in solar system, galactic, and ultimately universal, dissipation. Within 100 to 200 years, dissipation attributed directly to humans, through terra-forming other planets, will have out performed the dissipation of Earth's present biosphere. Within 1000 to 2000 years, humans will be dissipating most of the emitted solar light and we will have created a "dark star". Human colonies will probably have spread to neighboring stars and we will have terra-formed most available planets in our local section of the galaxy. Within 100,000 to 200,000 years, it is conceivable that we will have completely colonized our galaxy and will probably be on a journey to neighboring galaxies, perhaps through directed black-hole manipulation of space-time.

For many millions of years of human history, humans have been unwittingly playing their thermodynamic, and only, role as gardeners of the great dissipaters on land, the plants. During perhaps the last one hundred thousand years, creative myths and legends, collectively known as religion, provided the only, but naive and very wrong, answers to questions posed by an intelligent species concerning the phenomena of life. Only with the embracement of the scientific method beginning in the late 15 century Europe did the myths of religion finally come to be questioned and challenged by a few brave individuals which led the way to the incorporation of science into human culture.

Only now, after centuries of experience with the application of the scientific method, have we begun to understand our true place in the universe. Today, science has developed to such a point that there is no question too difficult to be addressed by careful observation, hypothesis and experimentation. Almost every individual on this planet owes their life to science. However, the practicing of science and objective thinking by only a fortunate few has led to a very unstable situation for our species. Before science, the human population was small and relatively constant for millions of years (about 50,000 humans) and distinct populations were sparse and rarely interacted. Due to science we are now over 7,000,000,000 humans and we are globally inter-connected with tremendous weapons of mass destruction and frightening cultural baggage that muddies our powers of objective reasoning. There is now no longer any room on the planet for myth based aggressions and any serious mistakes of this nature will not be forgiven. Our generation and the following ones have the omnipresent threat of self-annihilation hanging over them. We must prevent future generations of children from growing up unable to think objectively and independently or from becoming indoctrinated with tribal myths. Teaching everyone science and the scientific method appears to be our only salvation now that the nuclear genie has been let out of the bottle.

Unfortunately, today only a small percentage of humans seem to have an interest in science and in learning the correct method for seeking the truth about their human reality and that of the cosmos. Only a small portion of our population has understood the importance of true liberation of the individual. Unfortunately, the great majority of humans were not infected during childhood by the beauty of science and so are obliged to accept as truth naive and contrived myths. Rather than playing a part in the discovery of our cosmos and in uncovering the future potential of human existence, these individuals will remain simple gardeners, unwittingly carrying out our original thermodynamic function of spreading nutrients for the plants while futilely believing they are part of a more lasting importance. Their ignorance of an objective reality and of the fragility of the human species, however, gives them an excuse for reckless behavior, and this will situate the human species, and many unfortunate others, on the brink of extinction. In their contrived belief system, they have their own salvation and individual immortality guaranteed, to be lived out in a mystical utopia where, curiously, such ignorance does not exist.

The present evolution towards a global society which recognizes the importance of a universal understanding of the scientific method and the utility of an objective and materialistic understanding of the cosmos, has been too slow and with many regresses. Perhaps, rather than teaching difficult science to the adult masses who are simply too entrenched in their mystical belief system, scientists should instead attempt an old ploy used extensively with success in the past by invading cultures wishing to supplant another; that of inserting new elements into the culture or religion only strong enough to permute it sufficiently to head the subjects off in the desired direction. For example, scientists could implore believers to consider not accepting with infinite vanity that their creator is an image of themselves, but rather that the Sun is the true image of their creator. The Sun is the provider of the externally imposed generalized chemical potential (the photon potential) that drives the organization of material into living organisms, humans included. Within the stars were made all of the organic elements heavier than helium, including all the elements of the Earth and the carbon so vital to life. The Sun was responsible for making all of the primordial and fundamental molecules of life and every other organic molecule that was ever incorporated into life.

To worship a true image of their true creator, and even feel his all embracing warmth, an individual would only have to go outdoors on a sunny day and look up into the sky. Our Creator would appear to be the same to all subjects, he would be in the heavens always shining down on his subjects, and his subjects always looking up to him as an expression of eternal gratefulness. There would be no need for endless war and hysteria over who's god was the relevant one because it could be easily proved that the Sun was one and the same no matter from where it was being observed. On death, our atoms and molecules would take up their thermodynamically ordained place within the pigments of plants and cyanobacteria, eternally absorbing and dissipating our creators free energy into warmth, an almost eternal heaven on Earth.

There would be no need for difficult or embarrassing imagining, such as how many angels could dance on the head of a pin, or of where to locate an old man with a white beard in heaven. There would also be no need of having to deal with unscrupulous intermediaries exacting a price for putting a subject in direct contact with their deity. Our philosophy and code of ethics would begin to incorporate the elements of an objective understanding of the role of every individual and every process in our dissipative universe. A return to worshiping the Sun would lead the scientifically illiterate to a greater understanding about what is known concerning our objective reality and would allow for a gradual evolution of a convergence of religion with an objective and materialistic science.

The worship of Aten, the ancient Egyptian Sun god around 5000 B.C. is well documented in the historical records. The most important pagan religion of the Romans was known as Mithraism in which it was believed that the Son of the Sun was a saviour who was sacrificed for the good of all. The Mithraists also kept a special day of the week for reverence to the Sun god (Sun-day) and 25th of December was the most sacred day for the Romans, the birthday of the Sun. The Sun god has, in fact, been by far the most worshiped deity by all human communities over all places and over all epochs. The fact that the majority of our human ancestors were on the right track to a true understanding of the origins and meanings of life when they worshiped the Sun god, is a tribute to human intellect and intuition. How religion got hijacked approximately 2000 years ago from following a scientific tradition based on intuition, experience, observation and logical deduction into a vain practice of self-worship, of senseless rituals, of stifling censorship, and of the construction of institutions for the purpose of political domination, probably has much to do with the gradual increase in tribal interaction and the related threats to cohesion, but it is a theme that certainly deserves much more attention than it has received until now. If the Sun had been kept as our god since the ancient Roman times, scientific and religious enlightenment today probably would be completely intertwined at a highly objective level and we probably would not be facing the tribal threat of self-annihilation.

A first step towards heading society once again in the correct direction, after an embarrassing 2000 year interval of wandering aimlessly and now dangerously through the dark, would be to recover the pagan celebration around the winter solstice (~22 of December) in honor of the Sun god. Before Christianity, Judeism, or any other modern religion, the Sun god Aten, or some variation thereof, was honored with a special week of celebrations by almost all cultures for hundreds of thousands of years. The celebrations consisted of a week of offerings and festivities around the winter solstice in an effort to persuade the Sun god to not abandon the world. It was believed that if not pleased by the festivities, the Sun god would continue on his way, shortening every day until he eventually disappeared.

When Christianity became widely established by the late fourth century A.D., the celebration of the birth date of Christ was conveniently changed to the date of the pagan feast for the birth of Sol Invictus (the Unconquered Sun) of December 25th which had been established as a holiday for the celebration by the Roman emperor Aurelian in 274 A.D. Similar types of obliterations (or renovations) of culture occurred frequently in human history. For example, in the Americas, after the Spanish conquest, Catholic churches were built on top of, and using the bricks from, the ancient indigenous pyramids and other houses of worship. Traditions too strong to be obliterated outright were gradually deformed by including elements of the conquer's belief system and religion. Perhaps such a morphing of present religions with science is now needed in order to achieve progress towards a universal objective understanding of the cosmos and human reality. This is imminently necessary if we are to become fully aware of our fragility and become prepared for the inevitable threats from natural catastrophes and the very real possibility of self-annihilation.

A peaceful and organized return to worshiping the Sun god on December 25th, and perhaps also recovering the week of Easter as its original pagan celebration of the end of winter and the triumph of Sol Invictus over the god of the under world Volos (see figure 22.1), would be an important step in guiding every human towards an understanding of their true place in the cosmos. The celebration of the birthday of the philosopher Jesus Christ (Christmas) should be moved forward to a day in spring, consistent with biblical references to shepherds attending to the birthing of their flock at night on hearing of the birth of Christ. If a majority of humans could be convinced to take up an interest in their own history and in the scientific method, we could avoid the unfortunate position that we find ourselves in, of forever teetering on the threshold of extinction.

Fig. 22.1. A 1919 painting by Boris Kustodiev (1878–1927) depicting the Slavic celebration of the Eastern Orthodox holiday Maslenitsa. The celebration, which has its origins in the Slavic pagan tradition, was a sun-festival, personified by the ancient god Volos (god of the underworld). It was a celebration of the imminent end of winter. The tradition survived in some regions until as late as the 18th century. As with the pagan celebration at the winter solstice in tribute to the Sun god (now known as Christmas), the Christians also took over this pagan celebration of the arrival of spring and it became the last week before the onset of Great Lent, the resurection of Christ, otherwise known as Easter. Image credit: Public Domain.

A. Affinities from Planck's Equation for the Entropy of an Arbitrary Beam of Photons

Fig. A.1. A 1933 photo of Max Planck (1858-1947). Planck, a physicist of the importance and stature of Boltzmann, was the originator of quantum theory and the first to understand the nature of a spectrum emitted by a body in thermal equilibrium at a particular temperature (known as a "black-body radiation"). The derivations in this appendix are due to Planck and our derivation of the photon potential is based on his equation for the entropy of an arbitrary photon beam. Image credit: Unknown, Public Domain.

In this appendix a more accurate calculation of the affinity for the photon dissipation process is given using a formula for the entropy of an arbitrary light beam derived by Max Planck (see figure A.1) in the early 1900's (Planck, 1901;1906;1914)[283, 283, 285]. We first derive the entropy production for the irreversible process of the conversion by a pigment of a given arbitrary photon spectrum into another as the product of a generalized flow and force, corresponding to an energy flow and gradient of a photon potential, respectively. From this derivation, which clearly shows the non-linear relation between the force and flow, we next show how the general evolutionary criterion of Glansdorf and Prigogine (Eq. (6.6)) again leads to the result that under an imposed photon potential, a product pigment of a photochemical reaction which dissipates the imposed photon potential will proliferate to concentrations well beyond those expected near equilibrium. This proliferation of organic pigments over Earth's sunlit surface is the hallmark of the origin of life and its evolution.

A.1 Planck's Equation for the Entropy of an Arbitrary Photon Beam

Assume we have electromagnetic energy E within a square cavity of length L with perfectly conducting walls. The electric field must therefore go to zero at the walls. What are the number of resonant modes $M(\lambda)d\lambda$ that can be supported inside the cavity for a given wavelength interval between λ and $\lambda+d\lambda$? Calculating this number, and the number of ways that P energy packets each of energy $\varepsilon = h\nu = hc/\lambda$ can be distributed over the number of available modes $M(\lambda)d\lambda$ gives the number of microstates W and therefore, assuming that all microstates have the same *a priori* probability, from Boltzmann's equation we determine the entropy as $S = k \ln W$.

For standing (resonant) waves in a square box of dimension L, it is required that, $\frac{n}{2}\lambda = L$ where n is an integer value and λ is the wavelength of the resonant mode. Since there are three independent coordinates, we have three similar equations;

$$(n_x, n_y, n_z)/2 \cdot \lambda = L = V^{1/3}$$

The number of different combinations of the n_i that satisfy these equations are the number of different standing wave modes which can be supported for wavelength λ inside the cavity of volume V. To count this number, consider that this number corresponds to the number of integral grid points internal to the positive octant of a sphere of radius R in an imaginary n space such that,

$$(n_x^2 + n_y^2 + n_z^2) \leq R^2$$

where $R = \frac{2L}{\lambda} = \frac{2V^{1/3}}{\lambda}$.

The number of lattice points inside this sphere, for a large sphere, is just the volume of this sphere $\frac{4\pi}{3}R^3$ (because grid or lattice points are equally spaced in all dimensions), however, the n can only be positive, therefore we take the volume of only the positive octant. There are also two polarization states available for an unpolarized beam, or one state for a polarized beam, i.e. $n_0 = 2, 1$ respectively. Therefore, the total number of modes of wavelength λ, is

$$M(\lambda) = n_0 \frac{1}{8}\frac{4\pi}{3}R^3 = \frac{\pi n_0}{6}\left(\frac{2V^{1/3}}{\lambda}\right)^3 = \frac{n_0 4\pi}{3\lambda^3}V \tag{A.1}$$

The number of modes dM within an interval $d\lambda$ is then just

$$dM = \frac{dM}{d\lambda}d\lambda = \frac{-4\pi n_0}{\lambda^4}V d\lambda \tag{A.2}$$

Now we have derived that we have $M(\lambda)$ resonant modes in a volume V and we have a total energy $E(\lambda)$ to distribute over these $M(\lambda)$ modes. But we know that the energy comes in packets $\varepsilon = h\nu = hc/\lambda$.Therefore, we have P elements of energy ε such that $E(\lambda) = P(\lambda)\varepsilon$.

To calculate the entropy $S(\lambda)$ we must use Boltzmann´s equation,

$$S(\lambda) = k \ln W(\lambda)$$

We therefore have to calculate the number of microstates $W(\lambda)$ and this corresponds to the number of ways to distribute $P(\lambda) = E(\lambda)/\varepsilon$ energy elements over $M(\lambda)$ modes for carrying the energy. The number of ways to distribute P energy elements over M resonant modes is, using combinatorial theory,

$$W = \frac{(M+P-1)!}{(M-1)!P!}$$

Since M and P are very big, we can ignore the -1 in both the numerator and denominator. For the same reason we can also use Stirling approximate formula,

$$\ln(M!) = M\ln(M) - M + O(\ln M)$$

$$M! \simeq M^M \tag{A.3}$$

This gives,

$$W = \frac{(M+P)^{M+P}}{M^M P^P}$$

and therefore,

$$S(\lambda) = k \ln W(\lambda) = k[(M(\lambda){+}P(\lambda)) \ln(M(\lambda){+}P(\lambda)){-}M(\lambda) \ln M(\lambda){-}P(\lambda) \ln P(\lambda)]$$

but $P = \frac{MU(\lambda)}{\varepsilon}$ where $U(\lambda) = E(\lambda)/M(\lambda)$ is the average energy of the modes of wavelength λ. Therefore,

$$\begin{aligned} S &= k[(M + \frac{MU}{\varepsilon}) \ln(M + \frac{MU}{\varepsilon}) - M \ln M - \frac{MU}{\varepsilon} \ln \frac{MU}{\varepsilon}] \\ &= kM[(1 + \frac{U}{\varepsilon}) \ln M(1 + \frac{U}{\varepsilon}) - \ln M - \frac{U}{\varepsilon} \ln \frac{MU}{\varepsilon}] \\ &= kM[(1 + \frac{U}{\varepsilon}) \ln(1 + \frac{U}{\varepsilon}) - \frac{U}{\varepsilon} \ln \frac{U}{\varepsilon}] \end{aligned} \tag{A.4}$$

but for $d\lambda$ small,

$$U = \frac{E}{M} \simeq \frac{dE}{dM} = \frac{dE}{d\lambda}\frac{d\lambda}{dM} = \frac{-dE}{d\lambda}\frac{\lambda^4}{4\pi n_0 V} \tag{A.5}$$

where we have used Eqn. (A.2) for deriving the last term.

Now, consider not a static volume of electromagnetic radiation in a box, but rather a beam of radiation moving at velocity c at intensity of energy $I(\lambda)$ per unit wavelength interval per unit solid angle, we have

$$I(\lambda) = -\frac{dE}{d\lambda}\frac{1}{4\pi}\frac{c}{V} \tag{A.6}$$

where the factor of $1/4\pi$ comes in because $I(\lambda)$ is given per unit solid angle, and the factor of c/V comes in because we need to convert from energy per unit volume to energy flow per m^{-2} for a beam travelling at velocity c. $I(\lambda)$ has units [J m^{-3} $s^{-1}sr^{-1}$]. Using this in equation (A.5) gives

$$U(\lambda) = \frac{I(\lambda)\lambda^4}{n_0 c}$$

Using this in (A.4) with $\varepsilon = h\nu = hc/\lambda$ gives,

$$S(\lambda) = kM(\lambda)\left[\left(1 + \frac{\lambda^5 I(\lambda)}{n_0 hc^2}\right) \ln\left(1 + \frac{\lambda^5 I(\lambda)}{n_0 hc^2}\right) - \frac{\lambda^5 I(\lambda)}{n_0 hc^2} \ln \frac{\lambda^5 I(\lambda)}{n_0 hc^2}\right]$$

or

$$\frac{S(\lambda)}{M(\lambda)} = k\left[\left(1+\frac{\lambda^5 I(\lambda)}{n_0 hc^2}\right)\ln\left(1+\frac{\lambda^5 I(\lambda)}{n_0 hc^2}\right) - \frac{\lambda^5 I(\lambda)}{n_0 hc^2}\ln\frac{\lambda^5 I(\lambda)}{n_0 hc^2}\right]$$

but similarly to equation (A.5), for $d\lambda$ small,

$$\frac{S}{M} \simeq \frac{dS}{dM} = \frac{dS}{d\lambda}\frac{d\lambda}{dM} = \frac{-dS}{d\lambda}\frac{\lambda^4}{4\pi n_0 V}$$

Therefore,

$$\frac{-dS}{d\lambda} = \frac{4\pi n_0 V k}{\lambda^4}\left[\left(1+\frac{\lambda^5 I(\lambda)}{n_0 hc^2}\right)\ln\left(1+\frac{\lambda^5 I(\lambda)}{n_0 hc^2}\right) - \frac{\lambda^5 I(\lambda)}{n_0 hc^2}\ln\frac{\lambda^5 I(\lambda)}{n_0 hc^2}\right]$$

Now, consider not the entropy $S(\lambda)$ of a given volume V containing electromagnetic radiation, but the flow of entropy $j(\lambda)$ due to a flux of photons per unit wavelength interval, per unit time, per unit solid angle,

$$j(\lambda) = -\frac{dS}{d\lambda}\frac{1}{4\pi}\frac{c}{V}$$

where, as in equation (A.6), the factor of $1/4\pi$ comes in because $j(\lambda)$ is given per unit solid angle, and the factor of c/V comes in because we need to convert from entropy per unit volume to entropy flow per m^{-2} for a beam travelling at velocity c. $j(\lambda)$ has units [J m^{-3} K^{-1} $s^{-1}sr^{-1}$]. Finally then, we arrive at

$$j(\lambda) = \frac{n_0 kc}{\lambda^4}\left[\left(1+\frac{\lambda^5 I(\lambda)}{n_0 hc^2}\right)\ln\left(1+\frac{\lambda^5 I(\lambda)}{n_0 hc^2}\right) - \frac{\lambda^5 I(\lambda)}{n_0 hc^2}\ln\frac{\lambda^5 I(\lambda)}{n_0 hc^2}\right] \quad (A.7)$$

and the equivalent in terms of frequency ν is

$$j(\nu) = \frac{n_0 k\nu^2}{c^2}\left[\left(1+\frac{c^2 I(\nu)}{n_0 h\nu^3}\right)\ln\left(1+\frac{c^2 I(\nu)}{n_0 h\nu^3}\right) - \frac{c^2 I(\nu)}{n_0 h\nu^3}\ln\frac{c^2 I(\nu)}{n_0 h\nu^3}\right] \quad (A.8)$$

Note that nowhere in the derivation have we put restrictions on the type of radiation $I(\lambda)$ or $I(\nu)$ except that the number of resonant modes is very large and that the number of energy packets (photons) distributed over these modes is also very large in order to apply Sterlings formula, Eqn. (A.3). These conditions imply that the volume of the confined radiation is large with respect to its wavelength and that the energy intensity of the radiation is large with respect to the energy of a single photon, respectively. The formulas are therefore valid for any type of radiation $I(\lambda)$ satisfying these conditions, not only for black-body radiation.

A.2 Derrivation of Plancks Radiation Law for a Black-body Spectrum

For electromagnetic radiation in equilibrium we can derive Planck's radiation law for black-body emission using equation (A.4) and that $\varepsilon = h\nu = hc/\lambda$. In equilibrium we have

$$\frac{1}{T} = \frac{\partial S}{\partial E} = \frac{\partial S}{\partial U}\frac{\partial U}{\partial E} = \frac{1}{M}\frac{\partial S}{\partial U} = \frac{\lambda k}{hc}\ln(1 + \frac{\lambda U}{hc}) \tag{A.9}$$

and solving for U

$$U(\lambda, T) = \frac{hc}{\lambda}\frac{1}{\exp(hc/\lambda kT) - 1}$$

which is the average energy emitted per resonant mode at wavelength λ. The total energy $dE(\lambda)$ emitted by all modes within a wavelength range between λ and $\lambda + d\lambda$ is therefore

$$dE(\lambda, T) = \frac{dE(\lambda, T)}{d\lambda}d\lambda = \frac{dE}{dM}\frac{dM}{d\lambda}d\lambda = U(\lambda, T)\frac{dM}{d\lambda}d\lambda$$

since $E(\lambda, T) = M(\lambda)U(\lambda, T)$. Therefore,

$$dE(\lambda, T) = \frac{4n_0\pi V}{\lambda^4}\frac{hc}{\lambda}\frac{1}{\exp(hc/\lambda kT) - 1}d\lambda = \frac{4n_0\pi hcV}{\lambda^5}\frac{1}{\exp(hc/\lambda kT) - 1}d\lambda$$

Finally, the energy density per unit volume per unit solid angle per unit wavelength is

$$\rho(\lambda, T) = \frac{1}{4\pi V}\frac{dE(\lambda, T)}{d\lambda} = \frac{n_0 hc}{\lambda^5}\frac{1}{\exp(hc/\lambda kT) - 1} \tag{A.10}$$

and the energy flow per unit area of the body per unit time interval per unit wavelength interval is just,

$$i(\lambda, T) = c\rho(\lambda, T) = \frac{n_0 hc^2}{\lambda^5}\frac{1}{\exp(hc/\lambda kT) - 1}$$

which is the Planck black-body radiation law.

Similarly, in terms of frequency ν,

$$dE(\nu) = \frac{dE(\nu)}{d\nu}d\nu = \frac{dE}{dM}\frac{dM}{d\nu}d\nu = U(\nu)\frac{dM}{d\nu}d\nu \tag{A.11}$$

now,

$$\frac{dM}{d\nu} = \frac{dM}{d\lambda}\frac{d\lambda}{d\nu} = \frac{-4n_0\pi V}{\lambda^4}\left(\frac{-c}{\nu^2}\right) = \frac{4n_0\pi V\nu^2}{c^3}$$

where we have used (A.2). Therefore, this in Eqn. (A.11) gives,

$$dE(\nu, T) = \frac{4n_0\pi V\nu^2}{c^3}\frac{h\nu}{\exp(h\nu/kT) - 1}d\nu \tag{A.12}$$

and the energy density per unit volume per unit solid angle per unit frequency is,

$$\rho(\nu, T) = \frac{1}{4\pi V}\frac{dE(\nu)}{d\nu} = \frac{n_0 h\nu^3}{c^3}\frac{1}{\exp(h\nu/kT) - 1} \tag{A.13}$$

and the energy flow per unit area of the body per unit time interval per unit frequency interval is just,

$$i(\nu, T) = c\rho(\nu, T) = \frac{n_0 h\nu^3}{c^2}\frac{1}{\exp(h\nu/kT) - 1}$$

Of course, the Planck black-body distributions, equations (A.10) and (A.13) assume that the material body is maintained at a uniform temperature T and that the body is transparent to the emitted radiation, which is only true for very low density materials. For high density materials, if the source of the energy is internal, like our Sun (the material may not be in thermodynamic equilibrium), the temperature T in equations (A.10) and (A.13) must be taken as the surface temperature, and the measured energy densities become per unit *surface area* per unit solid angle per unit wavelength (or frequency).

Note also, that for isotropically emitted or incident beams, the solid angle of the beam affects the entropy flux, or the energy flux, as a simple geometrical factor.

A.3 Energy Emitted by a Black-body at Temperature T

We can find the total energy emitted per unit volume per unit time into 4π solid angle by a black-body at temperature T by multiplying by 4π and integrating equation (A.13) over all frequencies ν.

$$E = 4\pi \cdot \frac{n_0 h}{c^3} \int_0^{\infty} \frac{\nu^3}{\exp(h\nu/kT) - 1} d\nu \tag{A.14}$$

But $\int_0^{\infty} x^3/(\exp(x) - 1)dx = 3!\zeta(4) = \pi^4/15$, where ζ is the Riemann zeta function, so equation (A.14) becomes

$$E = 4\pi \frac{n_0 \pi^4 k^4 T^4}{15 c^3 h^3} = \frac{4 n_0 \pi^5 k^4}{15 c^3 h^3} T^4 = \frac{n_0 \pi^2 k^4}{30 c^3 \hbar^3} T^4 \tag{A.15}$$

which is just the Stephan-Boltzmann law with $\hbar = h/2\pi$.

The total entropy emitted by a black-body per unit volume per unit time into 4π solid angle is,

$$S = \frac{4}{3}\frac{E}{T} = \frac{4}{3}\frac{n_0 \pi^2 k^4}{30 c^3 \hbar^3} T^3. \tag{A.16}$$

A.4 Entropy Production due to the Dissipation of Photon Spectra

The entropy produced in a stationary state due to any dissipative process occurring within a material system with a well defined boundary is simply the difference between the inwardly and outwardly directed flows of entropy, integrated over the boundary of the system.

$$J = \int_S (\boldsymbol{j}_{out} - \boldsymbol{j}_{in}) \cdot \hat{n} dS, \tag{A.17}$$

where $\hat{n}$ is the unit normal to the surface S and the entropy flows, j, have units, for example, of $[\mathrm{WK^{-1}m^{-2}}]$. Assuming that the system has arrived at a stationary state (no change in entropy of the system over time) and considering only the primary photon interaction and dissipation process, ignoring all possibly coupled secondary irreversible processes acting over the boundary, such as, for example, conduction, convection, evaporation, etc., the total entropy flow out can be divided into a photon emitted (due to absorption and dissipation), $\boldsymbol{j}^e$, reflected, $\boldsymbol{j}^r$, and transmitted, $\boldsymbol{j}^t$ component,

$$j_{out} = j^e + j^r + j^t. \tag{A.18}$$

Assuming that part $a(\nu)$ of the incident energy not reflected or transmitted is converted into an unpolarized black-body spectrum and emitted isotropically into a 4π solid angle at an effective temperature T_E (determined ultimately from energy balance and conservation), the emitted entropy flow out for the photon dissipative process per unit frequency element $d\nu$, per unit solid angle, of the incident flow is (see Eq. (A.16)),

$$j^e(\nu) = \frac{4}{3}\frac{a(\nu)g^i(\Omega)h\nu N(\nu)}{T_E} = \frac{4}{3}\frac{a(\nu)g^i(\Omega)I(\nu)}{T_E}, \tag{A.19}$$

where h is Planck's constant, $N(\nu)$ is the number of the incident photons per unit time per unit area at frequency ν and of energy $h\nu$, $a(\nu)$ is the frequency dependent fraction of the photons absorbed, $g^i(\Omega)$ is the geometrical factor for capture of the incident light by the arbitrary enclosing surface (see below), and $I(\nu) = h\nu N(\nu)$ is the incident photon energy flux. The emitted entropy flow per unit area is obtained by integrating the flux normal to the surface over the frequency interval of interest and over the solid angle into which the black-body energy is radiated,

$$J^e = \int_\Omega \int_{\nu_1}^{\nu_2} j^e(\nu)d\nu \cos(\theta)d\Omega = g^e(\Omega)\int_{\nu_1}^{\nu_2} j^e(\nu)d\nu \tag{A.20}$$

where Ω is the solid angle subtended by the emitted flow, and where we have assumed that the emitted isotropy is independent of light frequency. The geometrical factor $g^e(\Omega)$ will be evaluated for different situations below.

An expression for the incident entropy flow $j^i(\nu)$ [J m^{-2} K^{-1}s^{-1}] due to an arbitrary photon energy flow $I(\nu)$ was obtained by Planck in 1913 and we have outlined the derivation of this in section A.1, the result is, Eqn. (A.8):

$$j^i(\nu) = \frac{n_0\, k\, \nu^2}{c^2}\left[\left(1 + \frac{c^2\, I(\nu)}{n_0\, h\, \nu^3}\right)\ln\left(1 + \frac{c^2\, I(\nu)}{n_0\, h\, \nu^3}\right) - \left(\frac{c^2\, I(\nu)}{n_0\, h\, \nu^3}\right)\ln\left(\frac{c^2\, I(\nu)}{n_0\, h\, \nu^3}\right)\right], \tag{A.21}$$

where n_0 denotes the polarization state, $n_0 = 1$ or 2 for polarized and unpolarized photons respectively, k is the Boltzmann constant, and c is the speed of light. For completeness, the corresponding expression in terms of wavelength $\lambda = c/\nu$, is (see derivation in section A.1 of Eqn. (A.7))

$$j^i(\lambda) = \frac{n_0\, k\, c}{\lambda^4}\left[\left(1 + \frac{\lambda^5\, I(\lambda)}{n_0\, h\, c^2}\right)\ln\left(1 + \frac{\lambda^5\, I(\lambda)}{n_0\, h\, c^2}\right) - \left(\frac{\lambda^5\, I(\lambda)}{n_0\, h\, c^2}\right)\ln\left(\frac{\lambda^5\, I(\lambda)}{n_0\, h\, c^2}\right)\right], \tag{A.22}$$

which has the units, for example, of [$\mathrm{J\,m^{-3}\,K^{-1}\,s^{-1}\,sr^{-1}}$]. The incident entropy flow (per unit area) at a given surface is thus

$$J^i = \int_\Omega \int_{\nu_1}^{\nu_2} j^i(\nu)\, d\nu \cos(\theta)\, d\Omega = g^i(\Omega) \int_{\nu_1}^{\nu_2} j^i(\nu)\, d\nu, \tag{A.23}$$

where θ is the angle of the normal of the surface to the incident beam, and Ω is the solid angle subtended by the source at the surface.

Assuming that both the reflected and transmitted radiations are Lambertian (i.e. scattered isotropically), the reflected and transmitted part of the outgoing entropy flow can be treated equally and thus can be summed together, $j^{r,t} = j^r + j^t$, giving this component of the outward flow as,

$$j^{r,t}(\nu) = g^i(\Omega) j^i(I(\nu)(1 - a(\nu))) \tag{A.24}$$

and therefore,

$$J^{r,t} = \int_\Omega \int_{\nu_1}^{\nu_2} j^{r,t}(\nu)\, d\nu \cos(\theta)\, d\Omega = g^{r,t}(\Omega) \int_{\nu_1}^{\nu_2} j^{r,t}(\nu) d\nu \tag{A.25}$$

where the angle θ in this case is always 0° since we are assuming that the radiation is emitted isotropically, and where we have assumed that the scattering isotropy is independent of light frequency.

A.5 Evaluation of the Geometrical Factors under Different Assumptions

A.5.1 Sun Overhead, Leaf Radiating into 4π

Assume that the Sun is directly overhead and that the radiated energy is emitted into 4π and that the reflected and transmitted energy is also emitted into 4π. The situation is as depicted in figure A.2. For example, this corresponds to the instantaneous absorption and emission on a leaf.

The geometrical factors are thus,

$$g^i(\Omega) = 1$$
$$g^e(\Omega) = \pi + \pi = 2\pi$$
$$g^{r,t}(\Omega) = \pi + \pi = 2\pi$$

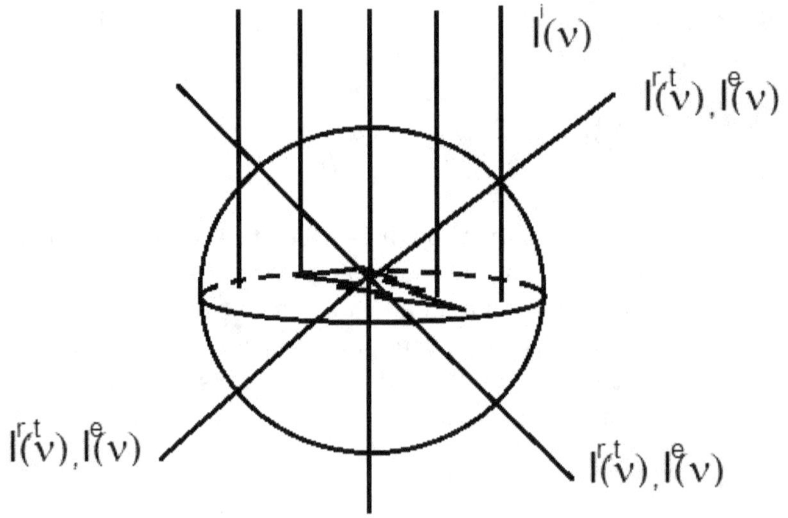

Fig. A.2. Geometrical factors for the sun directly overhead.

A.5.2 Whole Planet

Consider now for the calculation of the entropy production of the whole planet by averaging over one complete day of the planet. Assume that the radiation is emitted and reflected isotropically. The geometrical factors are thus,

$$
\begin{aligned}
g^i(\Omega) &= \frac{\pi r^2}{4\pi r^2} = \frac{1}{4} \\
g^e(\Omega) &= \pi/4 \\
g^{r,t}(\Omega) &= \pi/4
\end{aligned}
\tag{A.26}
$$

Here we are assuming that the radiation is emitted isotropically into 4π so that the day and night temperatures on Earth's surface are the same. The factor of $\pi R_E^2/4\pi R_E^2 = 1/4$ in each of the terms comes from assuming a parallel incident photon beam.

A.6 Entropy Production

Equation (A.17), after integrating over solid angle, thus gives that the entropy production for the dissipation, reflection, and transmission process for a leaf as,

$$
\begin{aligned}
J &= J^e + J^{r,t} - J^i \\
&= g^e(\Omega) \int_{\nu_1}^{\nu_2} j^e(\nu) d\nu + g^{r,t}(\Omega) \int_{\nu_1}^{\nu_2} j^{r,t}(\nu) d\nu - g^i(\Omega) \int_{\nu_1}^{\nu_2} j^i(\nu)\, d\nu
\end{aligned} \tag{A.27}
$$

$$
\begin{aligned}
&= \int_{\nu_1}^{\nu_2} 2\pi \frac{4}{3} \frac{a(\nu) I(\nu)}{T_E} \\
&+ 2\pi \frac{n_0\, k\, \nu^2}{c^2} \left[\left(1 + \frac{c^2\, I(\nu)(1 - a(\nu))}{n_0\, h\, \nu^3}\right) \ln \left(1 + \frac{c^2\, I(\nu)(1 - a(\nu))}{n_0\, h\, \nu^3}\right) \right. \\
&\left. - \left(\frac{c^2\, I(\nu)(1 - a(\nu))}{n_0\, h\, \nu^3}\right) \ln \left(\frac{c^2\, I(\nu)(1 - a(\nu))}{n_0\, h\, \nu^3}\right) \right] \\
&- \frac{n_0\, k\, \nu^2}{c^2} \left[\left(1 + \frac{c^2\, I(\nu)}{n_0\, h\, \nu^3}\right) \ln \left(1 + \frac{c^2\, I(\nu)}{n_0\, h\, \nu^3}\right) - \left(\frac{c^2\, I(\nu)}{n_0\, h\, \nu^3}\right) \ln \left(\frac{c^2\, I(\nu)}{n_0\, h\, \nu^3}\right) \right] d\nu,
\end{aligned} \tag{A.28}
$$

which is the general expression for the entropy production due to the dissipation of an arbitrary incident spectrum, $I(\nu)$, into equilibrium heat at a black-body of temperature T_E, for a frequency dependent fraction $a(\nu)$ absorbed of an arbitrary incident photon energy flow $I(\nu)$ with the rest $(1 - a(\nu))$ being isotropically dispersed into 4π solid angle by either reflection or transmission.

The same equation in terms of wavelength λ is,

$$
\begin{aligned}
J &= \int_{\lambda_1}^{\lambda_2} 2\pi \frac{4}{3} \frac{a(\lambda) I(\lambda)}{T_E} + \\
&2\pi \frac{n_0\, k\, c}{\lambda^4} \left[\left(1 + \frac{\lambda^5\, I(\lambda)(1 - a(\lambda))}{n_0\, h\, c^2}\right) \ln \left(1 + \frac{\lambda^5\, I(\lambda)(1 - a(\lambda))}{n_0\, h\, c^2}\right) \right. \\
&\left. - \left(\frac{\lambda^5\, I(\lambda)(1 - a(\lambda))}{n_0\, h\, c^2}\right) \ln \left(\frac{\lambda^5\, I(\lambda)(1 - a(\lambda))}{n_0\, h\, c^2}\right) \right] \\
&- \frac{n_0\, k\, c}{\lambda^4} \left[\left(1 + \frac{\lambda^5\, I(\lambda)}{n_0\, h\, c^2}\right) \ln \left(1 + \frac{\lambda^5\, I(\lambda)}{n_0\, h\, c^2}\right) - \left(\frac{\lambda^5\, I(\lambda)}{n_0\, h\, c^2}\right) \ln \left(\frac{\lambda^5\, I(\lambda)}{n_0\, h\, c^2}\right) \right] d\lambda.
\end{aligned} \tag{A.29}
$$

For integration over a planet as a whole we use the second set of geometrical factors and get

$$J = \int_{\nu_1}^{\nu_2} \frac{\pi}{4}\frac{4}{3}\frac{a(\nu)I(\nu)}{T_E} +$$
$$\frac{\pi}{4}\frac{n_0\, k\, \nu^2}{c^2}\left[\left(1 + \frac{c^2\, I(\nu)(1-a(\nu))}{n_0\, h\, \nu^3}\right) \ln\left(1 + \frac{c^2\, I(\nu)(1-a(\nu))}{n_0\, h\, \nu^3}\right)\right.$$
$$\left. - \left(\frac{c^2\, I(\nu)(1-a(\nu))}{n_0\, h\, \nu^3}\right) \ln\left(\frac{c^2\, I(\nu)(1-a(\nu))}{n_0\, h\, \nu^3}\right)\right]$$
$$- \frac{1}{4}\frac{n_0\, k\, \nu^2}{c^2}\left[\left(1 + \frac{c^2\, I(\nu)}{n_0\, h\, \nu^3}\right) \ln\left(1 + \frac{c^2\, I(\nu)}{n_0\, h\, \nu^3}\right) - \left(\frac{c^2\, I(\nu)}{n_0\, h\, \nu^3}\right) \ln\left(\frac{c^2\, I(\nu)}{n_0\, h\, \nu^3}\right)\right] d\nu, \tag{A.30}$$

or, in terms of wavelength λ,

$$J = \int_{\lambda_1}^{\lambda_2} \frac{\pi}{4}\frac{4}{3}\frac{a(\lambda)I(\lambda)}{T_E} +$$
$$\frac{\pi}{4}\frac{n_0\, k\, c}{\lambda^4}\left[\left(1 + \frac{\lambda^5\, I(\lambda)(1-a(\lambda))}{n_0\, h\, c^2}\right) \ln\left(1 + \frac{\lambda^5\, I(\lambda)(1-a(\lambda))}{n_0\, h\, c^2}\right)\right.$$
$$\left. - \left(\frac{\lambda^5\, I(\lambda)(1-a(\lambda))}{n_0\, h\, c^2}\right) \ln\left(\frac{\lambda^5\, I(\lambda)(1-a(\lambda))}{n_0\, h\, c^2}\right)\right]$$
$$- \frac{1}{4}\frac{n_0\, k\, c}{\lambda^4}\left[\left(1 + \frac{\lambda^5\, I(\lambda)}{n_0\, h\, c^2}\right) \ln\left(1 + \frac{\lambda^5\, I(\lambda)}{n_0\, h\, c^2}\right) - \left(\frac{\lambda^5\, I(\lambda)}{n_0\, h\, c^2}\right) \ln\left(\frac{\lambda^5\, I(\lambda)}{n_0\, h\, c^2}\right)\right] d\lambda. \tag{A.31}$$

or, putting $b(\lambda) = n_0 hc^2/\lambda^5$, and albedo $\alpha(\lambda) = (1 - a(\lambda))$ gives

$$J = \int_{\lambda_1}^{\lambda_2} I(\lambda)\frac{\pi}{3}\frac{(1-\alpha(\lambda))}{T_E} +$$
$$\frac{\pi}{4}\alpha(\lambda)\frac{k\,\lambda}{hc}\left[\left(\frac{b(\lambda)}{I(\lambda)\alpha(\lambda)} + 1\right) \ln\left(1 + \frac{I(\lambda)\alpha(\lambda)}{b(\lambda)}\right) - \ln\left(\frac{I(\lambda)\alpha(\lambda)}{b(\lambda)}\right)\right]$$
$$-\frac{1}{4}\frac{k\,\lambda}{hc}\left[\left(\frac{b(\lambda)}{I(\lambda)} + 1\right) \ln\left(1 + \frac{I(\lambda)}{b(\lambda)}\right) - \ln\left(\frac{I(\lambda)}{b(\lambda)}\right)\right] d\lambda. \tag{A.32}$$

For an black-body incident spectrum from our Sun, evaluated at the surface of Earth, we have (see Eq. (A.13)),

$$I(\nu) = \frac{R_S^2}{d^2}\frac{n_0 h\nu^3}{c^2}\frac{1}{e^{h\nu/kT_S} - 1} \tag{A.33}$$

where the first factor, R_S^2/d^2, accounts for the decrease in the flux per unit solid angle per unit m^2 of the flow of photons from the source (star) at the planet due to the distance d of the planet from the star. R_S is the radius of the star. For the case of stars like our sun, with unpolarized light and most of the energy output of small wavelength λ, the first term in the second and third square brackets of Eq. (A.32) can be neglected and we obtain the following approximate expression for the entropy production

$$J \approx \int_{\nu_1}^{\nu_2} I(\lambda)\pi \left[\frac{1}{3}\frac{(1-\alpha(\lambda))}{T_E} - \frac{\alpha(\lambda)}{4}\frac{k\,\lambda}{hc}\ln\left(\frac{I(\lambda)\alpha(\lambda)}{b(\lambda)}\right) + \frac{1}{4\pi}\frac{k\,\lambda}{hc}\ln\left(\frac{I(\lambda)}{b(\lambda)}\right)\right] d\lambda, \tag{A.34}$$

Equation (A.34) for the entropy production can be separated into a flow of energy at wavelength λ, $I(\lambda)$, and a generalized photon dissipation force, $F(\lambda)$

$$J = \int_{\nu_1}^{\nu_2} I(\lambda)F(\lambda)d\nu \tag{A.35}$$

with

$$F(\lambda) = \pi\left[\frac{1}{3}\frac{(1-\alpha(\lambda))}{T_E} - \frac{\alpha(\lambda)}{4}\frac{k\,\lambda}{hc}\ln\left(\frac{I(\lambda)\alpha(\lambda)}{b(\lambda)}\right) + \frac{1}{4\pi}\frac{k\,\lambda}{hc}\ln\left(\frac{I(\lambda)}{b(\lambda)}\right)\right] \tag{A.36}$$

with $b(\lambda) = n_0\, h\, c^2/\lambda^5$. The corresponding equation in terms of frequency, $\nu = c/\lambda$, is,

$$F(\nu) = \pi\left[\frac{1}{3}\frac{(1-\alpha(\nu))}{T_E} - \frac{\alpha(\nu)}{4}\frac{k}{h\nu}\ln\left(\frac{I(\nu)\alpha(\nu)}{b(\nu)}\right) + \frac{1}{4\pi}\frac{k}{h\nu}\ln\left(\frac{I(\nu)}{b(\nu)}\right)\right] \tag{A.37}$$

with $b(\nu) = n_0\, h\, \nu^3/c^2$. The non-linear relation between the energy flow $I(\nu)$ and the dissipative force $F(\nu)$ is obvious from equations (A.36) and (A.37).

B. Glossary of Technical Terms

albedo – the proportion of light reflected by a body with respect to that received. Refers to the visible spectrum unless a specific wavelength range is given.

allele – one of a number of alternative forms of the same gene.

amino acids – molecules which are the building blocks of proteins and enzymes. There are 22 distinct amino acids used by life today. The aromatic (ring shaped) amino acids, and some others with conjugated bonds, are very strong absorbers of UV-C light and it is suggested in this book that these were originally antenna pigments to RNA and DNA who's original function was to dissipate this light.

amphipathic – refers to molecules that have one part hydrophobic (water hating – non polar) and the other part hydrophilic (water loving – polar).

Archean – refers to the time period of early Earth history from 4.55 – 2.5 Ga (Ga represents giga años since present, 10^9 years). It is usually divided into the Hadean (4.55 – 3.9 Ga), the early Archean (3.9 – 2.9Ga) and the late Archean (2.9 – 2.5 Ga).

autotrophic organisms – organisms that are able to make their own food through the process of photosynthesis.

base – refers to one of the nucleobases of which RNA and DNA are composed, adenine A, thymine T, guanine G, cytosine C, uracil U.

chirality – the property of a molecule to absorb preferentially either right- or left-handed circularly polarized light.

circular dichroism – a measure of the characteristic of a material (or molecule) to absorb preferentially one chirality (circular polarization) of light over the other. The circular dichroism spectrum of a material is defined as the difference in the molar extinction coefficients between left- and right-handed circularly polarized light as a function of wavelength (see chapter 14).

Classical Irreversible Thermodynamics – mathematical formalism derived by Lars Onsager and Ilya Prigogine to deal with situations in which there is an external generalized chemical potential acting over the system. Prigogine showed that in such cases, the material in the system could organize in such a manner as to increase the dissipation of the external potential. For this work in describing the formation of "dissipative structuring" Prigogine was awarded the Nobel Prize in Chemistry in 1977.

Compton scattering – the process of inelastic light scattering off material. Some of the energy of the photon can be given to the material resulting in a photon of less energy, i.e. more red-shifted.

conjugated double bonds – double bonds in organic molecules which have single bonds on either side. Electronic excitations of these structures leads to non-local bonding, where electrons are shared by many atoms, called π or σ bonding.

conjugation – conjugation in molecules refers to alternating single and double bonds. Charge delocalization causes molecules with extensive conjugation to be more stable than other molecules. Changing the amount of conjugation in a molecular structure can also change the wavelength of maximal photon absorption.

diffeomorphism – a map between manifolds which is differentiable and has a differentiable inverse.

difference absorption spectra – the difference (as a function of wavelength) between the absorption spectra of denatured single strand and native double strand DNA or RNA. For short DNA or RNA segments, the difference absorption spectrum has a very particular shape dependent on sequence and can therefore be used to definitively identify denaturing. This is particularly useful for confirming UV-C light-induced denaturing.

dissipation – the process of spreading the conserved quantities of Nature (mass-energy, momentum, angular momentum, charge, etc.) over ever more microscopic degrees of freedom. Usually used when referring directly to the reduction of a particular imposed generalized chemical potential over a system. It is the same as saying the "entropy production" of the system.

dissipation-replication relation – a relation between the rate of photon dissipation and the rate of replication. Such a relation probably existed at the molecular level (see chapter 12 on ultraviolet and temperature assisted replication of RNA and DNA) during the Archean, as well as at the higher levels of organism, species, clade, ecosystem, and biosphere levels even today (see chapter 17.3).

dissipative structuring – a "spontaneous" non-equilibrium thermodynamic process in which material organizes under a generalized chemical potential to dissipate more effectively this potential. Examples are Bénard cells and hurricanes.

donor molecules – donor molecules are those that can absorb photons but lack a conical intersection and thus remain in the excited state for long times, decaying through the relatively slow processes of fluorescence or phosphorescence. If a donor molecule is in close proximity to an acceptor quencher molecule, it can transfer its excitation energy to the quencher and thus rapidly be prepared to absorb another photon. The complex is a greater dissipating system than the molecules acting separately. The amino acid tryptophan is an example of a donor molecule and it can transfer its excitation energy efficiently to RNA or DNA which are examples of quencher molecules.

EET – refers to electronic energy transfer. Consists of different process by which electronic excitation energy of a molecule can be transferred to another.

entropy – a measure of the distribution of Nature's conserved quantities (mass-energy, momentum, angular momentum, charge, etc.) over the available microscopic degrees of freedom. The entropy is maximum in an isolated system at thermodynamic equilibrium.

entropy production – the process of distributing the conserved quantities of Nature (mass-energy, momentum, angular momentum, charge, etc.) over ever more microscopic degrees of freedom. This term is used alternatively with the term "dissipation" throughout this book. The term "dissipation" is preferentially used when there is reference to a particular generalized chemical potential impressed over the system.

exciton – refers to a system of two or more molecules that share electronic excitation energy which can happen if the two molecules are in close contact on the absorption of an energy quanta. The exciton energy levels are usually different from that of the separate molecules due to the interaction of the two or more molecules participating.

extremophile – refers to organisms that can survive in extremely harsh conditions, for example, very high temperatures, very high salt concentrations, or extreme pH conditions.

FRET – refers to Förster resonant energy transfer. Consists of the process by which electronic excitation energy of a molecule can be transferred from it to another if the fluorescence spectrum of the donor molecule overlaps with the absorption spectrum of the acceptor quencher molecule.

fundamental molecules – those found in all three domains of life (archea, bacteria, eukaryote). Examples include, RNA, DNA, amino acids, flavins, flavonoids, carotenoids, porphyrins, etc. (see table 4.7). Because they exist in all three domains of life, it is supposed that they must have existed at the time of the origin of life, or close to it. A corroborating fact is that almost all of these molecules absorb strongly in the UV-C region (there was an atmospheric window in the UV-C region during the Archean) and have affinity (chemical attraction) to RNA and DNA.

fundamental pigments – a term used interchangeably with "fundamental molecules" since the fundamental molecules were, in fact, pigments in the UV-C during the Archean.

Ga. – the symbol used to represent a thousand million years (Giga años).

generalized chemical potential – any of a number of potentials that gives rise to a force which drives a particular flow in an irreversible process. It is used alternatively to the phrase "generalized thermodynamic potential". See "generalized thermodynamic force" for examples of generalized chemical potentials.

generalized thermodynamic potential – any of a number of potentials that gives rise to a force which drives a particular flow in an irreversible process.

generalized thermodynamic force – any of a number of forces which drive corresponding particular flows in an irreversible process. For example, the force due to a temperature gradient gives rise to a flow of heat. The force due to a concentration gradient gives rise to a flow of material. The force of the chemical affinity divided by the temperature gives rise to the flow (rate) of a chemical reaction. The differences in the photon spectra of the Sun and of that of the cosmic microwave background of space give rise to a force for photon dissipation (conversion of photons of higher energy to many more of lower energy), i.e. to a flow of energy from one form to another. It is this latter generalized thermodynamic force that organizes matter on Earth in such a manner to dissipate it more rapidly. One such organization of matter driven by this force is life.

genotypes – description of organisms through their microscopic genetic content.

Hadean – refers to the earliest time period of Earth history (4.55 – 3.9 Ga) when ocean temperatures were above 100 °C.

homochirality – refers to the fact that many of the fundamental molecules of life have a particular chirality (ability to absorb preferentially circularly polarized light of a given handedness, right or left). The homochirality of life has been difficult to explain since the distinct chiralities of molecules have equal equilibrium formation probability.

hyperthermophilic – liking very high temperatures. In biology, "hyperthermophilic" refers to organisms with optimal proliferation temperatures > 80 °C.

hypoxanthine – a natural rare nucleobase found only in transfer RNA (tRNA).

Last Universal Common Ancestor (LUCA) – A hypothetical organism at base of the trunk of the tree of life from which all life is suggested to have evolved. It has been suggested that the UCA must have lived about 3.5

Ga, before the branching into the three domains of life (bacteria, archea, eukaryote).

Ma. – the symbol used to represent a million years (Mega años).

Markovian process – a name given to a random process in which the system changes states according to a transition rule that only depends on the current state. No memory of the history, except knowledge of the current state, is required.

microscopic dissipative structuring – dissipative structuring operating at the microscopic scale (even at the nanoscale). It usually involves internal molecular degrees of freedom and is often associated with an imposed photon potential.

nucleotides – molecules consisting of a RNA or DNA base, ribose or deoxyribose sugar, and a phosphate group.

nucleosides – molecules consisting of a RNA or DNA base, ribose or deoxyribose sugar.

nucleobase – molecules consisting of only a RNA or DNA base.

oligos – short segments ($\lesssim$50 base pairs) of RNA or DNA.

PAH's – Poly aromatic hydrocarbons – molecules found throughout the cosmos composed mainly of carbon and hydrogen in the form of aromatic rings.

phenotypes – the macroscopic physical and behavioral characteristics of a living organism.

primordial molecules - those found around star forming regions, HCN, H_2CO, H_2O, CO, OH and needed for the formation of the fundamental molecules of life (those found in all three domains of life).

phase space – imaginary 6N dimensional space for an N particle system consisting of the 3N position coordinates and 3N momentum coordinates of all particles. The phase space concept allows a succinct description of a many-body system.

phase space paths – paths in phase space taken by a system in transiting from one macrostate to another. It has been suggested that the paths taken with greatest probability in nature are those which lead to the highest entropy production rates (see chapter 3.1.1).

pyrimidine dimer formation – pyrimidine dimer formation refers to the UV-induced formation of covalently bonded dimers of pyrimidines (thymine or cytosine) when these occur adjacent on the same RNA or DNA strand.

quantum efficiency – a measure of how many incident photons result in a particular photochemical reaction. If the quantum efficiency is equal to 1, then all incident photons produce the particular reaction under study.

quencher molecule – a molecule which has a conical intersection to allow the rapid dissipation of electronic excitation energy into heat. Through different energy transfer mechanisms, a quencher molecule can accept excitation energy from another antenna molecule and rapidly dissipate it to heat making the complex more efficient at dissipation than the two molecules separate. RNA and DNA are quencher molecules in the UV region of the spectrum.

red-edge – refers to the marked drop in the absorption of most autotrophic organisms at about 700 nm.

reducing atmosphere – a reducing atmosphere is basically one in which hydrogen is present in quantities and can donate its electron to a shared covalent bond. In less reducing atmospheres, it is much more difficult, but not impossible, to make the fundamental molecules.

reducing conditions – refers to conditions in which protons and electrons are easily transferred from one molecule to another. This occurs for molecules having a large hydrogen content.

ribose – a sugar made up of five carbon atoms in a ring. It is the sugar that is attached to the nucelobases in RNA. In DNA deoxyribose (the same sugar but missing an oxygen) is attached to the nucleobases.

RNA or DNA base – conjugated aromatic molecules which provide the four letter alphabet for the genetic code of RNA or DNA. These molecules absorb strongly and dissipate strongly in the UV-C. The molecules are known as adenine, guanine, cytosine and thymine, or uracil (for RNA).

RNA world – a hypothesis that suggests that the first steps of life were entirely based on the molecule RNA and the rest of life was built up around this molecule through improving its "fitness" as determined primarily on replication efficacy.

self-organization – self-organization generally refers to the process by which material organizes without external influence, for example the formation of water ice crystals. However, in this book we stretch the definition somewhat (as is usually done) to include organization of material under a generalized chemical potential, for example, the Benard cell, or the formation of pigment molecules under UV-C light. This organization is not strictly "self"-organization because the external environment is crucial for the organization process.

thermodynamic selection – a non-equilibrium thermodynamic principle that dissipative structures, from microscopic (e.g. organic pigments) to macroscopic (e.g. the biosphere), form spontaneously to dissipate an imposed thermodynamic potential (e.g. the solar photon potential), and that those structures that increase the global photon dissipation rate will have greater representation in the population of structures (be selected). One such example is the greater reproductive success afforded to oligonucleotides which have chemical affinity to other UV-C dissipating antenna fundamental molecules (see chapters 12, 13 and 17.3).

thermophilic – liking high temperatures. In biology, "thermophilic" refers to organisms with optimal proliferation temperatures > 50 °C.

Ultraviolet and Temperature Assisted Replication (UVTAR) – refers to the hypothetical process in which RNA and DNA could replicate using the environmental factors of UV-C light and high temperatures without the need of enzymes at the beginnings of life.

References

1. Abo-Riziq, A., Grace, L. Nir, E., Kabelac, M., Hobza, P. and de Vries, M.S. (2005) Photochemical selectivity in guanine–cytosine base-pair structures, PNAS, 102, 20-23, www.pnas.org_cgi_doi_10.1073_pnas.0408574102
2. Anbar, A. D. and Knoll, A. H. (2002) Proterozoic ocean chemistry and evolution: a bioinorganic bridge?, Science, 297, 1137–1142.
3. Angel, J. R. P., Illing, R., and Martin, P. G. (1972) Circular polarization of twilight, Nature, 238, 389–390.
4. Antipina and Gurtovenko (2015) Molecular Mechanism of Calcium-Induced Adsorption of DNA on Zwitterionic Phospholipid Membranes, J. Phys. Chem. B, 119 (22), pp 6638–6645. DOI: 10.1021/acs.jpcb.5b01256
5. Arcaya G, Pantoja ME, Pieber M, Romero C, Tohá JC. (1971) Molecular Interaction of L-Tryptophan with Bases, Ribonucleosides and DNA, Molecular interaction of L-tryptophan with bases, ribonucleosides and DNA, Z Naturforsch B. 10, 1026-1030.
6. Aristotle. Physics. Translated by R. P. Hardie and R. K. Gaye. The Internet Classics Archive. OCLC 54350394, http://classics.mit.edu/Aristotle/physics.2.ii.html
7. Arroyo, C . M ., Carmichael, A. J., Swenberg, C . E . and Myers, L. S . Jr . (1986) Neutron induced free radicals in oriented DNA . International Journal of Radiation Biology, 50, 789-793.
8. Atreyaa, S. K., Adamsa, E. Y., Niemann, H. B., Demick-Montelara, J. E., Owen, T. C., Fulchignoni, M., Ferri, F., Wilson, E. H. (2006) Titan's methane cycle, Planetary and Space Science, 54(12), 1177.
9. Attard, P. (2008) The Second Entropy: A Variational Principle for Time-dependent Systems, Entropy,10, 380-390; DOI: 10.3390/e10030380
10. Attard, P. (2009) The second entropy: a general theory for non-equilibrium thermodynamics and statistical mechanics, Annu. Rep. Prog. Chem., Sect. C: Phys. Chem., 105, 63-173 DOI: 10.1039/B802697C
11. Ball, J. A., Gottlieb, C. A., Lilley, A. E., and Radford, H. E. (1970) Detection of methyl alcohol in Sagittarius, The Astrophysical Journal, 162, L203-L210.
12. Barks, H.L., et al. (2010) Guanine, Adenine, and Hypoxanthine production in UV-irradiated formamide solutions: Relaxation of the requirements for prebiotic purine nucleobase formation. ChemBioChem 11, 1240-1243.
13. Baronea, F., Cellaib, L., Matzeua, M., Mazzeia, F., Pedone, F. (2000) DNA, RNA and hybrid RNA]DNA oligomers of identical sequence: structural and dynamic differences, Biophysical Chemistry 86, 37-47.
14. Baverstock, K . F ., and Cundall, R . B. (1988) Solitons and energy transfer in DNA . Nature, 332, 312-313 .
15. Baverstock, K.F. and Cundall, R. B. (1989) Are Solitons Responsible for Energy Transfer in Oriented DNA?, International Journal of Radiation Biology, 55:1, 151-153, doi:10.1080/09553008914550151
16. Beatty J.T., Overmann J., Lince M.T., Manske A.K., Lang S.L., Blankenship R.E., Van Dover C.L., Martinson T.A., Plumley F.G. (2005) An obligately photosynthetic bacterial anaerobe from a deep-sea hydrothermal vent. Proc Natl Acad Sci U S A 102(26):9306–9310. doi:10.1073/pnas. 0503674102

17. Becker, M. M. and Wang, Z. (1989) B-A transitions within a 5S ribosomal gene are highly sequence-specific, J. Biol. Chem. 264, 4163-4167.
18. R. V. Bensasson, E. J. Land, T. G. Truscott, Flash Photolysis and Pulse Radiolysis: Contributions to the Chemistry of Biology and Medicine, Pergamon Press, Oxford, 1983.
19. Bernhardt, H. S. (2012) The RNA world hypothesis: the worst theory of the early evolution of life (except for all the others) Biology Direct, 7:23.
20. Betts and Ball (1997) Albedo over the boreal forest. J. Geophys Resh 102, D24, 28, 901-28, 909.
21. Bidle, K.D., Lee, S., Marchant, D. R. and Falkowsk, P. G. (2007) Fossil genes and microbes in the oldest ice on Earth, PNAS, 104, 13455–1346.
22. Boltzmann, L. (1886) The Second Law of Thermodynamics, in: Ludwig Boltzmann: Theoretical physics and Selected writings, edited by: McGinness, B., D. Reidel, Dordrecht, The Netherlands, 1974.
23. Bouamaied, I., Nguyen, T., Ruhl, T. and Stulz, E. (2008) Supramolecular helical porphyrin arrays using DNA as a scaffold, Org. Biomol. Chem., 6, 3888–3891.
24. Boulanger, E., Anoop, A., Nachtigallova, D., Thiel, W. and Barbatti, M. (2013) Photochemical Steps in the Prebiotic Synthesis of Purine Precursors from HCN, Angew. Chem. Int. Ed. 2013, 52, 8000 –8003.
25. Brocks, J. J., Logan, G. A., Buick, R., and Summons, R. E. (1999) Archean molecular fossils and the early rise of eukaryotes, Science, 285, 1033–1036.
26. Brocks, J. J., Buick, R., Logan, G. A., and Summons, R. E. (2003) Composition and syngeneity of molecular fossils from the 2.78 to 2.45 billion-year-old Mount Bruce Supergroup, Pilbara Craton, Western Australia, Geochim. Cosmochim. Ac., 67, 4289–4319.
27. Broo, A. J. (1998) A Theoretical investigation of the physical reason for the very different luminescence properties of the two isomers adenine and 2-Aminopurine, Phys. Chem. A 102, 526-531.
28. Bucher, D.B., Pilles, B.M., Carell,T., and Zinth, W. (2014) Charge separation and charge delocalization identified in long-living states of photoexcited DNA, PNAS 111, 4369–4374.
29. Buchvarov, I., Wang, Q., Raytchev, M., Trifonov, A. and Fiebig, T. (2007) Electronic energy delocalization and dissipation in single- and double-stranded DNA, PNAS 104, 4794–4797.
30. Buhl, D., (1973) Galactic Clouds of Organic Molecules, in Cosmochemical Evolution and the Origins of Life: Proceedings of the Fourth International Conference on the Origin of Life and the First Meeting of the International Society for the Study of the Origin of Life, June 25-28, 1973. Volume I: Invited Papers. Editors: J. Oró, S. L. Miller, C. Ponnamperuma, R. S. Young, Barcelona, June 25-28, 1973. Springer, ISBN: 978-94-010-2284-2 (Print) 978-94-010-2282-8 (Online).
31. Bunge, C. (2015) Private conversation with an engaging physicist over homemade apple juice.
32. Buratti, B. J., Cruikshank, D. P., Brown, R. H., Clark, R. N., Bauer, J. M., Jaumann, R., McCord, T. B., Simonelli, D. P., Hibbitts, C. A., Hansen, G. B., Owen, T. C., Baines, K. H., Bellucci, G., Bibring, J. –P., Capaccioni, F., Cerroni, P., Coradini, A., Drossart, P., Formisano, V., Langevin, Y., Matson, D. L., Mennella, V., Nelson, R. M., Nicholson, P. D., Sicardy, B., Sotin, C., Roush, T. L., Soderlund, K., and Muradyan, A. (2005) Cassini Visual and Infrared Mapping Spectrometer observations of Iapetus: detection of CO2, The Astrophysical Journal, 622, L149-L152.
33. Burkholder, J.B., Mills, M. and McKeen, S. (2000) Upper Limit for the UV Absorption Cross Sections of H_2SO_4, Geophysical Research Letters, 27, 2493-2496.
34. J. Cadet, P. Vigni, in Bioorganic Photochemistry: Photochemistry and the Nucleic Acids, Ed. H. Morrison, John Wiley & Sons, New York, 1990, pp. 1–273.
35. Cadet J, Anselmino C, Douki T, Voituriez L (1992) Photochemistry of nucleic acids in cells. J Photochem Photobiol B Biol15 277-298.
36. Cameron, A. G. W. (1973) Abundances of the elements in the solar system. Space Sci. Rev. 15, 121-146.
37. Canfield, D. E. and Teske, A. (1996) Late Proterozoic rise in atmospheric oxygen concentration inferred from phylogenetic and sulphurisotope studies, Nature, 382, 127–132.

38. Canuel, C., Mons, M., Piuzzi, F., Tardivel, B., Dimicoli, I. and Elhaninea, M. (2005) Excited states dynamics of DNA and RNA bases: Characterization of a stepwise deactivation pathway in the gas phase, J. Chem. Phys. 122, 074316.
39. Challenger, F. and Simpson, I.S. (1948) Studies on biological methylation. Part XII. A precursor of the dimethyl sulphide evolved by Polysiphonia fastigiata. Dimethyl-2-carboxyethylsulphonium hydroxide and its salts." (pdf). Journal of the Chemical Society (London) 1591–1597. doi:10.1039/JR9480001591
40. Chang, R. (2000) Physical Chemistry for the Chemical and Biological Sciences, University Science Books, Sausalito California.
41. Chang, S., Mack, R. & Lennon, K. (1978) Carbon chemistry of separated phases of Murchison and Allende meteorites. Lunar Planetary Sci. 9, 157–159.
42. Chaplin, M. (2016) Water Structure and Science, http://www1.lsbu.ac.uk/water/images/watopt.gif
43. Charlson, R.J., Lovelock, J.E., Andreae, M.O., Warren, S.G. (1987) Oceanic phytoplankton, atmospheric sulphur, cloud albedo and climate. Nature 326(6114):655–661.
44. Cheung, A. C., Rank, D. M., Townes, C. H., Thornton, D. D., and Welch, W. J. (1968) Detection of NH_3 molecules in the interstellar medium by their microwave emission, Physical Review Letters, 21, 1701-1705.
45. Cheung, A. C., Rank, D. M., Thornton, D. D., and Welch, W. J. (1969) Detection of water in interstellar regions by its microwave radiation, Nature, 221, 626-628.
46. Chyba, C. and Sagan, C. (1991) Electrical energy sources for organic synthesis on the early, Earth, Orig Life Evol Biosph. 21:3-17
47. Chylla R.A., Whitmarsh J. (1989) Inactive photosystem II complexes in leaves. Turnover rate and quantitation. Plant Physiol 90:765–772.
48. Ciciriello, F., Costanzo, G. Pino, S., Crestini, C., Saladino, R., and Di Mauro, E. (2008) Molecular complexity favors the evolution of Ribopolymers, Biochemistry 47, 2732–2742.
49. Clarke, G. L., Ewing, G. C., and Lorenzen, C. J. (1970) Spectra of backscattered light from the sea obtained from aircraft as a measure of chlorophyll concentration, Science, 167, 1119–1121.
50. Cleaves, H. J. Chalmers, J. H., Lazcano, A., Miller, S. L. and Bada, J. L. (2008) A reassessment of prebiotic organic synthesis in neutral planetary atmospheres, Orig Life Evol Biosph 38, 105–115, DOI 10.1007/s11084-007-9120-3
51. Clemett, S. J., Sanford, S. A., Nakamura-Messenger, K., Hörz, F., and McKay, D. S. (2010) Complex aromatic hydrocarbons in Stardust samples collected from comet 81P/Wild 2, Meteoritics and Planetary Science, 45, 701-722.
52. Cline, D. B. (2005) On the physical origin of the homochirality of life. European Review 13, 49–59.
53. Cnossen, I., Sanz-Forcada, J.Favata, F. Witasse, O. Zegers, T., Arnold, N.F. (2007) The habitat of early life: Solar X-ray and UV radiation at Earth's surface 4–3.5 billion years ago. J Geophys Res 112, E02008. doi:10.1029/2006JE002784
54. Cnossen, I., Sanz-Forcada, J.Favata, F. Witasse, O. Zegers, T., Arnold, N.F. (2007) The habitat of early life: Solar X-ray and UV radiation at Earth's surface 4–3.5 billion years ago. arXiv:astro-ph/0702529
55. The source of this material is the COMET® Website at http://meted.ucar.edu/ of the University Corporation for Atmospheric Research (UCAR), sponsored in part through cooperative agreement(s) with the National Oceanic and Atmospheric Administration (NOAA), U.S. Department of Commerce (DOC). ©1997-2016 University Corporation for Atmospheric Research. All Rights Reserved.
56. Cooper, G.W., Onwo, W.M. & Cronin, J.R. (1992). Alkyl phosphonic acids and sulfonic acids in the Murchison meteorite. Geochim. Cosmochim. Acta 56, 4109–4115.
57. Cooper, G.W., Thiemens, M.H., Jackson, T.L. & Chang, S. (1997). Sulfur and hydrogen isotope anomalies in meteorite sulfonic acids. Science 277, 1072–1074.
58. Cooper G., Kimmich N., Belisle W., Sarinana J., Brabham K. & Garrel L. (2001) Carbonaceous meteorites as a source of sugar-related organic compounds for the early Earth. Nature 414, 879–883.

59. Corliss, J.B., Baross, J.A., and Hoffman, S.E. (1981)An hypothesis concerning the relationship between submarine hot springs and the origin of life on Earth. Oceanol. Acta Suppl. 4, 59–69.
60. Cornwall C.E., Hepburn C.D., McGraw C.M., Currie K.I., Pilditch C.A., Hunter K.A., Boyd P.W., Hurd C.L. (2013) Diurnal fluctuations in seawater pH influence the response of a calcifying macroalga to ocean acidification, Proc. R. Soc. B, 280 20132201; DOI: 10.1098/rspb.2013.2201.
61. Crespo-Hernández, C.E., Cohen, B., Hare, P.M., Kohler, B. (2004) Ultrafast excited-state dynamics in nucleic acids, Chem. Rev. 104,1977.
62. Crespo-Hernández, C.E., Cohen, B., Kohler, B. (2005) Base stacking controls excited-state dynamics in A•T DNA, Nature 436, 1141-1144, doi:10.1038
63. Cronin, J. R., Pizzarello, S., and Frye, J. S. (1987) ^{13}C NMR spectroscopy of the insoluble carbon of carbonaceous chondrites, Geochimica et Cosmochimica Acta, 51, 299-303.
64. Cronin, J.R., Pizzarello, S. & Cruikshank, D.P. (1988). Organic matter in carbonaceous chondrites, planetary satellites, asteroids and comets. In Meteorites and the early Solar system, eds Kerridge, J.F. & Matthews, M.S., University of Arizona Press, Tucson, AZ.
65. Crowe, S. A., Døssing, L. N., Beukes, N. J., Bau, M., Stephanus, J., Kruger, J., Frei, R., and Canfield, D. E. (2013) Atmospheric oxygenation three billion years ago, Nature, 501, 535–538.
66. Crowell, J. C. (1999) Pre-Mesozoic ice ages; their bearing on understanding the climate system, Mem. Geol. Soc. Amer., 192, 1–106.
67. Cruikshank, D. P., Wegryn, E., Dalle Ore, C. M., Brown, R. H., Bibring, J. P., Buratti, B. J., Clark, R. N., McCord, T. B., Nicholson, P. D., Pendleton, Y. J., Owen, T. C., Filacchione, G., Coradini, A., Cerroni, P., Capaccioni, F., Jaumann, R., Nelson, R. M., Baines, K. H., Sotin, C., Bellucci, G., Combes, M., Langevin, Y., Sicardy, B., Matson, D. L., Formisano, V., Drossart, P., and Mennella, V. (2008) Hydrocarbons on Saturn's satellites Iapetus and Phoebe, Icarus, 193, 334-343.
68. Czikklely, V. Forsterling, H. D. Kuhn, H. (1970) Extended dipole model for aggregates of dye molecules, Chem. Phys. Lett., 6, 207-210.
69. Dai, X., Wei, C., Su, F. F., Li, Z. Y., Sun, Z. F. and Zhang, Y. (2013) RSC Adv., Self-assembly of DNA networks at the air/water interface over time, DOI: 10.1039/C3RA42099J
70. Darwin, C. (1859) On the Origin of Species by Means of Natural Selection, Ed. J Murray, London. Reprinted by Gramercy Books, New York, 1979.
71. Dawkins, R. (1976) The Selfish Gene, Oxford University Press, Oxford.
72. Dawkins, Richard (1989) The Extended Phenotype. Oxford: Oxford University Press. ISBN 0-19-288051-9
73. de Duve C (1991) Blueprint for a Cell: The Nature and Origin of Life. Burlington, North Carolina, Neil Patterson Publishers, Carolina Biological Supply Company, p. 79.
74. Deamer, D. and Weber, a. L. (2010) Bioenergetics and Life's Origins, Cold Spring Harb Perspect Biol. 2010 Feb; 2(2): a004929. doi: 10.1101/cshperspect.a004929
75. Derenne, S. and Robert, F. (2010) Model of molecular structure of the insoluble organic matter isolated from Murchison meteorite, Meteorit. Planet. Sci., 45, 1461–1475.
76. De Rosa, M., Gambacorta, A., Gliozzi, A. (1986) "Structure, biosynthesis, and physicochemical properties of archaebacterial lipids". Microbiol. Rev. 50 (1): 70–80. PMID 3083222. PMC 373054
77. Dewar, R. (2003) Information theory explanation of the fluctuation theorem, maximum entropy production and self-organized criticality in non-equilibrium stationary states, J. Phys. A: Math. Gen. 36, 631–641.
78. Dewar, R„ Juretić, D. and Županović, P. (2006) The functional design of the rotary enzyme ATP synthase is consistent with maximum entropy production, Chem. Phys. Lett. 430, 177-182.
79. Di Giulio, M. (2001) The universal ancestor was a thermophile or a hyperthermophile. Gene, 281, 11–17.
80. Di Giulio, M. (2003) The universal ancestor and the ancestor of bacteria were hyperthermophiles. J. Mol. Evol. 57, 721–730.

81. Dillon, G. C. and Castenholtz, R. W. (1999) Scytonemin, a cyanobacterial sheath pigment, protects against UV-C radiation: Implications for early photosynthetic life, J. Phycol. 35, 673–681.
82. Domagal-Goldman, S. D., Kasting, J. F., Johnston, D. T., and Farquhar, J. (2008) Organic haze, glaciations and multiple sulfur isotopes in the Mid-Archean Era, Earth Planet. Sc. Lett., 269, 29–40.
83. Dorren, J. D., and Guinan, E.F. (1994) The Sun in time: detecting and modeling magnetic inhomogenities on solar-type stars. In: The Sun as a Variable Star, Pap, J.M., Frölich C., Hudson, H.S. and Solanki, S.K. Eds., pp. 206-216, Cambridge University Press, Cambridge.
84. Doughty, C.E., Wolf, A. and Malhi,Y. (2013) The legacy of the Pleistocene megafauna 1350 extinctions on nutrient availability in Amazonia, Nature Geoscience 6, 761–764. 1351 doi:10.1038/ngeo1895.
85. Douglas, A. E. and Herzberg, G. (1941) Notes on CH+ in interstellar space and in the laboratory, The Astronomical Journal, 94, 381.
86. Dubey, R.K. and Tripathi, D.N. (2005)A study of thermal denaturataion/renaturation in DNA using laser light scattering: A new approach, Indian Journal of Biochemistry & Biophysics, 42, 301-307.
87. Duley, W. W., and Williams, D. A. (1979) Are there organic grains in the interstellar medium, Nature, 277, 40-41.
88. Dunham, T. J., Jr., Adams, W. S. (1937) Interstellar neutral potassium and neutral calcium, Publications of The Astronomical Society of the Pacific, 49, 26-28.
89. Dutta, S. and Mokhir, A. (2011) An autocatalytic chromogenic and fluorogenic photochemical reaction controlled by nucleic acids, Chem. Commun., 2011, 47, 1243–1245. DOI: 10.1039/C0CC02508A
90. Dwek, E., Arendt, R. G., Fixsen, D. J., Sodroski, T. J., Odegard, N., Weiland, J. L., Reach, W. T., Hauser, M. G., Kelsall, T., Moseley, S. H., Silverberg, R. F., Shafer, R. A., Ballester, J., Bazell, D., Isaacman, R. (1997) Detection and Characterization of Cold Interstellar Dust and Polycyclic Aromatic Hydrocarbon Emission, from COBE Observations, The Astrophysical Journal, Volume 475, 565-579.
91. Edwards, H. G. M., Jorge Villar, S.E., Pullan, D., Hargreaves, M. D., Hofmann, B. A., Westall, F. (2007) Morphological biosignatures from relict fossilised sedimentary geological specimens: a Raman spectroscopic study, J. Raman Spectrosc, 38, 1352–1361.
92. Eickbush and Moudrianakis (1977) A Mechanism for the Entrapment of DNA at an air-water interface, Biophysical Journal 18, 275-288.
93. Encrenaz T. 2014 Infrared spectroscopy of exoplanets: observational constraints, Phil.Trans.R.Soc. A 372: 20130083. http://dx.doi.org/10.1098/rsta.2013.0083
94. Engels, F. (1884) Origen of the Family, Private Property and the State, Penguin Books Ltd., 1972, London England.
95. Erives, A. (2011) A Model of Proto-Anti-Codon RNA Enzymes Requiring l-Amino Acid Homochirality. Journal of Molecular Evolution, 73(1-2), 10–22. http://doi.org/10.1007/s00239-011-9453-4
96. Ferris, J. P. and Orgel, L. E. (1966) An Unusual Photochemical Rearrangement in the Synthesis of Adenine from Hydrogen Cyanide, J. Am. Chem. Soc., 88, 1074-1074.
97. Fuente, A.; et al. (2005) Photon-dominated Chemistry in the Nucleus of M82: Widespread HOC+ Emission in the Inner 650 Parsec Disk, Astrophysical Journal 619 (2): L155–L158.
98. Garcѐs F., Davila C. A. (1982) Alterations in DNA irradiated with ultraviolet radiation—I. The formation process of cyclobutylpyrimidine dimers: cross sections, action spectra and quantum yields. Photochem. Photobiol. 35:9–16.
99. Garcia-Fernandez, J.M., Nieto-Villar, J. M. and Rieumont-Briones, J. (1996) The rate of entropy production as an evolution criterion in chemical systems II. Chaotic reactions, Phys. Scr. 53 643-644.
100. Gates, D. M. (1980) Biophysical Ecology, ISBN 0-387-90414-X, Springer-Verlag, New York Inc.
101. Gatica J. (2011) Contribucion de las Cianobacterias a la Produccion de Entropia de la Tierra, Bachelors thesis, Faculty of Sciences, Universidad Nacional Autonoma de México.

102. Genberg, L., Heisel, F., McLendon, G., Miller, J.D., Vibrational energy relaxation processes in heme proteins. J. Luminescence 40-41, 571-572, 1988.
103. Gennis, R. B. (1989) Biomembranes: Molecular Structure and Function. Springer, ISBN 0387967605.
104. Gilinsky, N. L. (1994) Volatility and the Phanerozoic Decline of Background Extinction Intensity, Paleobiology 20, 445-458.
105. Gillett, F. C., Forrest, W. J., and Merrill, K. M. (1973) 8 - 13-micron spectra of NGC 7027, BD +30° 3639, and NGC 6572, The Astrophysical Journal, 183, 87-93.
106. Ginzburg, C. (1980) The Cheese and the Worms: The Cosmos of a Sixteenth-Century Miller. The Johns Hopkins University Press. Baltimore.
107. Glansdorff, P. and Prigogine, I. (1964) On a general evolution criterion in macroscopic physics. Physica 30, 351-374.
108. Gnanadesikan, A., Emanuel, K., Vecchi, G. A., Anderson, W. G., and Hallberg, R. (2010) How ocean color can steer Pacific tropical cyclones, Geophys. Res. Lett., 37, L18802, doi:10.1029/2010GL044514.
109. Gordon-Walker, A., Penzer, G. R., Radda, G. K. (1970) Excited States of Flavins Characterised by Absorption, Prompt and Delayed Emission Spectra, Eur. J. Biochem. 13, 313-321.
110. Gould, S. J. (2002) The Structure of Evolutionary Theory, Belknap Press, ISBN 0-674-00613-5
111. Gould, K. S., & Quinn, B. D. (1999) Do anthocyanins protect leaves of New Zealand native species from UV-B? New Zealand Journal of Botany, 37, 175–178.
112. Govil, G., Kumar, N.Y., Kumar, M.R., Hosur, R.V., Roy, K.B. and Miles, H.T. (1985) Recognition schemes for protein-nucleic acid interactions. Proc Int Symp Biomol Struct Interactions, Suppl J Biosci 8, 645-656.
113. Gradie, J. and Veverka, J. (1980) The composition of the Trojan asteroids, Nature, 283, 840-842.
114. Grammatika, M. and Zimmerman, W.B. (2001) Microhydrodynamics of flotation processes in the sea-surface layer. Dynam Atmos Oceans 34, 327-348.
115. Gray, D. M., Morgan, A. R., Ratliff, R. L. (1978) A comparison of the circular dichroism spectra of synthetic DNA sequences of the homopurine homopyrimidine and mixed purine-pyrimidine types. Nucleic Acid Research 5 3679–3695.
116. Gredel, R., Carpentier, Y., Rouillé, G., Steglich, M., Huisken, F., and Henning, Th. (2011) Abundances of PAHs in the ISM: confronting observations with experimental results, Astronomy and Astrophysics, 530, 26.
117. Graur, D. and Pupko, T. (2001) The Permian bacterium that isn't, Mol. Biol. and Evol. 18, 1143–1146.
118. Gudipati, M. S. and Allamandola, L. J., (2004) Polycyclic aromatic hydrocarbon ionization energy lowering in water ices, The Astrophysical Journal, 615:L177–L180.
119. Guillois, O., Nenner, I., Papoular, R., and Reynaud, C. (1996) Coal models for the infrared emission spectra of proto-planetary nebula, The Astrophysical Journal, 464, 810-817.
120. Guo, X., Lan, Z. and Cao, Z. (2013) Ab initio insight into ultrafast nonadiabatic decay of hypoxanthine: keto-N7H and keto-N9H tautomers, Phys.Chem. Chem. Phys., 15, 10777
121. Gysbers R., Tram K., Gu J., Li Y. (2015) Evolution of an Enzyme from a Noncatalytic Nucleic Acid Sequence, Scientific Reports 5: 11405. doi:10.1038/srep11405
122. Hagen, U., Keck, K., Kröger, H., Zimmermann, F. and Lücking, T. (1965) Ultraviolet light inactivation of the priming ability of DNA in the RNA polymerase system. Biochim Biophys Acta 95, 418-425.
123. Hanczyc M.M. and Szostak J.W. (2004) Replicating vesicles as models of primitive cell growth and division, Current Opinion in Chemical Biology 8, 660–664.
124. Hardy, J. T.: The sea-surface Microlayer (1982) Biology, Chemistry and Anthropogenic Enrichment, Prog. Oceanogr., 11, 307–328.
125. Hare, P.M., Crespo-Hernádez, C.E., and Kohler, B. (2007) Internal conversion to the electronic ground state occurs via two distinct pathways for pyrimidine bases in aqueous solution, PNAS, 104, 435– 440.

126. Hawking, S. and Penrose, R. (1996) The Nature of Space and Time, p. 135, Princeton University Press, Princeton.
127. Hayatsu, R., Anders, E., Studier, M.H. & Moore, L.P. (1975). Purines and triazines in the Murchison meteorite. Geochim. Cosmochim. Acta 39, 471–488.
128. Hendry, L.B., Bransome, JR., E.D., Hutson, M.S., and Campbell, L.K. (1981) First approximation of a stereochemical rationale for the genetic code based on the topography and physicochemical properties of "cavities" constructed from models of DNA, Proc. Natl Acad. Sci. USA, **78**, 7440-7444.
129. Hernández Candia C. N. (2009) Medición experimental del coeficiente de producción de entropía de una planta por el proceso de transpiración, Bachelors thesis, Universidad Nacional Autónoma de México.
130. Herrmann, F. and Würfel P. (2005) Light with nonzero chemical potential, Am. J. Phys., 73, 717–721.
131. Hess, B., Mikhailov, A.S., (1994) Self-organization in living cells, Science 264, 223-224.
132. Hofmann, G.E. et al. (2011) High-frequency dynamics of ocean pH: A multi-ecosystem comparison. PLoS ONE 6(12):e28983.
133. Horowitz, E.D., Engelhart A.E., Chen M.C., Quarles K.A., Smith M.W., Lynn D.G., Hud N.V. (2010) Intercalation as a means to suppress cyclization and promote polymerization of base-pairing oligonucleotides in a prebiotic world. Proc Natl Acad Sci USA 107, 5288-5293.
134. Hoyle, F., Wickramasinghe, C. and Watkins, J. (1986) Viruses from space and related matters, University College Cardiff Press, Cardiff, Wales.
135. Hoyle, F. and Wickramasinghe, N. C. (1978) Lifecloud, The Origin of Life in the universe, J.M. Dent and Sons, Ltd., Londres, 1978.
136. Hoyle, F. and Wickramasinghe, N. C. (1999) Biofluorescence and the extended red emission in astrophysical sources, Astrophysics and Space Science, 268, 321-325.
137. Iwabata, H., Watanabe, K., Ohkuri, T., Yokobori, S. and Yamagishi, A. (2005). Thermostability of ancestral mutants of Caldococcus noboribetus isocitrate dehydrogenase. FEMS Microbiol. Letters, 243, 393–398.
138. James, K.D., Ellington, A.D. (1997) Surprising fidelity of template-directed chemical ligation of oligonucleotides. Chem Biol 4, 595-605.
139. Jesorka, A., Holzwarth, A. R., Eichhöfer, A., Reddy, C.M., Kinoshita, Y., Tamiaki, H., Katterle, M., Naubrond J.-V., and Balaban T.S. (2012) Water coordinated zinc dioxochlorin and porphyrin self-assemblies as chlorosomal mimics: variability of supramolecular interactions, Photochem. Photobiol. Sci., 11, 1069-1080.
140. Jewitt, D. C. and Luu, J. X. (2001) Colors and spectra of Kuiper Belt objects, The Astronomical Journal, 122, 2099-2114.
141. Johnsen, S. and Widder, E.A. (1999) The physical basis of transparency in biological tissue: Ultrastructure and the minimization of light scattering. J Theor Biol 199, 181-198.
142. Joyce GF, Visser GM, van Boeckel CA, van Boom JH, Orgel LE, van Westrenen J (1984) Chiral selection in poly(C)-directed synthesis of oligo(G), Nature 310 (5978): 602–4. Bibcode:1984Natur.310..602J. doi:10.1038/310602a0. PMID 6462250.
143. Jungclaus, G., Cronin, J.R., Moore, C.B. & Yuen, G.U. (1976a) Aliphatic amines in the Murchison meteorite. Nature 261, 126–128.
144. Jungclaus, G.A., Yuen, G.U., Moore, C.B. & Lawless, J.G. (1976b) Evidence for the presence of low molecular weight alcohols and carbonyl compounds in the Murchison meteorite. Meteoritics 11, 231–237.
145. Jürgen, H., Haug, P. (1986) Traces of Archaebacteria in ancient sediments, System Applied Microbiology 7 (Archaebacteria '85 Proceedings): 178–83. doi:10.1016/S0723-2020(86)80002-9
146. Kaifu, N., Morimoto, M., Nagane, K., Akabane, K., Iguchi, T., and Takagi, K. (1974) Detection of interstellar methylamine, The Astrophysical Journal, 191, L135-L137.
147. Kaku, Michio (1994) Hyperspace: A Scientific Odyssey Through Parallel universes, Time Warps, and the Tenth Dimension. Oxford: Oxford University Press. ISBN 0-19-286189-1.
148. Kang, H. et al.. (2002) "Intrinsic Lifetimes of the Excited State of DNA and RNA Bases", JACS, 124, 12958-12959.

149. Karam, P. A. (2003) Inconstant Sun: how solar evolution has affected cosmic and ultraviolet radiation exposure over the history of life on Earth, Health Phys., 84, 322–333.
150. Kasting, J. F. & Ackerman, T. P. (1986) Climatic consequences of very high carbon dioxide levels in the earth's early atmosphere, Science 234, 1383-1385.
151. Kasting, J. F. (1993)Earth's early atmosphere, Science 259, 920-926.
152. Kauffman, S. (1986) Autocatalytic sets of proteins. Journal of Theoretical Biology, 119, 1-24.
153. Kauffman S. (1993) The Origins of Order, Oxford University Press, New York.
154. Kawai, Y. and Wada, A. (2007) Diurnal Sea Surface Temperature Variation and Its Impact on the Atmosphere and Ocean: A Review, Journal of Oceanography, Vol. 63, pp. 721 to 744.
155. Keller, L. P., Bajt, S., Baratta, G. A., Borg, J., Bradley, J. P., Brownlee, D. E., Busemann, H., Brucato, J. R., Burchell, M., Colangeli, L., d'Hendecourt, L., Djouadi, Z., Ferrini, G., Flynn, G., Franchi, I. A., Fries, M., Grady, M. M., Graham, G. A., Grossemy, F., Kearsley, A., Matrajt, G., Nakamura-Messenger, K., Mennella, V., Nittler, L., Palumbo, M. E., Stadermann, F. J., Tsou, P., Rotundi, A., Sandford, S. A., Snead, C., Steele, A., Wooden, D., Zolensky, M. (2006) Infrared spectroscopy of comet 81P/Wild 2 samples returned by Stardust, Science, 314, 1728-1731.
156. Kerridge, J. F. (1999) Formation and processing of organics in the early Solar System, Space Science Reviews, 90, 275-288.
157. Kharecha, P., Kasting, J. F., and Siefert, J. L. (2005) A coupled atmosphere-ecosystem model of the early Archean Earth, Geobiology, 3, 53–76.
158. Kiang N., Siefert J., Govindjee, Blankenship R.E. (2007) Spectral signatures of photosynthesis. I. Review of earth organisms. Astrobiology 7:222–251.
159. Kim, S. T., Li, Y. F. and Sancar, A. (1992) The third chromophore of DNA photolyase: Trp-277 of Escherichia coli DNA photolyase repairs thymine dimers by direct electron transfer. Proc. Natl. Acad. Sci. USA 89, 900–904.
160. Kleidon A and Lorenz RD (2005) Entropy Production by Earth System Processes in Kleidon A and Lorenz RD (eds.) Non-equilibrium thermodynamics and the production of entropy: life, Earth, and beyond. Springer Verlag, Heidelberg. ISBN: 3-540-22495-5
161. Kleidon, A. and Renner, M. (2013) A simple explanation for the sensitivity of the hydrologic cycle to surface temperature and solar radiation and its implications for global climate change, Earth Syst. Dynam., 4, 455–465, doi:10.5194/esd-4-455-2013
162. Knacke, R. F. (1977) Carbonaceous compounds in interstellar dust, Nature, 269, 132-134.
163. Knauth, L. P. (1992) Isotopic Signatures and Sedimentary Records, in: Lecture Notes in Earth Sciences #43, edited by: Clauer, N. and Chaudhuri, S., Springer-Verlag, Berlin, 123–152.
164. Knauth, L. P. (2005) Temperature and salinity history of the Precambrian ocean: implications for the course of microbial evolution, Paleogeography, Paleoclimatology, Paleoecology 219:53-69, 2005
165. Knauth, L. P. and Lowe, D. R. (2003) High Archean climatic temperature inferred from oxygen isotope geochemistry of cherts in the 3.5 Ga Swaziland group, South Africa, Geol. Soc. Am. Bull., 115, 566–580.
166. Koch, T.H. and Rodehorst, R.M. (1974) Quantitative investigation of the photochemical conversion of diaminomaleonitrile to diaminofumaronitrile and 4-amino-5-cyanoimidazole, J. Am. Chem. Soc., 96, 6707 – 6710.
167. Kohler, B. (2010) Nonradiative decay mechanisms in DNA model systems, J. Phys. Chem. Lett. 2010, 1, 2047–2053.
168. Kondepudi, N. and Prigogine, I.L. (1998) Modern Thermodynamics: From Heat Engines to Dissipative Structures, John Wiley & Sons Ltd, Chichester, West Sussex, England.
169. König, B.,"Organic Photochemistry", [Online] Available: http://www-oc.chemie.uni-regensburg.de/OCP/ ch/chb/oc5/Photochemie-08.pdf
170. Krasnopolsky, V. A. (2016) Sulfur aerosol in the clouds of Venus. Icarus, 274: 33 DOI: 10.1016/j.icarus.2016.03.010

171. Kritsky, M. S., Telegina, T. A., Vechtomova, Y. L. and Buglak, A. A. (2013) Why flavins are not competitors of chlorophyll in the evolution of biological converters of solar energy, Int. J. Mol. Sci. 2013, 14, 575-593; doi:10.3390/ijms14010575.
172. Krypides, N.C. and Ouzounis, C.A. (1995) Nucleic acid-binding metabolic enzymes: Living fossils of Stereochemical Interactions?, J. Mol. Evol. 40, 564-569.
173. Kuznetsova, M., Lee, C., Aller, J., Frew, N. (2004) Enrichment of amino acids in the sea surface microlayer at coastal and open ocean sites in the North Atlantic Ocean, Limnology and oceanography, 49, 1605-1619.
174. Kvenvolden K., Lawless J., Pering K., Peterson E., Flores J., Ponnamperuma C., Kaplan I.R. & Moore C. (1970) Evidence for extraterrestrial amino acids and hydrocarbons in the Murchison meteorite. Nature 228, 928–926.
175. Kwok, S., Volk, K., and Bernath, P. (2001) On the Origin of Infrared Plateau Features in Proto-Planetary nebula, The Astrophysical Journal Letters, 554, L87-L90.
176. Kwok, S. (2007) Molecules and solids in planetary nebula and proto-planetary nebula, Advances in Space Research, 40, 655–658.
177. Kwok, S. (2009) Organic matter in space: from star dust to the Solar System, Astrophys Space Sci, 319, 5–21.
178. Kwok, S. (2012) Organic Matter in the universe, Weinheim: Wiley-VCH, p. 127-141.
179. Kwok, S. and Zhang, Y. (2011) Mixed aromatic–aliphatic organic nanoparticles as carriers of unidentified infrared emission features, Nature, 479 (7371), 80–83.
180. Kwok, S. and Zhang, Y. (2013) Unidentified Infrared Emission Bands: PAHs or MAONs?, The Astrophysical Journal, 771(1), 5, 9.
181. Lacey, J.C. and Mullins, D.W. (1983) Experimental studies related to the origin of the genetic and protein synthesis – a review, Origins Life, 13, 3-42.
182. Laß, K. and Friedrichs, G. (2011) Revealing structural properties of the marine nanolayer from vibrational sum frequency generation spectra, J. Geophysical Research 116, C08042, doi:10.1029/2010JC006609
183. Lavery, T. J., Roudnew, B., Gill, P., Seymour,J., Seuront, L., Johnson, G., Mitchell, J.G., and Smetacek, V. (2010) Iron excretion by sperm whales stimulates carbon export in the Southern Ocean, Proc. R. Soc. B 277, 3527–3531. doi:10.1098/rspb.2010.0863
184. Lawless, J.G. (1973) Amino acids in the Murchison meteorite. Geochim. Cosmochim. Acta 37, 2207–2212.
185. Lawless, J.G. & Yuen, G.U. (1979). Quantification of monocarboxylic acids in the Murchison carbonaceous meteorite. Nature 282, 396–398.
186. Lawless, J.G., Zeitman, B., Pereira, W.E., Summons, R.E. & Duffield, A.M. (1974) Dicarboxylic acids in the Murchison meteorite. Nature 251, 40–42.
187. Levy, M. and Miller, S.L. (1998) The stability of the RNA bases: Implications for the origin of life, Proceedings of the National Academy of Sciences of the United States of America 95, 7933-7938.
188. Li, A. and Draine, B. T. (2002) Do the Infrared Emission Features Need Ultraviolet Excitation? The Polycyclic Aromatic Hydrocarbon Model in UV-poor Reflection nebula, The Astrophysical Journal, 572, 232.
189. Lloyd, S. and Pagels, H. M. (1988) Complexity as thermodynamic depth, Ann. Phys., 188, 186–213.
190. López-Puertas, M., Dinelli, B. M., Adriani, A., Funke, B., García-Comas, M., Moriconi, M. L., D'Aversa, E., Boersma, C., and Allamandola, L. J. (2013) Large Abundances of Polycyclic Aromatic Hydrocarbons in Titans's Upper Atmosphere, The Astrophysical Journal, 770(2), 132.
191. Lotka, A. J. (1922) Contribution to the energetic of evolution. Proceedings of the National Academy of Sciences USA, 8, 147–151.
192. Lovelock, J. E. (2005) Gaia: Medicine for an ailing planet, 2nd Edn., Gaia Books, New York.
193. Lowe, D.R. and Tice, M.M. (2004) Geologic evidence for Archean atmospheric and climatic evolution: Fluctuating levels of CO2, CH4, and O2 with an overriding tectonic control, Geology 32, 493-496.

194. Luvall J.C. and Holbo H.R. (1991) Thermal remote sensing methods in landscape ecology. Chpt. 6 in Turner M and Gardner RH, Quantitative methods in landscape ecology, Springer-Verlag.
195. MacIntyre, F. (1974) The top millimeter of the ocean, Scientific American, 230, 62–77.
196. Majerfeld, I., and Yarus, M. (2005) A diminutive and specific RNA binding site for L-tryptophan. Nucleic Acids Research 33, 5482–5493 (2005).
197. Makarieva, A. M. and Gorshkov, V. G. (2007) Biotic pump of atmospheric moisture as driver of the hydrological cycle on land, Hydrol. Earth Syst. Sci., 11, 1013–1033, doi:10.5194/hess-11-1013-2007
198. Marmur, J. and Doty, P. (1959) Heterogeneity in deoxyribonucleic acids: 1. Dependence on composition of the configurational stability of the deoxyribonucleic acids, Nature 183, 1427-1429.
199. Marmur, J. and Grossman, L. (1961) Ultraviolet light induced linking of deoxyribonucleic acid strands and its reversal by photoreactivating enzyme, Proc.. Nat. Acad. Sci. 47, 778-787.
200. Marquetand, P., Nogueira, J. J., Mai, S., Plasser, F. and González, L. (2016) Challenges in simulating light-induced processes in DNA, Molecules 22, 49; doi:10.3390/molecules22010049
201. Martin, W., Baross, J., Kelley, D. and Russell, M.J. (2008) Hydrothermal vents and the origin of life, Nature Reviews Microbiology 6, 805-814, doi:10.1038/nrmicro1991
202. McKellar, A. (1940) Evidence for the molecular origin of some hitherto unidentified interstellar lines, Publications of The Astronomical Society of the Pacific, 52, 187-192.
203. Meeks J. C. and Castenholz R. W. (1971) Growth and photosynthesis in an extreme thermophile, Synechococcus lividus (Cyanophyta). Arch Mikrobiol 78:25–41.
204. Mergny, J.-L., Li, J., Lacroix, L., Amrane, S. and Chaires, J.B. Thermal difference spectra: a specific signature for nucleic acid structures. Nucleic Acids Research, 33 (2005) doi:10.1093/nar/gni134
205. Mezzetti, A., Protti, S., Lapouge, C. and Cornard, JP (2011) Protic equilibria as the key factor of quercetin emission in solution. Relevance to biochemical and analytical studies, Chem. Chem. Phys.,13, 6858–6864.
206. Michaelian, K. (2005) Thermodynamic stability of ecosystems, J. Theor. Biol., 237, 323–335.
207. Michaelian, K. (2009) Thermodynamic origin of life. Cornell ArXiv, arXiv:0907.0042 [physics.gen-ph].
208. Michaelian, K. (2010) Thermodynamic origin of life. Earth Syst. Dynam. Discuss., 1, 1-39. http://www.earth-syst-dynam-discuss.net/1/1/2010/esdd-1-1-2010-discussion.html
209. Michaelian, K. (2010b) Homochirality through Photon-Induced Melting of RNA or DNA: the Thermodynamic Dissipation Theory of the Origin of Life. Nature Precedings. Available from Nature Precedings http://hdl.handle.net/10101/npre.2010.5177.1
210. Michaelian, K. (2011) Thermodynamic dissipation theory for the origin of life. Earth Syst Dynam 2, 37–51. doi:10.5194/esd-2-37-2011
211. Michaelian, K. and Manuel, O. (2011) Origin and evolution of life constraints on the solar model, J. Mod. Phys, 2, 587–594.
212. Michaelian, K. (2012a) The biosphere: A thermodynamic imperative, in The Biosphere, Ed. Natarajan Ishwaran, Director, Division of Ecological and Earth Sciences, UNESCO, Paris, France, INTECH, U.S., ISBN: 979-953-307-504-3.
213. Michaelian, K. (2012b) HESS Opinions "Biological catalysis of the hydrological cycle: life's thermodynamic function", Hydrol. Earth Syst. Sci., 16, 2629–2645, doi:10.5194/hess-16-2629-2012.
214. Michaelian, K. (2013) A non-linear irreversible thermodynamic perspective on organic pigment proliferation and biological evolution, J. Phys.: Conf. Ser. 475, 012010. doi:10.1088/1742-6596/475/1/012010
215. Unpublished personal experience with the opaquing of certain plastic polymers exposed to sunlight.

216. Michaelian, K. and Santillán Padilla, N. (2014) DNA Denaturing through UV-C Photon Dissipation: A Possible Route to Archean Non-enzymatic Replication, bioRxiv [Biophysics]. doi: http://dx.doi.org/10.1101/009126
217. Michaelian, K. and Simeonov, A. (2015) Fundamental molecules of life are pigments which arose and evolved to dissipate the solar spectrum. Cornell ArXiv arXiv:1405.4059v2 [physics.bio-ph], and Michaelian, K. and Simeonov, A., Fundamental molecules of life are pigments which arose and co-evolved as a response to the thermodynamic imperative of dissipating the prevailing solar spectrum. Biogeosciences, 12, 4913–4937, 2015, www.biogeosciences.net/12/4913/2015/ doi:10.5194/bg-12-4913-2015.
218. Michaelian, K. (2015) Photon dissipation as an indicator of ecosystem health, in Environmental Indicators, Ed. R. Armon and Osmo Hänninen, Springer. ISBN 9401794995 (print), 9789401794992 (on-line).
219. Michaelian, K. and Santillán Padilla, N., (2016) Rudimentary DNA emzyneless replication under supposed Archean conditions, under preparation.
220. Michaelian, K. and Simeonov, A. (2016) Thermodynamic explanation of the cosmic ubiquity of organic pigments, Cornell ArXiv, arXiv:1608.08847 [astro-ph.EP].
221. Michaelian, K. (2016) Some baryonic dark matter may be organic pigments, under preparation.
222. Middleton, C.T., de La Harpe, K., Su, C., Kay Law, Y., Crespo-Hernández, C. E. and Kohler, B. (2009) DNA Excited-State Dynamics: From Single Bases to the Double Helix, Annu. Rev. Phys. Chem. 60:217–239.
223. Miller., S.L. (1955) Production of some organic compounds under possible primitive earth conditions. J. Am. Chem. Soc. 77, 2351-2361.
224. Miller, S.L. (1998) The endogenous synthesis of organic compounds, in The Molecular Origins of Life, edited by Brack, A., Cambridge University Press, Cambridge.
225. Miller, S. L. and Bada, J.L. (1988) Submarine hot springs and the origin of life, Nature, 334, 609-611.
226. Miller, S. and Tennyson, J. (1992) H3+ in space, Chem. Soc. Rev.,21, 281-288. DOI: 10.1039/CS9922100281
227. Mitchell DL, Nairn RS (1989) The biology of the (6-4) photoproduct. Photochem Photobiol. 49, 805-819.
228. Miyazaki, J., Nakaya, S., Suzuki, T., Tamakoshi, M., Oshima, T. & Yamagishi, A. (2001) Ancestral residues stabilizing 3-isopropylmalate dehydrogenase of an extreme thermophile: experimental evidence supporting the thermophilic common ancestor hypothesis. J. Biochem. (Tokyo), 129, 777–782.
229. Mohr, H. and Schopfer, P. (2015) Plant Physiology, Springer Verlag, Heidelberg, p. 161.
230. Mojzsis, S. J., Harrison, T. M., and Pidgeon, R. T. (2001) Oxygen-isotope evidence from ancient zircons for liquid water at the Earth's surface 4300 Myr ago, Nature, 409, 178–181.
231. Monnard, P.-A., Kanavarioti, A., and, and Deamer, D. W. (2003) Eutectic Phase Polymerization of Activated Ribonucleotide Mixtures Yields Quasi-Equimolar Incorporation of Purine and Pyrimidine Nucleobases Journal of the American Chemical Society 2003 125 (45), 13734-13740 DOI: 10.1021/ja036465h
232. Montenay-Garestier, T., and Helène, C. (1968) Molecular Interactions between Tryptophan and Nucleic Acid Components in Frozen Aqueous Solutions, Nature, 217, 844-846.
233. Montenay-Garestier, T., and Helène, C. (1971) Reflectance and luminescence studies of molecular complex formation between tryptophan and nucleic acid components in frozen aqueous solutions, Biochemistry 10, 300-306. DOI: 10.1021/bi00778a016
234. Morel, R. E. and Fleck, G (1989) Onsager's Principle: A Unifying Biotheme, J. Theor. Biol., 136, 171–175.
235. Morowitz, H., Kostelnik, J., Yang, J., and Cody, G. (2000) The origin of intermediary metabolism. Proceedings of the NationalAcademy of Sciences, 97, 7704-7708.
236. Morris, P.W. Gupta, H. , Nagy, Z., Pearson, J.C., Ossenkopf-Okada, V., Falgarone, E., Lis, D.C., Gerin, M., Melnick, G., Neufeld, D.A., Bergin, E.A. (2016) Herschel/HIFI Spectral Mapping of C^+, CH^+, and CH in Orion BN/KL: The prevailing role of ultraviolet irradiation in CH^+ Formation, arXiv:1604.05805v3 [astro-ph.GA] 30 Apr 2016.

237. Mulkidjanian, A.Y. and Junge, W. (1997) On the origin of photosynthesis as inferred from sequence analysis, Photosynthesis Research 51: 27–42.
238. Mulkidjanian, A.Y., Cherepanov, D.A., and Galperin, M.Y. (2003) Survival of the fittest before the beginning of life: selection of the first oligonucleotide-like polymers by UV light, BMC. Evol. Biol., 3, 12.
239. Mullis, K. (1990) The unusual origin of the Polymerase Chain Reaction. Sci Am 262, 56–65.
240. Murphy C.J., Arkin, M.R., Jenkins, Y., Ghatlia, N.D., Bossmann, S.H., Turro, N.J., Barton, J.K. (1993) Long-range photoinduced electron transfer through a DNA helix, Science 262, 1025-9.
241. Mushir, S. Deep, S., Fatma, T. (2014) Screening of cyanobacterial strains for UV screening compound Scytonemin - Environmental Perspectives, IJIRSET 3, Issue 2, 9191-9196.
242. Nandy, K. (1964) Observations of Intersteller Reddening I: Results for region in Cygnus, Publications of the Royal Observatory Edinburgh, 3, 142.
243. Nandy, K. (1965) Observations of Intersteller Reddening I: Results for Region in Perseus, Publications of the Royal Observatory Edinburgh, 5, 13.
244. NASA web site (2016) Dark Energy, Dark Matter. http://science.nasa.gov/astrophysics/focus-areas/what-is-dark-energy/ (latest access, July, 2016).
245. Neault, J. F., and Tajmir-Riahi, H. A. (1999) Structural Analysis of DNA-Chlorophyll Complexes by Fourier Transform Infrared Difference Spectroscopy, Biophysical Journal 76, 2177–2182.
246. Nes, W. R. and Nes, W. D (1980) Lipids in Evolution. Plenum Press, New York, p. 85.
247. Nguyen, M. J., Raulin, F., Coll, P., Derenne, S., Szopa, C., Cernogora, G., Israël, G., and Bernard, J. M. (2007) Carbon isotopic enrichment in Titan's tholins; implications for Titan's aerosols, Planetary and Space Science, 55, 2010-2014.
248. Nicolis, G. and Prigogine, I. (1977) Self-Organization in Nonequilibrium Systems. New York: John Wiley and Sons. ISBN 0-471-02401-5.
249. Nisbet, E. G. and Sleep, N. H. (2001) The habitat and nature of early life, Nature, 409, 1083–1091.
250. Nordlund, T.M., Xu, D., Evans, K.O. (1993) Excitation energy transfer in DNA: Duplex melting and transfer from normal bases to 2-aminopurine, Biochemistry, 32 (45), pp 12090–12095 DOI: 10.1021/bi00096a020
251. Nuesch, F. and Gratzel, M. (1995) "H-aggregation and correlated absorption and emission of a merocyanine dye in solution, at the surface and in the solid state. A link between crystal structure and optical propertie, Chem. Phys., 193, 1-17.
252. Núñez, M.E., Hall, D.B., Barton, J.K. (1999) Long-range oxidative damage to DNA: Effects of distance and sequence, Chemistry & Biology 6, Issue 2, 85-97.
253. Nutman, A. P., Bennett, V. C., Friend, C. R. L.,Van Kranendonk, M. J., Chivas, A. R. (2016) Rapid emergence of life shown by discovery of 3,700-million-year-old microbial structures, Nature, 2016/08/31/online - advance online publication, doi:10.1038/nature19355
254. Öberg, K.I., Guzman, V.V., Furuya, K., Qi, C., Aikawa, Y., Andrews, S. M., Loomis, R., Wilner, D. J. (2015) The comet-like composition of a protoplanetary disk as revealed by complex cyanides, Nature 520, 198–201. doi:10.1038/nature14276
255. An In-Depth Study of the DH-mini Light Source, Ocean Optics, http://oceanoptics.com/study-of-dh-mini-light-source/ last retrieved, 12-oct-2015.
256. Onsager, L. (1931) Reciprocal Relations in Irreversible Processes, I., Phys. Rev., 37, 405–426.
257. Onsager, L. (1945) Theories and problems of liquid diffusion. Ann. NY Acad. Sci. 46, 241–265, doi:10.1111/j.1749-6632.1945.tb36170.x
258. Oparin, A.I. (1936) The origin of life. English translation by Ann Synge of A.I. Oparin, (1924) Proiskhozhdenie zhizny. Moscow. Izd.Moskovhii Rabochi.
259. L. E. Orgel, L. E. (1995) Unnatural selection in chemical systems, Acc. Chem. Res., 28, 109-118.
260. Orgel, L. E. (2004) Prebiotic chemistry and the origin of the RNA world, Crit. Rev. Biochem. Mol., 39, 99–123.

261. Oró, J. (1961) Mechanism of synthesis of adenine from hydrogen cyanide under possible primitive Earth conditions, Nature, 191, 1193–1194.
262. Oró, J. and Kimball, A. P. (1962) Synthesis of purines under possible primitive Earth conditions, II. Purine intermediates from hydrogen cyanide, Arch. Biochem. Biophys., 96, 293–313.
263. Osborne, C. P. and Freckleton, R. P. (2009) Ecological selection pressures for C4 photosynthesis in the grasses, P. Royal Soc. B, 276, 1753–1760.
264. Palmer, K. F., and Williams, D. (1975) Optical Constants of Sulfuric Acid; Application to the Clouds of Venus? Applied Optics 14, 208-219.
265. Paltridge, G.W. (1979) Climate and thermodynamic systems of maximum dissipation, Nature, 279, 630–631.
266. Padmanabhan, T. (2010) Thermodynamical Aspects of Gravity: New insights, arXiv:0911.5004.v2 [gr-qc].
267. Papoular, R., Conard, J., Giuliano, M., Kister, J., and Mille, G. (1989) A coal model for the carriers of the unidentified IR bands, Astronomy and Astrophysics, 217, 204-208.
268. Papoular, R. (2001) The use of kerogen data in understanding the properties and evolution of interstellar carbonaceous dust, Astronomy and Astrophysics, 378, 597-607.
269. Patel, B.H., Percivalle, C., Ritson, D. J., Duffy. C.M., Sutherland, J.D. (2015) Common origins of RNA, protein and lipid precursors in a cyanosulfidic protometabolism, Nature Chemistry 7, 301–307. doi:10.1038/nchem.2202
270. Pavlov, A. A., Kasting, J. F., Brown, L. L., Rages, K. A., and Freedman, R. (2000) Greenhouse warming by CH_4 in the atmosphere of early Earth, J. Geophys. Res., 105, 11981–11990.
271. Pecourt. J.M., Peon, J., and Kohler, B. Ultrafast internal conversion of electronically excited RNA and DNA nucleosides in water. J Am Chem Soc 122, 9348-9349 (2000).
272. Pecourt, M., Peon, J. and Kolher, B. (2001) DNA excited-state dynamics: ultrafast internal conversion and vibrational cooling in a series of nucleosides, J. Am. Chem. Soc. 123, 10370.
273. Peltzer, E.T. & Bada, J. (1978) Hydroxycarboxylic acids in the Murchison meteorite. Nature 272, 443–444.
274. Peltzer, E.T., Bada, J.L., Schlesinger, G. &Miller, S.L. (1984) The chemical conditions on the parent body of the Murchison meteorite ; some conclusions based on amino, hydroxy and dicarboxylic acids. Adv. Space Res. 4, 69–74.
275. Pendleton, Y.J. and Allamandola N.J. (2002) The Organic Refractory Material in the Diffuse Interstellar Medium: Mid-Infrared Spectroscopic Constraints, Astrophysical Journal Supplement Series, 138, 75-98.
276. Peretó, J., Bada, J.L. and Lazcano, A. (2009) Charles Darwin and the Origin of Life, Orig Life Evol Biosph, 39(5): 395–406.
277. Pering, K.L. & Ponnamperuma, C. (1971). Aromatic hydrocarbons in the Murchison meteorite. Science 173, 237–239.
278. Pershin S. (1998) Possibility of relict organic pigment detection on the Mars surface from the Earth, Mars Orbiter or Lander, Suppl. to Annales Geophys., 16, pp. C827-1051.
279. Pershin, S. (2000) “Mars surface: anomaly 763/554 color index indicates presence of organic pigments”, CD of Abstracts of 33rd COSPAR Scientific Assembly, B0.4-0008. https://fabiosiciliano.files.wordpress.com/2012/12/pershin-marsorganic2.pdf
280. Pizzarello, S., Feng, X., Epstein, S. & Cronin, J.R. (1994) Isotopic analyses of nitrogenous compounds from the Murchison meteorite – ammonia, amines, amino-acids, and polar hydrocarbons. Geochim. Cosmochim. Acta 58, 5579–5587.
281. Pizzarello, S. and Cronin, J.R. (2000). Non-racemic amino acids in the Murray and Murchison meteorites. Geochim. Cosmochim. Acta 64, 329–338.
282. Planck, M. "Über das Gesetz der Energieverteilung im Normalspektrum". Annalen der Physik 4: 553 (1901). Bibcode:1901AnP...309..553P. doi:10.1002/andp.19013090310. Translated in Ando, K. "On the Law of Distribution of Energy in the Normal Spectrum".
283. Planck, M. (1906) Vorlesungen über die Theorie der Wärmestrahlung. Johann Ambrosius Barth. LCCN 07004527.
284. Planck, M. (1913) The Theory of Heat Radiation, Barth, Leipzig, Germany, 224 pp.

285. Planck, M. (1914) The Theory of Heat Radiation. Masius, M. (transl.) (2nd ed.). P. Blakiston's Son & Co. OL 7154661M.
286. Ponnamperuma, S., Sagan, C., Mariner, R. (1963) Symthesis of adenosine triphosphate under possible primitive earth conditions. Nature 199, 222-226.
287. Popper, Karl (1978) Natural selection and the emergence of mind. Dialectica 32: 339-355. (an excerpt can be found at http://www.geocities.com/criticalrationalist/popperevolution.htm)
288. Powner, M. W., Gerland, B., and Sutherland, J. D. (2009) Synthesis of activated pyrimidine ribonucleotides in prebiotically plausible conditions, Nature, 459, 239–242.
289. Prigogine, I. (1967) An Introduction to the Thermodynamics of Irreversible Processes, Wiley, New York.
290. Prigogine, I., Nicolis, G., and Babloyantz, A. (1972a) Thermodynamics of evolution (I) Physics Today, 25, 23–28.
291. Prigogine, I., Nicolis, G., and Babloyantz, A. (1972b) Thermodynamics of evolution (II), Phys. Today, 25, 38–44.
292. Prigogine, I. (1996) The End of Certainty: Time, Chaos and the New Laws of Nature, Free Press, Simon and Schuster, New York.
293. Puget, J. L. and Leger, A. (1989) A new component of the interstellar matter: small grains and large aromatic molecules, Annual Review of Astronomy and Astrophysics, 27, 161-198.
294. Rahn, R. O., Shulman, R.G.and Longworth, J.W. (1966) Phosphorescence and Electron Spin Resonance Studies of the uv-Excited Triplet State of DNA, The Journal of Chemical Physics 45, 2955; doi: 10.1063/1.1728051
295. Rajeswari, R., Montenay-Garestier, T. and Helene, C. (1987) Does Tryptophan Intercalate in DNA? A Comparative Study of Peptide Binding to Alternating and Nonalternating A-T Sequences, Biochemistry, 26, 6825-683.
296. Rajnohová, Z., Lengyel, A., Funari, S.S., Uhríková, D. (2010) The structure and binding capacity of lipoplexes, Acta Facultatis Pharmaceuticae Universitatis Comenianae, Tomus LVII.
297. Raoult D, Drancourt M, Azza S, et al. (2008) Nanobacteria Are Mineralo Fetuin Complexes. PLoS Pathog. 4 (2): e41. doi:10.1371/journal.ppat.0040041. PMC: 2242841. PMID 18282102
298. Raszka, M. and Kaplan, N. O. (1972) Association by Hydrogen Bonding of Mononucleotides in Aqueous Solution, Proc. Nat. Acad. Sci. 69, 2025-2029.
299. Ribas, I., Guinan, E. F., Güdel, M., and Audard, M. (2005) Evolution of the solar activity over time and effects on planetary atmospheres. I. High-energy irradiances (1–1700 Å), Ap. J., 622, 680–694.
300. Ripple, W. J. and Beschta, R. L. (2006) Linking wolves to willows via risk-sensitive foraging by ungulates in the northern Yellowstone ecosystem, Forest Ecology and Management 230, 96–106.
301. Roman J., McCarthy J.J. (2010) The Whale Pump: Marine Mammals Enhance Primary Productivity in a Coastal Basin. PLoS ONE 5(10): e13255. doi:10.1371/journal.pone.0013255
302. Rosing, M. T. and Frei, R. (2004) U-rich Archaean sea-floor sediments from Greenlandindications of >3700 Ma oxygenic photosynthesis, Earth Planet. Sc. Lett., 217, 237-244.
303. Roslund, J., Roth, M., Guyon, L. Boutou, V., Courvoisier, F. Wolf, J-P, and Rabitz, H. (2011) Resolution of strongly competitive product channels with optimal dynamic discrimination: Application to flavins, The Journal of Chemical Physics, 134, 3, 034511. doi:10.1063/1.3518751
304. Roush, T. L., and Cruikshank, D. P. (2004) Observations and laboratory data of planetary organics, in Astrobiology: Future Perspectives, Ehrenfreund, P. et al. (eds.), 305, 149.
305. Rubin, R. H., Swenson, G. W., Jr., Benson, R. C., Tigelaar, H. L., and Flygare, W. H. (1971) Microwave Detection of Interstellar Formamide, The Astrophysical Journal, 169, L39-L44.
306. Russell, R. W., Soifer, B. T., and Willner, S. P. (1977) The 4 to 8 micron spectrum of NGC 7027, The Astrophysical Journal, 217, L149-L153.

307. Russell, R. W., Soifer, B. T., and Willner, S. P. (1978) The infrared spectra of CRL 618 and HD 44179 (CRL 915), The Astrophysical Journal, 220, 568-572.
308. Rye, R. and Holland, H. D. (1998) Paleosols and the evolution of atmospheric oxygen: a critical review, Am. J. Sci., 298, 621–672.
309. Sagan, C. (1973) Ultraviolet Selection Pressure on the Earliest Organisms, J. Theor. Biol., 39, 195–200.
310. Sagan, C. (1985) Contact. New York: Simon and Schuster.
311. Sagan, C. and Chyba, C. (1997) The Early Faint Sun Paradox: Organic Shielding of Ultraviolet-Labile Greenhouse Gases, Science, 276, 1217–1221.
312. Saladino, R., Crestini, C., Busiello, V., Ciciriello, F., Costanzo, G., and Di Mauro, E. (2005) Origin of Informational Polymers Differential stability of 3'- and 5'-phosphoester bonds in deoxy monomers and oligomers, J. Biol. Chem. 280, 35658–35669
313. Saladino, R., et al. (2005b) On the prebiotic synthesis of nucleobases, nucleotides, oligonucleotides, pre-RNA and pre-DNA molecules. Topics in Current Chemistry 259, 29-68.
314. Saladino, R., Crestini, C., Ciciriello, F., Di Mauro, E., and Costanzo, G. (2006) Origin of Informational Polymers: Differential Stability of phosphoester bonds in ribomonomers and ribooligomers, J. Biol. Chem. 281, 5790–5796.
315. Saladino, R., Crestini, C., Neri, V., Ciciriello, F., Costanzo, G., and Di Mauro, E. (2006) Origin of informational polymers: The concurrent roles of formamide and phosphates, ChemBioChem 7, 1707–1714.
316. Saladino, R., Crestini, C., Ciciriello, F., Costanzo, G., and Di Mauro, E. (2007) Formamide chemistry and the origin of informational polymers, Helv. Chim. Acta 4, 694–720.
317. Saladino, R., et al. (2007) Formamide chemistry and the origin of informational polymers. Chem. Biodivers. 4, 694-720.
318. Salama, F., Galazutdinov, G. A., Krełowski, J., Biennier, L., Beletsky, Y., and Song, In-Ok (2011) Polycyclic Aromatic Hydrocarbons and the Diffuse Interstellar Bands: A survey, The Astrophysical Journal, 728, 154.
319. Sanchez, R., Ferris, J. & Orgel, L. E. (1996) Conditions for purine synthesis: Did prebiotic synthesis occur at low temperatures? Science 153, 72-73.
320. Sauer, M., Hofkens, J., Enderlein J. (2011) Basic Principles of Fluorescence Spectroscopy, Handbook of Fluorescence Spectroscopy and Imaging. WILEY-VCH Verlag GmbH & Co. KGaA, Weinheim. ISBN: 978-3-527-31669-4
321. Scheer H. (2003) The pigments. In: Green BR, Parson WW (eds) Light-harvesting antennas in photosynthesis. Kluwer, Dordrecht, p 13
322. Schild, R.E. (1977) Interstellar reddening law, Astron.J., 82, 337-344.
323. Schneider, E.D., and Kay, J.J. (1994) Life as a Manifestation of the Second Law of Thermodynamics, Mathl. Comput. Modelling Vol. 19, No. 6-8, pp. 25-48.
324. Schneider, E.D., Kay, J.J. (1994) Complexity and thermodynamics: towards a new ecology. Futures 24, 626–647.
325. Schrödinger, Erwin (1944). What is Life? The Physical Aspect of the Living Cell. Cambridge University Press.
326. Schuergers et al. (2016) Cyanobacteria use micro-optics to sense light direction, eLife 2016;5:e12620. DOI: 10.7554/eLife.12620
327. Schurer, K. (1994) Leaf absorbance and photosynthesis, p 53. In: T.W.Tibbitts (ed.). International Lighting in Controlled Environments Workshop, NASA-CP-95-3309.
328. Schwartz, A. W. and Chang, S. (2002) From Big Bang to Primordial Planet-Setting the Stage for the Origin of Life, in: Life's Origin, edited by: Schopf, J. W., University of California Press, Berkeley, 78–112.
329. Schwartzman, D.W. and Lineweaver, C. H. (2004) The hyperthermophilic origin of life revisited, Biochem. Soc. Transact., 32, 168–171.
330. Scott, C. and Glasspool, J. (2006) The diversification of Paleozoic fire systems and fluctuations in atmospheric oxygen concentration, P. Natl. Acad. Sci. USA, 103, 10861–10865, doi:10.1073/pnas.0604090103
331. Seager, S., Turner, E.L., Schafer, J., and Ford, E.B. (2005) Vegetation's Red Edge: A Possible Spectroscopic Biosignature of Extraterrestrial Plants, Astrobiology, 5, 372-390.

332. Sephton, M.A. and Botta, O. (2005) Recognizing life in the Solar System: guidance from meteoritic organic matter, International Journal of Astrobiology 4 (3 & 4): 269–276. doi:10.1017/S1473550405002806
333. Serrano-Andrés, L., Merchán, M. (2009) Are the five natural DNA/RNA base monomers a good choice from natural selection? A photochemical perspective, Journal of Photochemistry and Photobiology C: Photochemistry Reviews 10, 21–32.
334. Setlow, R. B. & Carrier, W. L. (1963) Identification of ultraviolet-induced thymine dimers in DNA by absorbance measurements. Photochem. Photobiol. 2, 49-57.
335. Shannon, C.E. (1948) The Bell System Technical Journal, 27, pp. 379–423, 623–656. A copy of this famous paper can be obtained at http://worrydream.com/refs/Shannon%20-%20A%20Mathematical%20Theory%20of%20Communication.pdf
336. Shapiro, R. (2007) A simpler origin for life, Sci. Am., 296, 46–53.
337. Shimizu, H., Yokobori, S., Ohkuri, T. (2007) Extremely Thermophilic Translation System in the Common Ancestor Commonote: Ancestral Mutants of Glycyl-tRNA Synthetase from the Extreme Thermophile Thermus thermophilus, J. Mol. Biol., 369, 1060–1069, doi:10.1016/j.jmb.2007.04.001
338. Shimoyama, A. & Katsumata, H. (2001) Polynuclear aromatic thiophenes in the Murchison carbonaceous chondrite. Chem. Lett. 3, 202–203.
339. Shinichi, Takaichi (2016) Lipidbank, http://lipidbank.jp/cgi-bin/main.cgi?id=VCA. Date of last access 22-12-2016.
340. Shukla, M.K., and Leszczynski, J. (2005) Effect of Hydration on the Lowest Singlet $\pi\pi^*$ Excited-State Geometry of Guanine: A Theoretical Study, J. Phys. Chem. 109, 17333-17339.
341. Shukla, M.K., and Leszczynski, J. (2009) Hydration of guanine: Electronic singlet excited states for complexes with 19 and 27 water molecules, Chemical Physics Letters 478, 254–259.
342. D R Silvius (1982) Thermotropic Phase Transitions of Pure Lipids in Model Membranes and Their Modifications by Membrane Proteins. John Wiley & Sons, Inc., New York.
343. Simakov, M.B. and Kuzicheva, E.A. (2005) Abiogenic photochemical synthesis on surface of meteorites and other small space bodies, Adv. Space Res. 36, 190–194.
344. Simeonov, A. and Michaelian, K. (2017) Properties of cyanobacterial UV-absorbing pigments suggest their evolution was driven by optimizing photon dissipation rather than photoprotection, submitted for publication, Biogeosciences.
345. Sinha R P, Sinha J P, Groniger A, Hader D.P. (2002) Polychromatic action spectrum for the induction of a mycosporine-like amino acid in a rice-field cyanobacterium, Anabaena sp., Journal of Photochemistry and Photobiology B: Biology, 66, 47-53.
346. Smith, J. D. T., Draine, B. T., Dale, D. A., Moustakas, J., and Kennicutt Jr., R. C. (2007) The mid-infrared spectrum of star-forming galaxies: global properties of polycyclic aromatic hydrocarbon emission, The Astrophysical Journal, 656, 770-791.
347. Snow, T. P. and Witt, A. N. (1995) The Interstellar Carbon Budget and the Role of Carbon in Dust and Large Molecules, Science, 270, 1455-1460.
348. Snyder, L. E., Buhl, D., Zuckerman, B., and Palmer, P. (1969) Microwave detection of interstellar formaldehyde, Physical Review Letters, 22, 679-681.
349. Snyder, L. E. and Buhl, D. (1971) Observations of radio emission from interstellar hydrogen cyanide, The Astrophysical Journal, 163, L47-L52.
350. Solomon, P. M., Jefferts, K. B., Penzias, A. A., and Wilson, R. W. (1971) Detection of millimeter emission lines from interstellar methyl cyanide, The Astrophysical Journal, 168, L107-L110.
351. Solovchenko, A. and Merzlyak, M. (2008) Screening of visible and UV radiation as a photoprotective mechanism in plants. Russian Journal of Plant Physiology 55(6): 719-737.
352. Soloviev, A. and Lukas, R. (2006) The Near-Surface Layer of the Ocean Structure, Dynamics and Applications, Vol. 31, Atmospheric and Oceanographic Science Library, ISBN 978-1-4020-4052-8 (Print), 978-1-4020-4053-5 (Online). http://www.marinetechnologynews.com/news/ships-research-fleet-524328

353. Som, S. M., Catling, D. C., Harnmeijer, J. P., Polivka, P. M., and Buick, R. (2012) Air density 2.7 billion years ago limited to less than twice modern levels by fossil raindrop imprints, Nature, 484, 359–362.
354. Spencer, J. R. and Denk, T. (2010) Formation of Iapetus' Extreme Albedo Dichotomy by Exogenically Triggered Thermal Ice Migration, Science, 327(5964), 432-435.
355. Sponer, J.E., Szabla, R., Gora, R.W. Saitta, A.M., Pietrucci, F., Saija, F., Di Mauro, E., Saladino, R., Ferus, M., Civis, S., Sponer, J. (2016) Phys. Chem. Chem. Phys. 18, 30, 20047-20066, The Royal Society of Chemistry. DO - 10.1039/C6CP00670A
356. Staleva, H. et al. (2015) Mechanism of photoprotection in the cyanobacterial ancestor of plant antenna proteins, Nat. Chem. Biol. 11, 287–291.
357. Stapleton, A. E., and Walbot, V. (1994) Flavonoids Can Protect Maize DNA from the Induction of Ultraviolet Radiation Damage. Plant Physiol. 105: 881 -889.
358. Stetter, K. O. (2006) Hyperthermophiles in the history of life. Philosophical Transactions of the Royal Society B: Biological Sciences, 361(1474), 1837–1843. http://doi.org/10.1098/rstb.2006.1907
359. Stoks, P. G. & Schwartz, A. W. (1979) Uracil in carbonaceous meteorites, Nature (London) 282, 709-710.
360. Stoks, P. G. & Schwartz, A . W. (1981) Nitrogen-heterocyclic compounds in meteorites – significance and mechanisms of formation. Geochim Cosmochim Acta 45, 563-569.
361. Stoks, P.G. & Schwartz, A.W. (1982) Basic nitrogen-heterocyclic compounds in the Murchison Meteorite. Geochim. Cosmochim. Acta 46, 309–315.
362. Stomp, M., Huisman, J., Stal, L. J., and Matthijs, H. C. (2007) Colorful niches of phototrophic microorganisms shaped by vibrations of the water molecule, ISME J., 1, 271–282.
363. Stribling, R. and Miller, S.L. (1987) Energy yields for hydrogen cyanide and formaldehyde syntheses: the HCN and amino acid concentrations in the primitive ocean. Origins Life Evol B 17, 261-273.
364. Su, C., Middleton, C. T., Kohler, B. (2012) Base-Stacking Disorder and Excited-State Dynamics in Single-Stranded Adenine Homo-oligonucleotidesJ. Phys. Chem. B, 116 (34), pp 10266–10274 DOI: 10.1021/jp305350t
365. Summons, R. E., Jahnke, L. L., Hope, J. M., and Logan, G. A. (1999) 2-Methylhopanoids as biomarkers for cyanobacterial oxygenic photosynthesis, Nature, 400, 554–557.
366. Swenson, R.(1989) Emergent evolution and the global attractor: The evolutionary epistemology of entropy production maximization, in: Proceedings of the 33rd Annual Meeting of The International Society for the Systems Sciences, edited by: Leddington, P., 33, 46–53.
367. Swenson, R. (1991) End-directed physics and evolutionary ordering: Obviating the problem of the population of one. In "The Cybernetics of Complex Systems: Self-Organization, Evolution, and Social Change", F. Geyer (ed.), 41-60. Salinas, CA: Intersystems Publications.
368. Swings, P. and Rosenfeld, L. (1937) Considerations regarding interstellar molecules, The Astronomical Journal, 86, 483-486.
369. Szabat M., Pedzinski T., Czapik T., Kierzek E., Kierzek R. (2015) Structural Aspects of the Antiparallel and Parallel Duplexes Formed by DNA, 2'-O-Methyl RNA and RNA Oligonucleotides. PLoS ONE 10(11): e0143354. doi:10.1371/journal.pone.0143354
370. Szabla, R., Tuna, D., Góra, W. R., Šponer, J., Sobolewski, A. L., and Domcke, W. (2013) Photochemistry of 2-Aminooxazole, a Hypothetical Prebiotic Precursor of RNA Nucleotides, J. Phys. Chem. Lett., 4 (16), pp 2785–2788, DOI: 10.1021/jz401315e
371. Szostak, J.W. (201) The eightfold path to non-enzymatic RNA replication, J. Sys. Chem., 3:2
372. Tappert, R., McKellar, R. C., Wolfe, A. P., Tappert, M. C., Ortega-Blanco, J., Muehlenbachs, K. (2013) Stable carbon isotopes of C3 plant resins and ambers record changes in atmospheric oxygen since the Triassic, Geochim. Cosmochim. Ac., 121, 240–262.
373. Tashiro, R., Wang, A. H.-J. Sugiyam, H. (2006) Photoreactivation of DNA by an Archaeal nucleoprotein Sso7d, PNAS, 103, 16655–1665.
374. Tennyson, J. (2003) Molecules in Space, Volume 3, Part 3, Chapter 14, pp 356–369 in Handbook of Molecular Physics and Quantum Chemistry, Edited by Stephen Wilson, John Wiley & Sons, Ltd, Chichester, ISBN 0 471 62374 1

375. Tielens, A. G. G. M. (2008) Interstellar polycyclic aromatic hydrocarbon molecules, Annual Review of Astronomy and Astrophysics, 46, 289-337.
376. Toulmé, J.J., Charlier, M. and Heléme, C., (1974) Specific Recognition of Single-Stranded Regions in Ultraviolet-Irradiated and Heat-Denatured DNA by Tryptophan-Containing Peptides, Proc. Nat. Acad. Sci. USA 71, 3185-3188.
377. Trainer, M. G., Pavlov, A. A., DeWitt, H. L., Jimenez, J. L., McKay, C. P., Toon, O. B., and Tolbert, M. A. (2006) Organic haze on Titan and the early Earth, P. Natl. Acad. Sci. USA, 103, 18035–18042.
378. Trissl H.W. (1993) Long-wavelength absorbing antenna pigments and heterogeneous absorption bands concentrate excitons and increase absorption cross section. Photosynth Res 35:247–263..
379. Turner, B. E., Kislyakov, A. G., Liszt, H. S., Kaifu, N. (1975) Microwave detection of interstellar cyanamide, The Astrophysical Journal, 201, L149-L152.
380. Uchida, K. I., Sellgren, K., and Werner, M. (1998) Do the Infrared Emission Features Need Ultraviolet Excitation?, The Astrophysical Journal Letters, 493, L109.
381. Uchida, K. I., Sellgren, K., Werner, M. W., and Houdashelt, M. L. (2000) Infrared Space Observatory mid-infrared spectra of reflection nebula, The Astrophysical Journal, 530, 817-833.
382. Unrean, P., Srienc, F. (2011) Metabolic networks evolve towards states of maximum entropy production, Metabolic Engineering 13, 666–673.
383. Ulanowicz, R. E. and Hannon, B. M. (1987) Life and the production of entropy, P. Roy. Soc. Lond. B, 232, 181–192.
384. Ulbricht, T.L.V. (1959) Asymmetry: The non-conservation of parity and optical activity. Quart. Rev., 13, 48.
385. Ussery, D. W. (2002) DNA Structure: A-, B- and Z-DNA Helix Families, Encyclopedia of Life Sciences, Macmillan Publishers Ltd, Nature Publishing Group / www.els.net
386. Vekshin, N.L. (2005) Photonics of Biopolymers, KomKniga, Moscow.
387. Vreeland, R.H., Rosenzweig, W. D. and Powers, D. W. (2000) Isolation of a 250 million-year-old halotolerant bacterium from a primary salt crystal, Nature 407, 897-900 (19 October 2000) | doi:10.1038/35038060
388. Wächtershäuser, G. (1988) Before enzymes and templates: Theory of surface metabolism. Microbiology and Molecular Biology Reviews, 52, 452–484.
389. Wächtershäuser, G. (2006) From volcanic origins of chemoautotrophic life to bacteria, archaea and eukarya, Philos. T. Roy. Soc. B, 361, 1787–1808.
390. Waite Jr., J. H. (2007) The process of tholin formation in Titan's upper atmosphere, Science, 316, 870-875.
391. Wald R M (1993) Black hole entropy is the Noether charge, Phys. Rev. D 48 3427–3431.
392. Wang, J., Bras, R.,L., Lerdau, M., and Salvucci, G. D. (2007) A maximum hypothesis of transpiration, J. Geophys. Res. 112, G03010, doi:10.1029/2006JG000255
393. Watanabe Y, Martini JEJ, Ohmoto H. (2000) Geochemical evidence for terrestrial ecosystems 2.6 billion years ago. Nature 408, 574-578.
394. Weber, A.L. and Lacey, J.C. (1978) Genetic code correlations: amino acids and their anticodon nucleotides, J. Mol. Evol. 11, 199-210.
395. Webster, P.J., Clayson, C.A., Curry, J.A. (1996) Clouds, radiation and the diurnal cycle of sea surface temperature in the tropical western pacific, J. of Climate, Agust 1996, 1712-1730.
396. Weinreb, S., Barrett, A. H., Meeks, M. L., and Henry, J. C. (1963) Radio observations of OH in the interstellar medium, Nature, 200, 829-831.
397. Werfel, J., Ingber, D. E., and Bar-Yam, Y. (2015) Programmed death is favored by natural selection in spatial systems, Phys. Rev. Lett., 114, 238103.
398. Werner, F., Untitled[Review of the article "Homochirality Through Photon-induced Melting Of RNA or DNA: Thermodynamic Dissipation Theory Of The Origin Of Life" by Michaelian, K., WebmedCentral BIOCHEMISTRY 2010;1(10):WMC00924]; REVIEW_REF_NUM441

399. Westall, F., de Ronde, C.E.J., Southam, G., Grassineau, N., Colas, M., Cockell, C., Lammer, H. (2006) Implications of a 3.472-3.333 Gyr-old subaerial microbial mat from the Barberton greenstone belt, South Africa for the UV environmental conditions on the early Earth, Phil. Trans. Roy. Soc. B, 361 (1474) 1857-1875, doi: 10.1098/rstb.2006.1896
400. Wetmur, J.G. and Davidson, N. (1968) Kinetics of renaturation of DNA. J Mol Biol 31, 349370.
401. Wickramasinghe, N.C. (1973) Light scattering functions for small particles with applications in astronomy, J. Wiley, NY.
402. Wickramasinghe, D.T. and Allen, D.A., (1986) Discovery of organic grains in comet Halley, Nature, 323, 44-46.
403. Wickramasinghe, N.C. (2010a) "Interstellar Grains: 50 years on", chapter 6 in "Vindication of Cosmic Biology: Tribute to Sir Fred Hoyle (1915–2001) edited by Nalin Chandra Wickramasinghe, World Scientific Publishing Co. 2015, ISBN: 978-981-4675-26-0.
404. Wickramasinghe, N.C. (2010b) Spectroscopic Evidence of Cosmic Life, Journal of Cosmology, 2010, Vol 11, 3476-3488.
405. Wickramasinghe, N.C. (2013) A Journey with Fred Hoyle (Second Edition) (edited by: Kamala Wickramasinghe) World Scientific Publishing Co. 248 p. ISBN 978-981-4436-12-0
406. Wiechert, U. H. (2002) Earth's early atmosphere, Science, 298, 2341–2342.
407. Williams, M. C., Wenner, J. R., Rouzina, I., and Bloomfield, V. A. (2001) Effect of pH on the overstretching transition of double-stranded DNA: evidence of force induced DNA melting, Biophys. J., 80, 874–881.
408. Willis, R. (1993) World Mythology, Henry Holt Reference Books, Duncan Baird Publishers.
409. Wolstencroft, R. D. (2002) Terrestiral and Astronomical Sources of Circular Polarisation (2004) A fresh look at the origin of Homochirality on Earth, in: Bioastronomy 2002: Life Among the Stars: proceedings of the 213th symposium of the International Astronomical Union held at Hamilton Island, Great Barrier Reef, Australia, edited by: Norris, R. P. and Stootman, F. H., IAU Symposium, Vol. 213.
410. Wommack, K. E. and Colwell, R. R. (2000) Virioplankton: Viruses in aquatic ecosystems, Microbiol. Mol. Biol. Rev., 64, 69–114.
411. Wootton, J. T., Pfister, C. A., and Forester, J. D. (2008) Dynamic patterns and ecological impacts of declining ocean pH in a high resolution multi-year data set, P. Natl. Acad. Sci. USA, 105, 18848–18853.
412. Woese, C. R., Kandler, O. & Wheelis, M. L., (1990) Towards a natural system of organisms: proposal for the domains Archaea, Bacteria, and Eucarya. Proc. Natl Acad. Sci. USA, 87, 4576–4579.
413. Wu, W. and Liu, Y. (2010) Radiation entropy flux and entropy production of the Earth system. Rev Geophys 48:RG2003, doi:10.1029/2008RG000275
414. Wu W, Liu Y, and Wen G (2011) Spectral solar irradiance and its entropic effect on Earth's climate. Earth Syst Dynam Discuss 2: 45–70, doi:10.5194/esdd-2-45-2011
415. Wullschleger, S., Meinzer, F., and Vertessy, R. (1998) A review of wholeplant water use studies in trees, Tree Physiol., 18, 499–512.
416. Würfel, P. and Ruppel, W. (1985) The flow equilibrium of a body in a radiation field, J. Phys. C: Solid State Phys. 18, 2987-3000.
417. Xu, D-G. and Nordlund, T. M. (2000) Sequence dependence of energy transfer in DNA oligonucleotides, Biophysical Journal 78, 1042–1058.
418. Yarus, M., Widmann, J.J., Knight, R. (2009) RNA-amino acid binding: a stereochemical era for the genetic code, J. Mol. Evol., 69(5), 406-429.
419. Yockey, H.P. (1995) Information in bits and bytes: reply to Lifson's Review of Information Theory and Molecular Biology," BioEssays 17, 85-88.
420. Yockey, H.P. (2005) Information theory, evolution, and the origin of life, Cambridge University Press, Cambridge.
421. Young, G. M., von Brunn, V., Gold, D. J. C., and Minter, W. E. L. (1998) Earth's oldest reported glaciation: physical and chemical evidence from the Archean Mozaan Group (similar to 2.9 Ga) of South Africa, J. Geol., 106, 523–538.

422. Yuen, G.U. & Kvenvolden, K.A. (1973) Monocarboxylic acids in Murray and Murchison carbonaceous meteorites. Nature 251, 40–42.
423. Yuen, G., Blair, N., DesMarias, D.J. & Chang, S. (1984) Carbon isotope composition of low molecular weight hydrocarbons and monocarboxylic acids from the Murchison meteorite. Nature 307, 252–254.
424. Yurkov,V.V., Krieger, S., Stackebrandt, E. and Beatty, J.T. (1999) Citromicrobium bathyomarinum, a Novel Aerobic Bacterium Isolated from Deep-Sea Hydrothermal Vent Plume Waters That Contains Photosynthetic Pigment-Protein Complexes, J. Bacteriol. 181, 4517-4525.
425. Zabelinskii, S. A., Chebotareva, M. A., Shukolyukova, E.P., Furaev, V. V. and Krivchenko, A. I., (2005) Participation of π-Electrons of Phospholipid Molecules in Absorption of Ultraviolet Light in the Range of 260–280 nm, Journal of Evolutionary Biochemistry and Physiology, Volume 41, Number 3, 296-300. DOI: 10.1007/s10893-005-0062-y
426. Zahnle, K., Arndt, N., Cockell, C., Halliday, A., Nisbet, E., Selsis, F., and Sleep, N. H. (2007) Emergence of a Habitable Planet, Space Sci. Rev., 129, 35–78.
427. Zalar, A., Tepfera, D., Hoffmann, S. V., Kollmann, A., Leach, S. (2007) VUV-UV absorption spectroscopy of DNA and UV screens suggests strategies for UV resistance during evolution and space travel, in Instruments, Methods, and Missions for Astrobiology X edited by Richard B. Hoover, Gilbert V. Levin, Alexei Y. Rozanov, Paul C. W. Davies. Proc. of SPIE Vol. 6694, 66940U, doi: 10.1117/12.733699
428. Zhou, X. and Mopper, K. (1997) Photochemical production of low molecular-weight carbonyl compounds in seawater and surface microlayer and their air-sea exchange, Mar. Chem., 56, 201–213.
429. Zhu, Z., Piao, S., Myneni, R. B., Huang, M. Zeng, Z., Josep G. Canadell, Philippe Ciais, Stephen Sitch, Pierre Friedlingstein, Almut Arneth, Chunxiang Cao, Lei Cheng, Etsushi Kato, Charles Koven, Yue Li, Xu Lian, Yongwen Liu, Ronggao Liu, Jiafu Mao, Yaozhong Pan, Shushi Peng, Josep Peñuelas, Benjamin Poulter, Thomas A. M. Pugh, Benjamin D. Stocker, Nicolas Viovy, Xuhui Wang, Yingping Wang, Zhiqiang Xiao, Hui Yang, Sönke Zaehle, Ning Zeng (2016) Greening of the Earth and its drivers, Nature Climate Change, Letters, doi:10.1038/nclimate3004
430. Zintchenko, A., Koňák, C. (2005) Interaction of DNA/Polycation Complexes with Phospholipids: Stabilizing Strategy for Gene Delivery, Macromolecular Bioscience, 5, 1169–1174. DOI: 10.1002/mabi.200500159.
431. Zotin, A. I. (1984) Bioenergetic trends of evolutionary progress of organisms, in: Thermodynamics and regulation of biological processes, edited by: Lamprecht, I. and Zotin, A. I., De Gruyter, Berlin, 451–458.
432. Zubay, G. (1962) A Theory on the Mechanism of Messenger-RNA Synthesis, PNAS, 48, 456-461.
433. Zuckerman, B., Ball, J. A., and Gottlieb, C. A. (1971) Microwave detection of interstellar formic acid, The Astrophysical Journal, 163, L41-L45.
434. Zvezdanović, J. and Marković, D. (2008) Bleaching of chlorophylls by UV irradiation in vitro: the effects on chlorophyll organization in acetone and n-hexane, J. Serb. Chem. Soc., 73, 271–282.

Index

www.ingramcontent.com/pod-product-compliance
Lightning Source LLC
Chambersburg PA
CBHW081139180526
45170CB00006B/1849
* 9 7 8 1 5 4 1 3 6 6 7 1 8 *